RELIABILITY TECHNOLOGY
Theory and Applications

RELIABILITY TECHNOLOGY
Theory and Applications
Third Edition

Dr. Deepika Garg
Associate Professor
Department of Mathematics
G D Goenka University, Gurugram (Haryana, India)
Ph.D. (Applied Mathematics), NIT Kurukshetra (Haryana, India)

Dr. Jai Singh Gurjar
Formerly Professor
NIT Kurukshetra (Haryana)
Principal Punjab College of Engineering and Technology Mohali (Punjab, India)
Principal Modern Institute of Engineering and Technology Ambala (Haryana, India)

Mr. Arun Kumar
Research Scholar
G D Goenka University Gurugram (Haryana, India)
M.Sc. (Applied Mathematics) Kurukshetra University, Kurukshetra (Haryana, India)

TECHSAR™
NEW DELHI

Published and Printed by
TechSar Pvt. Ltd.
4435–36/7, Ansari Road, Daryaganj
New Delhi-110 002 (India)
E-mail: tech_info@techsarworld.com
Website: www.techsarworld.com

ISBN 978-93-90620-47-0

Reprint 2025

Preface to the Third Edition

We feel quite happy to present the third edition of *Reliability Technology: Theory and Applications*. This edition is an improvement over the second edition. Minor corrections in the second edition and suggestions by students and faculty have been taken care of. The techniques mentioned in the book are useful for reliability engineers without bothering about equations and we obtain many parameters related with reliability, availability and maintainability analysis. Also, some new references are added to the list.

The book is of unique nature. We may claim without any doubt that there is no other book in the market at present which has a treatment like this and also such applications of reliability technology. Thus, this third edition will add significantly to this feature of the book. We are thankful to all our colleagues, friends and students for the continues support and encouragement.

Suggestions for further improvement of the text are welcome.

Dr Deepika Garg
Dr Jai Singh Gurjar
Arun Kumar

Preface to the Second Edition

Nowadays for manufacturing industries to be successful in market, a lot of efforts are required. Success depends on many factors and two most important factors are reliability and reliable production processes. From one side it is affected by external factors like increase in customer demands for high quality and reliable products. On the other hand by internal factors, for example, unreliable production processes. Process reliability is a method for identifying internal problems, which provide significant ways to cost reduction for profit optimization. Most of the manufacturing industries discussed in this book have been personally visited at the sight either by our research scholars or myself. Since we have just tried to change the existing trend of writing texts on reliability from simple cases to practical complex cases, we hope that the text will open some new channels in the field of reliability.

This second edition is an improvement over the first edition. Moreover, there is one important addition in chapter 4—a newly tested technique—Gupta and Singh Graphical Technique. When we use this method we don't need to develop any complex mathematical equations. Only we need transition diagram and formulae which we have derived. This technique is useful to the reliability engineers without bothering about equations and one can obtain more than one parameter related to the system performance simultaneously. Also, some new references are added to the list. In this book, the aim of the current research is to develop reliability assessment tools which must help engineers to estimate system parameters quickly and with great precision.

We are thankful to all my friends and students for the support and encouragement they rendered in the preparation of the book.

Dr Deepika Garg
Dr Jai Singh Gurjar
Arun Kumar

Preface to the First Edition

The words 'reliable' and 'reliability' are generally used in our daily life indicating some degree of confidence in a person/thing/system. The word reliability (= re + liability) means repeated liability because of various breakdowns and failures of a device/system. When we sell or purchase something we talk about its reliability. At present, we are living in industrial era, so reliable operation of industrial systems is the prime interest. Not only the industries, reliable functioning of other systems also is the intention of everybody. In mathematical terms, our present life is the function of reliability in one way or the other. A number of daily problems can be easily solved if basic reliability concepts are known.

Several texts on reliability theory are available which deal with simple electrical/electronics and mechanical systems as examples. This text deals with simple as well as complex systems taken from electrical/electronics, mechanical, agricultural, industrial and other daily life problems demonstrating the use of various techniques. The text is self-contained. The text provides an insight that how to apply reliability theory to solve daily life problems and industrial problems. It may be mentioned here confidently that this text is quite different and a new one providing answers to the problems of workers working in the field of reliability. Most of the problems discussed in this text have been personally visited at the sight either by my research scholars or myself. Since I have just tried to change the existing trend of writing texts on reliability from simple cases to practical complex cases, suggestions for improvement in the text from readers will help. I hope that the text will open some new channels in the field of reliability.

The first half of the book (i.e., Chapters 1 to 4) provides basics about reliability theory, probability and distributions, system modeling and solution techniques while the second half (Chapters 5 to 8) contains the analysis of various important process industries, agro-based and other systems.

I thank all my friends and students for the help and encouragement they rendered in the preparation of the book.

Dr Deepika Garg
Dr Jai Singh Gurjar
Arun Kumar

Contents

Preface to the Third Edition *v*
Preface to the Second Edition *vi*
Preface to the First Edition *vii*

1. Introduction **1**

1.1 Historical Background 1
1.2 Scope of the Subject 1
1.3 Some Definitions 2
1.4 Various Systems 6
1.5 How to Analyse a System 7
Exercise 8

2. Probability, Probability Distributions and Stochastic Processes **9**

2.1 Probability 9
2.2 Conditional Probability 10
2.3 Theorems of Probability 10
2.4 Baye's Theorem 13
2.5 Random Variable 15
2.6 Some Discrete and Continuous Probability Distributions 15
2.7 A Stochastic Process is a Process which has Some Random or Stochastic Element Involved in Its Structure 24
Exercise 29

3. Important Concepts in Reliability Study **31**

3.1 Mean Time to System Failure (MTSF) 31
3.2 Mean Time to Repair (MTTR) 32
3.3 Mean Time Between Failures (MTBF) 32
3.4 System Availability 33
3.5 Mission Reliability 34
3.6 System Reliability 34
3.7 Reliability Allocation 37
3.8 Preventive Maintenance 39
3.9 Redundancies 39
3.10 Data Collection 39
3.11 Data Recovery 40
3.12 Data Analysis Procedures 40

3.13 Fault Tree Analysis Method 41
3.14 Reliability Testing 43
3.15 Risk Analysis and Management 45
3.16 Structural Redundancy 45
3.17 Extremal Distributions 46
3.18 Accelerated Tests 47
3.19 Reliability Enhancement Testing (RET) 49
3.20 Reliability and Its Equirement 50
3.21 Reliability vs Quality Control 50
Exercise 50

4. System Modeling and Solution Techniques **53**

4.1 System Description 53
4.2 Mathematical Modeling 54
4.3 Decomposition Method 54
4.4 Cut-Set Method 55
4.5 Tie-Set Method 56
4.6 Event Space Method 56
4.7 Path-Tracing Method 56
4.8 Boolean Table Method 57
4.9 Reduction Method 58
4.10 Renewal Theoretic Approach 58
4.11 Markov Method 61
4.12 Supplementary Variables Technique 63
4.13 Discrete Transform Method 65
4.14 Boolean Function Technique 68
4.15 Gupta-Singh Technique 72
4.16 The Sign Test 81
4.17 The Paired-sample Sign Test 82
4.18 A Rank Sum Test : The Mann-Whitney U Test 83
4.19 The One Sample Runs Test 85
Exercise 86

5. Reliability Analysis of Some Electrical-Electronics, Telecommunication and Mechanical Systems **87**

5.1 k-out-of-n Systems 87
5.2 On-Surface Transit System 96
5.3 Priority Systems 98
5.4 Mission Reliability 106
5.5 Container Manufacturing 111
Exercise 117

6. Reliability Technology in Process Industries **119**

6.1 The Paper Industry 119
6.2 The Cement Industry 131
6.3 The Utensils Industry 136
6.4 Availability Analysis of the Cattle Feed Plant 141
Exercise 154

7. Reliability Technology Applied to Agro-based Industries **155**

7.1 The Sugar Industry 155
7.2 The Butter-Oil (Ghee) Manufacturing Plant 163
7.3 The Fertilizer Plant 167
7.4 Availability Analysis of the Rice Plant 174
7.5 Decision Support System 195
Exercise 201

8. Availability Analysis of a Biogas Plant and a Plant Pathological System **203**

8.1 Biogas Plant (Non-conventional Energy System) 203
8.2 Plant Pathology Problem 206
8.3 Reliability Optimization 207
Exercise 208

Bibliography **209**
Index **218**

1

Chapter

Introduction

1.1 HISTORICAL BACKGROUND

The theory of reliability is a science which studies the laws of occurrence of failures in technical equipment. Some of the areas of reliability research are life testing, structural reliability (including redundancy considerations), machine maintenance problems (a part of queueing theory) and replacement problems (closely connected to renewal theory). Overlapping areas are quality control, extreme value theory, order statistics and censorship in sampling. Problems of machine maintenance were discussed by Khintchine (1932) and Palm (1947), of replacement by Lotka (1939), of renewal theory by Feller (1941, 49) and of extreme value theory by Weibull (1939), Gumbel (1935), Epstein (1948) [Ref. 1, 2, 3].

Reliability theory grew up in World War-II and since then this subject has been given a good attention by various countries. In advanced countries like the U.S.A., U.S.S.R., U.K. and others, various reliability research groups for industries and defence establishments were organised. In 1950, in U.S.A., Air Force formed an Adhoc Group on Reliability of Electronic Equipment (AGREE); AGREE published its first report in 1957. Thus, from an initiation by U.S.A. the subject progressed very much. Canada, Japan, Germany and France have also done a remarkable job in the area. Due to the fact that today, technology without reliability is unthinkable, the subject is progressing.

Due to the increasing importance of reliability theory in industry and defence, the two main organisations in every country, the subject must be given a due attention for development.

1.2 SCOPE OF THE SUBJECT

Every company or establishment should have a reliability group for the following purposes:

1. Reliability evaluation.
2. Reliability apportionment.
3. Design review.
4. Design control.
5. Specification, material and processing review.
6. Vendor control.
7. Test planning, operation and analysis.

8. Reliability knowledge.
9. Reliability and failure reporting systems.
10. Mathematical and statistical services for reliability problems.
11. Internal coordination of reliability activities.

To meet the above-mentioned needs, an establishment should have the following sections:

1. Management
2. Engineering
3. Testing
4. Manufacturing
5. Quality Control
6. Purchasing and Contracts

Viewing the above, we observe that in industry, reliability technology plays an important role. Also in defence establishments without reliability technology, we are unable to meet the requirements of our forces satisfactorily. Besides, industry and defence, reliability technology is also being used in biomedical, electronics, mining and electrical engineering. In our view reliability technology is applicable to almost all practical problems of our life. Some of the problems are discussed in this text.

Feeling the necessity of reliability technology advanced countries like U.S.A., U.S.S.R., U.K., France, Germany, Canada and Japan formed reliability groups in all major and medium industries. Government agencies also have full-fledged support groups. In fact, these countries have done a remarkable progress in this area. Owing to the huge importance of reliability technology, every developing country must follow the developed ones and should progress considerably in this area. Without diverting the attention towards reliability technology their progress is almost impossible or dim even if it is an oil producing or coal producing country. In U.K., at national level an organisation 'Safety and Reliability Directorate', was formed in 1959. In 1970, another organisation called 'Systems Reliability Service' was formed. There are journals on reliability from U.S.A., U.K. and other developed countries, namely, IEEE Transactions on Reliability (U.S.A.), Microelectronics and Reliability (U.K.) and Reliability Engineering & System Safety (U.K.), etc., disseminating knowledge and experiences in this technology. In U.K., universities and technical institutions have courses covering reliability and a few have a course leading to M.Sc. degree in Reliability. This pattern of advanced countries must be followed by developing countries for better development.

1.3 SOME DEFINITIONS

Before proceeding to further discussion on the subject, we mention here some definitions and basic concepts.

Reliability

Reliability of a device or system is defined as the probability that it performs its purpose within tolerance for the period of time intended under the given operating conditions.

The definition of reliability stresses four elements: probability, tolerance or adequate performance, time and operating conditions. It tells us how much reliable an equipment or a system is. Tolerance or adequate performance indicates that criteria must be established which clearly specify or define what is considered to be satisfactory operation. Time is very important for reliability concept because without a knowledge of the probability of a device functioning or surviving for a given time, there is no way of assessing the probability of completing a mission or task which is scheduled to last for a given period. The last element, operating conditions is also an important factor. A figure of reliability is meaningless unless the operating conditions are specified.

Availability

It is defined as the probability that the system will be able to operate within tolerance at a given instant of time.

Reliability can be obtained from availability, which requires calculation or probabilities of the system being in operable state at an instant of time, by integration.

Quality

The quality of a device is the degree of performance to applicable specification and workmanship standards.

To differentiate quality with reliability, it is to mention that a device which has undergone all quality tests may not necessarily be reliable. Quality is associated with manufacture whereas reliability superiors to that already built in design.

Maintainability

The probability that when maintenance action is initiated under given conditions, a failed system will be restored to operable condition within a specified total down time.

Repairability

The probability that a failed system will be restored to operable condition with a specified active repair time.

Serviceability

The degree of ease or difficulty with which a system can be repaired.

Mean time to failure

The total operating time of a number of items (or system) divided by the number of failures.

System effectiveness

The probability the system can successfully meet an operational demand within a given time under specified conditions.

Uptime

The total time during which the system is in acceptable operating condition.

Downtime

The total time during which the system is not in acceptable operating condition.

Redundancy

The existence of more than one means of performing a given function.

Stand by redundancy

The redundancy wherein the alternative means of performing the function is inoperative until needed, and is switched on upon failure of the primary means of performing the function.

Active redundancy

The redundancy wherein all redundant items are operating simultaneously rather than being switched on when needed.

Partial redundancy

The redundancy wherein two or more redundant items are required to perform the function.

K-out-of-m:G system

The system which is good if at least k of its m items are good.

K-out-of-m:F system

The system which is failed if at least k of its m items are failed.

Reliability function

Let N_o be the size of the population out of which N_s units survive the test while N_f fail, then the reliability function $R(t)$ is given by [2].

$$R(t) = \frac{N_s}{N_o} = \frac{N_o - N_f}{N_o} = 1 - \frac{N_f}{N_o} \tag{1.3.1}$$

giving

$$\frac{dR(t)}{dt} = -\frac{1}{N_o}\frac{dN_f}{dt} \quad \text{(Taking } N_o \text{ fixed)} \tag{1.3.2}$$

or,

$$\frac{dN_f}{dt} = -N_o \frac{dR(t)}{dt} \tag{1.3.3}$$

dN_f/dt is recognized as the rate at which units fail. Dividing both sides of (1.3.3) by $N_s(t)$, we obtain the instantaneous probability of failure $r(t)$ that is,

$$r(t) = \frac{1}{N_s}\frac{dN_f}{dt} = -\frac{N_o}{N_s}\frac{dR(t)}{dt} \tag{1.3.4}$$

Using (1.3.1) in (1.3.4), we get

$$r(t) = -\frac{1}{R(t)}\frac{dR(t)}{dt} \tag{1.3.5}$$

Integrating (1.3.5), we obtain

$$R(t) = \exp\left(-\int_{o}^{t} r(t)\,dt\right) \tag{1.3.6.}$$

The function $r(t)$ is called the failure rate function, hazard function, hazard rate.

Hazard rate Concept

The measure of an equipment reliability is the infrequency with which failures occur in time. A failure distribution represents an attempt to describe mathematically the length of the life of a material, a structure or a device. There are many physical causes that individually or collectively may be responsible for the failure of a device at any particular instant.

Non-symmetric distributions are importantly different at the tails and actual observations are sparse, particularly at the right-hand tail, because of limited sample size.

In view of these difficulties, it is necessary to appeal to a concept that makes it possible to distinguish between the different distribution functions on the basis of a physical consideration. Such a concept is based on the failure rate function which is known in the literature of reliability as the *hazard rate.* In actuarial statistics the hazard rate goes under the name of *force of mortality*, in extreme-value theory it is called the *intensity function* and in economics its reciprocal is called *Mill's Ratio.*

Let $F(x)$ be the distribution function of the time-to-failure of random variable X and let $f(x)$ be its probability density function. Then the hazard rate $h(x)$, is defined as

$$h(x) = \frac{f(x)}{1 - F(x)} \tag{1.3.7}$$

Here $1 - F(x)$ is called the reliability $R(x)$ at time x. The hazard rate, which is a function of time has a probabilistic interpretation; namely, $h(x)\,dx$ represents the probability that a device of age x will fail in the interval $(x, x + \Delta x)$, or

$$h(x) = \lim_{\Delta x \to 0} \frac{[P\{\text{a device of age } x \text{ will fail in the interval } (x, x + \Delta x) \text{ given that it has survived upto } x\}]}{\Delta x}$$

On the basis of physical considerations, one is at liberty to choose the functional form of $h(x)$ for a particular device. Once this is done, a differential equation in $h(x)$ is obtained, from which $f(x)$ and $F(x)$ can be recovered.

Let us assume the following:

$F(o^-) = 0,\ F(+\infty) = 1.$

Since $\frac{d}{dx}F(x) = f(x)$, we have from (1.3.7)

$$h(x)dx = \frac{dF(x)}{1-F(x)}$$

or,

$$\int_0^t h(x)dx = -[\log\{1 - F(x)\}]_0^t = -\log\frac{1-F(t)}{1-F(0)}$$

or,

$$1 - F(t) = \exp\left[-\int_0^t h(x)dx\right]$$

which gives on differentiation

$$f(t) = h(t)\exp\left[-\int_0^t h(x)dx\right] \qquad (1.3.8)$$

1.4 VARIOUS SYSTEMS

(i) ***Series Systems:*** It is a configuration of equipment such that the system is said to be in operation iff all the equipment in the configuration are operative. Such systems are also referred to as chain models (Fig. 1).

To evaluate the reliability induced by this type of systems, we need only to consider the state where all the equipment are operating. The probability that the system is operating at time $t + dt$ is the compound probability that it had been operating at time t and that none of the equipment failed in the interval $(t, t + dt)$.

(ii) ***Standby Redundant Systems:*** It is a configuration of equipment (say n) in which only one equipment is on-line at a time. When it fails, a standby equipment is immediately switched on-line and the failed equipment taken off-line. The process continues until all $n - 1$ (or a specified number) standby equipment have been exhausted (Fig. 1.1).

The system is capable of performing its function if at least one of the n equipment is operable. Thus, the reliability in this case is simply the sum of the probabilities that the system is in a state other than failed.

(iii) ***Parallel Redundant Systems:*** In this configuration we allow all n equipment to operate simultaneously. For the system to be in an operable state at time t, the system can be in any state other than n (when all equipment have failed). Such systems are also known as rope models, since the system fails when all the fibres break (Fig. 1.1).

In the standby case, since the off-line equipment cannot fail or will have a failure rate less than the on-line equipment, the system reliability will always be greater than for parallel redundancy. If the off-line equipment has the same failure rate as the on-line equipment, standby redundancy reduces to parallel redundancy.

(iv) ***Maintained Systems:*** These are the systems on which maintenance action is possible during a finite interval of time.

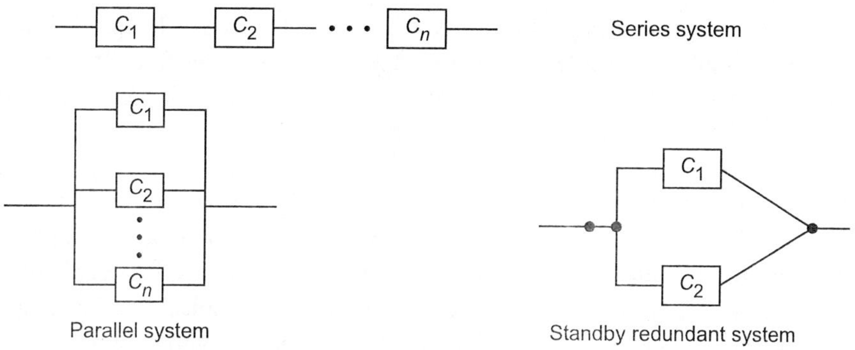

Fig. 1.1

(v) ***(K, m) systems:*** Suppose a system has m independent components and for successful operation of the system at least K components should operate. Such a system is known as (K, m) system.

(vi) ***Series-parallel Systems:*** A system consisting of m identical parallel systems of order n arranged in series is called series-parallel system of order (m, n) (Fig. 1.2).

(vii) ***Parallel-series Systems:*** A system consisting of m identical series systems of order n arranged in parallel is called a parallel-series system of order (m, n) (Fig. 1.2).

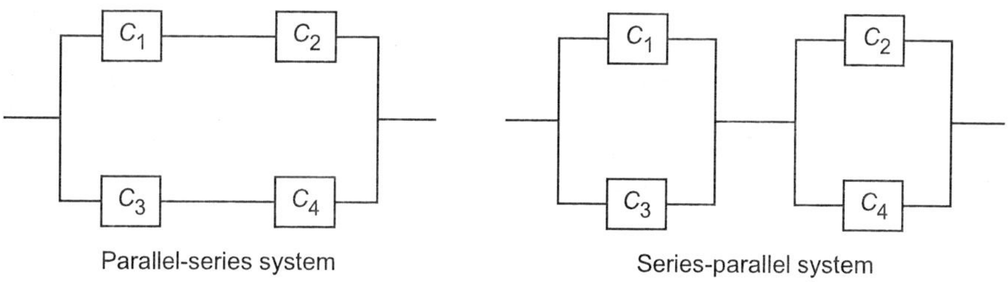

Fig. 1.2

(viii) ***Complex Systems:*** A combination of above types of systems or different from above categories having complex structures are called complex systems.

System subjected to two types of failure:

Many systems consist of components which can fail in two mutually exclusive ways, with the result that the system can fail in either of the two mutually exclusive ways. Systems with one proper mode of operation and two mutually exclusive failure modes are called trichotomous in contrast to the usual dichotomous of one operation and one failure mode.

1.5 HOW TO ANALYSE A SYSTEM

When we are to analyse a given system under certain relevant assumptions, first of all, define the various states of the system along with state probabilities. Next, form the governing equations/differential equations for the system along with initial and boundary conditions. Now the problem is to solve the equations/differential equations under the given conditions. There are techniques to solve differential equations. The more commonly used technique is by taking Laplace transforms of different equations and solving the resulting equations for probability transforms. Taking the inversions of probability transforms, various probabilities can be obtained. Sometimes, when we have a system of differential-difference equations, we use generating functions after taking Laplace transforms of the system of equations. The governing differential equations may also be analysed without making use of Laplace transforms and generating functions as we do in this text for some problems. Some problems may be solved with the help of regenerative point technique instead of forming differential equations.

Regenerative technique is also a combersome technique involving Laplace transforms and its inverse. To overcome this problem, Gupta-Singh introduced a new technique which is given in Chapter 4.

Solving the governing equations, we should evaluate the various parameters of the system, such as, MTSF, reliability function, etc. The limiting behaviour of the system should also be studied.

EXERCISE

1. Define the term reliability with suitable example.
2. Define a short note on sign test.
3. State k-out-of-m system redundancy.
4. Draw and explain hazard rate concept used in reliability.
5. What is reliability function?
6. Find out the system reliability for a serial and parallel configuration with 2 components.
7. Define complex system and series system.
8. Derive the expression for reliability evaluation of series and parallel systems.
9. Write short notes on:
   (a) Quality
   (b) Maintainability
   (c) Series-parallel system

2

Chapter

Probability, Probability Distributions and Stochastic Processes

In Chapter 1, we have mentioned the various aspects of the theory of reliability along with some definitions. It is obvious from Chapter 1 that the study of the theory of reliability requires a background of probability. Therefore, before proceeding to the study of various systems, we give here an outline of probability theory reliable for the study of reliability as well as queueing problems. The theory of probability is concerned with the study of those methods of analysis that are common to the study of random phenomena in all the fields in which they arise.

2.1 PROBABILITY

Before proceeding to the definition of probability, we explain certain terms to be used at subsequent occasions.

An experiment resulting in any of the possible outcomes is called a *trial* and the possible outcomes are known as the events or cases. The events are said to be *equally likely* when we have no reason to expect any one rather than the other. Events are called *mutually exclusive* or incompatible if the occurrence of one of them precludes the occurrence of all the others. On the contrary, events are compatible if it is possible for them to happen simultaneously. Moreover, events are said to be exhaustive when they include all possible ones. Also, we call the cases to be favourable to an event if they entail its happening.

We now consider the definitions of probability:

(a) ***Mathematical or Priori Definition:*** If there are n exhaustive, mutually exclusive and equally likely cases and m are favourable to an event A, the probability of the happening of A is defined as the ratio $\frac{m}{n}$.

(b) ***Statistical or Empirical definition:*** If trials be repeated a great number of times under essentially the same conditions then the limit of the ratio of the number of times that an event happens to the total number of trials as the number of trials increases indefinitely is called the probability of the happening of that event. It is assumed that the ratio approaches a finite and a unique limit.

The two definitions of probability are apparently different. In the mathematical definition, it is the relative frequency of favourable cases to the total number of cases while in statistical definition, it is the limit of the relative frequency of the happening of the event under consideration.

Denoting the probability of the happening of an event A by p and that of the non-happening by q, we have

$$p + q = 1, \text{ since } p = \frac{m}{n},\ q = \frac{n-m}{n} = 1 - \frac{m}{n} \tag{2.1.1}$$

Sometimes, we call the above saying that odds in favour of event A are m to n or odds against of event A are $n - m$ to n.

2.2 CONDITIONAL PROBABILITY

The probability of an event B occurring when it is known that some event A has already occurred is called the conditional probability and is denoted by $P(B|A)$. It is usually read as the probability that B occurs given that A has already occurred or simply as the probability of B given A. Symbolically, it is given by

$$P(B|A) = \frac{P(AB)}{P(A)}\ P(A) > 0 \tag{2.2.1}$$

where, $P(AB)$ denotes the probability that both the events A and B occur.

From (2.2.1), we have

$$P(AB) = P(A)\ P(B|A) \tag{2.2.2}$$

As a natural generalization of relation (2.2.2) for n events $A_1, A_2, A_3, \ldots\ldots\ldots, A_n$, we have

$$P(A_1A_2A_3\ldots A_n) = P(A_1)\ P(A_2|A_1)\ P(A_3|A_1A_2)\ldots P(A_n|A_1A_2A_3\ldots A_{n-1}) \tag{2.2.3}$$

For independent events, relations (2.2.2) and (2.2.3) reduce to the following forms:

$$P(AB) = P(A)\cdot P(B) \tag{2.2.4}$$

$$P(A_1A_2A_3\ldots A_n) = P(A_1)\cdot P(A_2)\cdot P(A_3)\ldots P(A_n) \tag{2.2.5}$$

2.3 THEOREMS OF PROBABILITY

Theorem 1: The probability that one of several mutually exclusive events $A_1, A_2, \ldots, A_n$ will happen is the sum of the probabilities of the separate events, i.e.

$$P(A_1 + A_2 + \ldots A_n) = P(A_1) + P(A_2) + \ldots P(A_n) \tag{2.3.1}$$

Proof: Let N be the total number of exhaustive, mutually exclusive and likely cases and out of these let $m_1, m_2, \ldots, m_n$ be favourable to the events $A_1, A_2, \ldots, A_n$ respectively.

Since the n events are mutually exclusive, these $m_1, m_2, \ldots, m_n$ cases are quite distinct and non-overlapping. Therefore, the total number of cases which are favourable to either A_1 or A_2 or…or A_n is $m_1 + m_2 + \ldots m_n$. Hence,

$$P(A_1 + A_2 + \ldots + A_n) = \frac{m_1 + m_2 + \ldots + m_3}{N}$$

$$= \frac{M_1}{N} + \frac{m_2}{N} + \ldots + \frac{m_n}{N}$$

$$= P(A_1) + P(A_2) + \ldots + P(A_n)$$

which proves the theorem. This theorem is called the theorem of total probability.

Theorem 2: If the events $A_1, A_2, \ldots, A_n$ are not mutually exclusive, then probability of at least one of the n events is given by

$$P(A_1 + A_2 + \ldots + A_n) = \sum_{i=1}^{n} P(A_i) - \sum_{\substack{i,j=1 \\ i \neq j}}^{n} P(A_i A_j) + \ldots + (-1)^{n-1} P(A_1 A_2 \ldots A_n) \quad (2.3.2)$$

Proof: Consider the events A and B only.

Since, AB and $A\bar{B}$ are two exhaustive and mutually exclusive forms in which A can occur, we have

$$P(A) = P(AB) + P(A\bar{B})$$

Similarly, $$P(B) = P(BA) + P(B\bar{A}) = P(AB) + P(\bar{A}B).$$

Adding, we get

$$P(A) + P(B) = P(AB) + [P(A\bar{B}) + P(\bar{A}B) + P(AB)]$$

By Theorem 1, the expression within the square brackets represents the probability $P(A + B)$ of the occurrence of at least one of the events A and B. Hence,

$$P(A + B) = P(A) + P(B) - P(AB). \quad (2.3.3)$$

Let B mean the occurrence of at least one of the events A_2 and A_3, then writing A_1 for A, (2.3.3) gives

$$P(A_1 + A_2 + A_3) = P(A_1) + P(A_2) + P(A_3) - P(A_1 A_2) - P(A_2 A_3) - P(A_3 A_1) + P(A_1 A_2 A_3)$$

$$= \sum_{i=1}^{3} P(A_i) - \sum_{\substack{i,j=1 \\ i \neq j}}^{3} P(A_i A_j) + P(A_1 A_2 A_3) \quad (2.3.4)$$

The general law for n events, which may be proved by mathematical induction, is

$$P(A_1 + A_2 + \ldots + A_n) = \sum_{i=1}^{n} P(A_i) - \sum_{\substack{i,j=1 \\ i \neq j}}^{n} P(A_i A_j) + \sum_{\substack{i,j,k=1 \\ i \neq j \neq k}}^{n} P(A_i A_j A_k) - \ldots + (-1)^{n-1} P(A_1 A_2 \ldots A_n), \quad (2.3.5)$$

where the second sum is overall combination of the numbers 1, 2, …, n taken two at a time, the third is overall combination of the numbers taken three at a time, and so forth.

Note: $\bar{A}$ and $\bar{B}$ denote the non-happening of events A and B, respectively.

Cor. If $A_1, A_2, \ldots, A_n$ are n mutually exclusive events, then $P(A_iA_j) = 0$, $P(A_iA_jA_k) = 0$, $P(A_1A_2 \ldots A_n) = 0$.

Consequently, (2.3.5) reduces to

$$P(A_1 + A_2 + \ldots + A_n) = \sum_{i=1}^{n} P(A_i) = P(A_1) + P(A_2) + \ldots + P(A_n)$$

which is the theorem of total probability.

Theorem 3: The probability of the simultaneous occurrence of two events A and B is equal to the probability of A multiplied by the conditional probability of B, given that A has occurred (or it is equal to the probability of B multiplied by the conditional probability of A given that B has occurred), i.e.,

$$P(AB) = P(A)\ P(B|A) = P(B)\ P(A|\mathrm{B}). \tag{2.3.6}$$

Proof: Let N denotes the total number of mutually exclusive and equally likely cases among which m cases are favourable to the event A. The cases favourable to both the events A and B are included in the m cases favourable to A. Let their number be m_1. Then the probability $P(AB)$ that both the events A and B will happen is given by

$$P(AB) = \frac{m_1}{N} = \frac{m}{N} \cdot \frac{m_1}{m}.$$

The ratio m/N is the probability of A, i.e., $m/N = P(A)$.

Assuming the occurrence of A there are only m equally likely cases left out of which m_1 are also favourable to B. Hence, the ratio m_1/m represents the conditional probability $P(B|A)$ of B, supposing that A has occurred. Hence,

$$P(AB) = P(A)\ P(B|A)$$

Since, the compound event AB involves A and B symmetrically, we shall have

$$P(AB) = P(B)\ P(A|B).$$

This theorem is called theorem of compound probability.

If A and B are independent events, then $P(B|A)$ is same as $P(B)$, and we have

$$P(AB) = P(A)\ P(B). \tag{2.3.7}$$

In general, for independent events, we have

$$P(A_1A_2 \ldots A_n) = P(A_1)\ P(A_2) \ldots P(A_n). \tag{2.3.8}$$

Problems

1. Three groups of children contain respectively 3 girls and 1 boy; 2 girls and 2 boys; 1 girl and 3 boys. One child is selected at random from each group. Show that the chance that the three selected, consist of 1 girl and 2 boys is $\frac{13}{32}$.

   One girl and two boys may be selected in the following ways:

   (i) Girl from the I group, boy from II group, boy from III group.

The probability of this event $= \frac{3}{4} \cdot \frac{2}{4} \cdot \frac{3}{4} = \frac{9}{32}$ (Theorem 3)

(ii) Boy from I group, girl from II group, boy from III group. The probability of this event

$\frac{1}{4} \cdot \frac{2}{4} \cdot \frac{3}{4} = \frac{3}{32}$.

(iii) Boy from I group, boy from II group, girl from III group. The probability of this event

$\frac{1}{4} \cdot \frac{2}{4} \cdot \frac{1}{4} = \frac{1}{32}$

All these events are mutually exclusive, hence the chance that any of these events happen

$= \frac{9}{32} + \frac{3}{32} + \frac{1}{32} = \frac{13}{32}$ (Theorem 1)

2. *A* can hit a target in 4 times in 5 shots; *B* 3 times in 4 shots; *C* twice in 3 shots. They fire a volley. What is the probability that two shots at least hit?

Chance of *A*'s hitting $= \frac{4}{5}$; Chance of *B*'s hitting $= \frac{3}{4}$;

Chance of *C*'s hitting $= \frac{2}{3}$.

For at least two hits, we may have,

(i) *A*, *B*, *C* all may hit with probability $= \frac{4}{5} \cdot \frac{3}{4} \cdot \frac{2}{3} = \frac{2}{5}$ (Th. 3)

(ii) *B*, *C* may hit and *A* may lose with probability $= \frac{1}{5} \cdot \frac{3}{4} \cdot \frac{2}{3} = \frac{1}{10}$

(iii) *C*, *A* may hit and *B* may lose with probability $= \frac{4}{5} \cdot \frac{1}{4} \cdot \frac{2}{3} = \frac{2}{15}$

(iv) *A*, *B* may hit and *C* may lose with probability $= \frac{4}{5} \cdot \frac{3}{4} \cdot \frac{1}{3} = \frac{1}{5}$

Since these are mutually exclusive events, the required probability $= \frac{2}{5} + \frac{1}{10} + \frac{2}{15} + \frac{1}{5}$

$= \frac{5}{6}$ (Theorem 1)

2.4 BAYE'S THEOREM

Statement: An event *A* can be explained by a set of exhaustive and mutually exclusive hypotheses B_1, B_2, …, B_n. Given

(i) 'a priori' probabilities $P(B_1)$, $P(B_2)$, …, $P(B_n)$ corresponding to a total absence of knowledge regarding the occurrence of *A* and

(ii) Conditional probabilities $P(A|B_1)$, $P(A|B_2)$, ..., $P(A|B_n)$,

(a) It is required to form the 'a posterion' probabilities $P(B_1|A)$, $P(B_2|A)$, ..., $P(B_n|A)$.

(b) Further, find the probabilities of materialisation of another event C, given the probabilities $P(C|AB_1)$, $P(C|AB_2)$, ..., $P(C|AB_n)$.

Proof:

(i) By the theorem of compound probability (Sec. 2.3),

$$P(AB_i) = P(B_i)\ P(A|B_i) = P(A)\ P(B_i|A)$$

$$\therefore \qquad P(B_i|A) = \frac{P(B_i)\,P(A|B_i)}{P(A)} \tag{2.4.1}$$

Since the event A can materialise in the mutually exclusive forms AB_1, AB_2, ..., AB_n, we have by the theorem of total probability (Theorem 1, Section 2.3)

$$P(A) = P(AB_1) + P(AB_2) + \ldots + P(AB_n)$$

$$= \sum_{i=1}^{n} P(B_i)\ P(A|B_i) \tag{2.4.2}$$

$$\therefore \qquad P(B_i|A) = \frac{P(B_i)\,P(A|B_i)}{\sum_{i=1}^{n} P(B_i)\,P(A|B_i)} \tag{2.4.3}$$

(ii)

$$P(C_1A) = \sum_{i=1}^{n} P(CB_i|A) = \sum_{i=1}^{n} P(B_i|A)\ P(C|AB_i)$$

$$= \frac{\sum_{i=1}^{n} P(B_i)\,P(A|B_i)\,P(C|AB_i)}{\sum_{i=1}^{n} P(B_i)\,P(A|B_i)} \quad \text{(using 2.4.3)} \tag{2.4.4}$$

Relation (2.4.3) is also known as the formula for 'a posterion' probability. This name is explained by the fact that this formula gives the probability of relation of the events B_i with respect to the occurrence of A. $P(B_i)$ is known as a priori probability.

EXAMPLE: There are five urns of the following compositions:

2 urns with 2 white and 3 black balls each,

2 urns with 1 white and 4 black balls each

1 urn with 4 white balls and 1 black ball.

A ball is chosen from one of the urns taken at random. It turned out to be white. What is the probability after the experiment (a posterion probability) that the ball was taken from the urn of the third composition

By hypothesis, we have

$$P(B_1) = \frac{2}{5},\ P(B_2) = \frac{2}{5},\ P(B_3) = \frac{1}{5}$$

$$P(A|B_1) = \frac{2}{5},\ P(A|B_2) = \frac{1}{5},\ P(A|B_3) = \frac{4}{5}$$

By Baye's formula, we have

$$P(B_3|A) = \frac{P(B_3)\,P(A|B_3)}{P(B_1)\,P(A|B_1) + P(B_2)\,P(A|B_2) + P(B_3)\,P(A|B_3)}$$

$$= \frac{\frac{1}{5}\cdot\frac{4}{5}}{\frac{2}{5}\cdot\frac{2}{5}+\frac{2}{5}\cdot\frac{1}{5}+\frac{1}{5}\cdot\frac{4}{5}} = \frac{4}{10} = \frac{2}{5}$$

In exactly the same way, we may find

$$P(B_1|A) = \frac{2}{5},\ P(B_2|A) = \frac{1}{5}.$$

2.5 RANDOM VARIABLE

Many applications of probability theory are based on the notion of a random variable. We use the term statistical experiment to describe any process by which several chance measurements are obtained. All the possible outcomes of a statistical experiment comprise a set which we call sample space. Often we are not interested in the details associated with each sample point but in some numerical description of the outcome for which we require the random variable concept. We define a random variable as follows.

A random variable is a variable quantity whose values depend on chance and for which a distribution function of probabilities has been defined or equivalently a quantity X is said to be a random variable (or, equivalently, X is said to be an observed value of a numerical valued random phenomenon) if for every real number x there exists a probability that X is less than or equal to x. Random variables are usually denoted by capital letters.

If the sample space contains a finite number of points or a sequence with as many elements as there are whole numbers, it is called a discrete sample space. On the other hand, if the elements of the sample space are infinite in number or as many as the number of points on a line segment, we say that we have continuous sample space. A random variable defined over a discrete sample space is called a discrete random variable while over a continuous sample space is called a continuous random variable. In most practical problems, continuous random variables represent measured data and discrete random variables represent count data such as the number of defectives in a sample of k items or the number of accidents per year.

2.6 SOME DISCRETE AND CONTINUOUS PROBABILITY DISTRIBUTIONS

The distribution function $F(.)$ of a numerical valued random phenomenon is defined as having its value, at any real number x, the probability that an observed value of the random phenomenon will be less than or equal to the number x. In symbols, for any real number x,

$$F(x) = P[\text{real numbers } x' : x' \le x].$$

If the probability function is specified by a probability mass function $p(.)$, then the distribution function $F(.)$ for any real number x, called discrete distribution function, is given by

$$F(x) = \Sigma\, p(x'), \text{ for any real number } x,$$

points $x' \le x$ such that $p(x') > 0$.

If the probability function is specified by a probability density function $f(.)$, then the distribution function $F(.)$ for any real number x, called continuous distribution function, is given by

$$F(x) = \int_{-\infty}^{x} f(x')dx', \text{ for any real number } x.$$

Most of the distribution functions arising in practice are either discrete or continuous. Also, there are distribution functions which are neither discrete nor continuous. Such distribution functions are called *mixed.* A distribution function $F(.)$ is called mixed if it can be written as a linear combination of two distribution functions, namely, discrete and continuous denoted by $F_d(.)$ and $F_c(.)$ respectively in the following way:

$F(x) = C_1F_d(x) + C_2F_c(x)$, for any real number x in which C_1 and C_2 are constants between 0 and λ, whose sum is one.

It is to be noticed that $p(x)$ is a probability mass function for a discrete random variable X if, for any real number x,

(i) $p(x) = p(X = x)$,

(ii) $p(x) \ge 0$, and

(iii) $\sum_x p(x) = 1$

Similarly, $f(x)$ is a probability density function for a continuous random variable X, if for any real number x

(i) $P(a < x < b) = \int_a^b f(x)dx$,

(ii) $f(x) \ge 0$, and

(iii) $\int_{-\infty}^{\infty} f(x)dx = 1.$

If instead of one random variable X, we have two random variables X and Y, the probability mass function for their simultaneous occurrence is represented by $p(x, y)$ and is called joint probability mass function of X and Y. $p(x, y)$ is called a joint probability mass function for discrete random variables X and Y, if, for all real numbers x and y,

(i) $p(x, y) = P[X = x, Y = y]$,

(ii) $p(x, y) \geq 0$, and

(iii) $\sum_x \sum_y p(x, y) = 1$,

and is called a joint probability density function $f(x, y)$ for continuous random variables X and Y, if, for all real numbers x and y,

(i) $P[a < x < b, c < y < d] = \int_c^d \int_a^b f(x, y)dx\, dy$

(ii) $f(x, y) \geq 0$, and

(iii) $\int_{-\infty}^{\infty} \int_{-\infty}^{\infty} f(x, y)dx\, dy = 1$.

We now discuss some important probability distributions which can describe the behaviour of most of the random variables encountered in practice.

(a) ***Uniform distribution:*** The simplest of all discrete probability distributions is one where the random variable assumes all its values with equal probability. This distribution is called uniform probability distribution.

If a random variable X assumes the values $x_1, x_2, \ldots x_k$ with equal probability, then the discrete uniform distribution has the probability mass function given by

$$p(x, k) = 1/k,\ x = x_1, x_2, \ldots, x_k. \tag{2.6.1}$$

(b) ***Binomial distribution:*** We call an experiment a binomial experiment if it has the following properties:

(i) It consists of n repeated trials.

(ii) Each trial results in an outcome that may be classified as a success or a failure.

(iii) The probability of success remains constant from trial to trial.

(iv) The repeated trials are independent.

The number X of successes in n trials of a binomial experiment is called a binomial random variable. The probability distribution of the binomial variable X is called binomial distribution, generally denoted by $b(x; n, p)$ since its value depends upon the number of trials and the probability of success on a given trial.

To derive an expression for probability mass function, we consider an experiment for exactly x successes and $n - x$ failures in a specified order. Each success occurs with probability p and each failure with probability $q = 1 - p$. Since all the trials are independent, the probability for x successes in a specified order from n trials is $p^x q^{n-x}$. The possible number of ways in which x successes can occur from n trials is $\binom{n}{x}$. Since all these possible ways of getting x successes are mutually exclusive, each having a probability $p^x q^{n-x}$, the probability of getting exactly x successes in a series of n independent trials is given by

$$b(x;\, n,\, p) = \begin{cases} \binom{n}{x}\, p^x\, q^{n-x}, \text{for } x = 0, 1, \ldots, n \\ 0, \text{otherwise} \end{cases} \tag{2.6.2}$$

$b(x;\, n,\, p)$ given by (2.6.2) is the probability mass function for binomial distribution. The sum of the probabilities

$$\sum_{x=0}^{n} b(x;\, n,\, p) = \sum_{x=0}^{n} \binom{n}{x}\, p^x\, q^{n-x} = (q + p)^n = 1.$$

The binomial distribution contains two independent constants, n and p(or q). These are called parameters of the binomial distribution. If $p = q = \frac{1}{2}$, the binomial distribution is symmetrical distribution and when $p \neq q$, it is a skew distribution.

(c) ***Poisson distribution:*** An experiment having the following properties is called a Poisson experiment.

(i) The number of successes occurring in one time interval or specified region are independent of those occurring in any other disjoint interval or region.
(ii) The probability of a single success occurring during a very short time interval or in a small region is proportional to the length of time interval or the size of the region and does not depend on the number of successes occurring outside this time interval or region.
(iii) The probability of more than one success occurring in such an interval or fall in such a small region is negligible.

The number X of successes in a Poisson experiment is called a Poisson random variable. The probability distribution of the Poisson variable X is called the Poisson distribution and is generally denoted by $p(x;\, \lambda)$.

Poisson distribution is the limiting form of binomial distribution when p (or q) is very small and n is large but the average number of successes np(or nq) $= \lambda$ is a finite constant. To derive probability mass function $p(x;\, \lambda)$ for Poisson distribution, we proceed as follows from binomial distribution.

The probability of x successes is given by

$$p(x) = \binom{n}{x}\, p^x\, q^{n-x}$$

which can be written as

$$p(x) = \frac{n!}{x!(n-x)!} \left(\frac{\lambda}{n}\right)^x \left(1 - \frac{\lambda}{n}\right)^{n-x} \quad [\because p = \frac{\lambda}{n}]$$

$$= \frac{\lambda^x}{x!}\left(1 - \frac{\lambda}{n}\right)^n \cdot \frac{\left(1 - \frac{\lambda}{n}\right)^{-x}}{n^x} \cdot \frac{n!}{(n-x)!}$$

Taking limit as $n \to \infty$, we have

$$\lim_{n\to\infty} p(x) = \frac{\lambda^x}{x!} e^{-\lambda} \lim_{n\to\infty} \left[\frac{1}{n^x} \cdot \frac{n!}{(n-x)!}\right]$$

Using Stirling's formula for $n!$,

$$n! \sim \sqrt{2\pi}\, e^{-n}\, n^{n+\frac{1}{2}}$$

where, e is the Naperian base, 2.71828 ..., we have

$$\lim_{n\to\infty} p(x) = \frac{\lambda^x e^{-\lambda}}{x!} \lim_{n\to\infty} \left[\frac{\sqrt{2\pi}\, e^{-n}\, n^{n+\frac{1}{2}}}{n^x \sqrt{2\pi}\, e^{-n+x} \cdot (n-x)^{n-x+\frac{1}{2}}} \right]$$

$$= \frac{\lambda^x e^{-\lambda}}{x!e^x} \lim_{n\to\infty} \left[\frac{1}{\left(1-\frac{x}{n}\right)^n \left(1-\frac{x}{n}\right)^{-x+\frac{1}{2}}} \right]$$

$$= \frac{\lambda^x e^{-\lambda}}{x!} \left(\because \lim_{n\to\infty} \left(1-\frac{x}{n}\right)^n = e^{-x} \right)$$

Hence, $$p(x;\lambda) = \frac{\lambda^x e^{-\lambda}}{x!}, \text{ for } x = 0, 1, 2, \ldots \qquad (2.6.3)$$

$$= 0, \text{ otherwise}$$

The Poisson distribution involves only one parameter, λ.

The sum of the probabilities $= \sum_{x=0}^{\infty} \frac{\lambda^x e^{-\lambda}}{x!} = e^{-\lambda} \cdot e^{\lambda} = 1.$

(d) ***Geometric distribution:*** Consider a sequence of Bernoulian trials with probability of success being p. Let x be the number of failures preceding the first success then x is a random variable with probability distribution given by

$$p(x) = q^x p,\ x = 0, 1, 2, \ldots,\ q = 1 - p \qquad (2.6.4)$$

since the first x trials are to be failures clearly $\sum_{x=0}^{\infty} p \quad \sum_{x=0}^{\infty} q^x p = \frac{p}{1-q} = 1$

Note: An experiment having only two possible outcomes, success and failure, is called a Bernouli trial.

(e) ***Negative Binomial distribution:*** Consider an experiment in which we require exactly $x + k$ trials to produce k successes in an indefinite series of Bernoulian trials. In this experiment, clearly the last trial must be a success with probability p. Among the other $(x + k - 1)$ trials, there must be $k - 1$ successes having the probability.

$${}^{x+k-1}C_{k-1}\, p^{k-1}\, q^x,\ q = 1 - p.$$

Multiplying the two probabilities, we get

$$p(x = k) = {}^{x+k-1}C_{k-1}\, p^k q^x,\ x = 0, 1, 2, \ldots, \text{ and } x > 0 \tag{2.6.5}$$

Clearly,

$$\sum_{x=0}^{\infty} p(x=k) = p^k \sum_{x=0}^{\infty} {}^{x+k-1}C_{k-1}\, q^x p^k (1-q)^{-k} = 1.$$

Since

$${}^{x+k-1}C_{k-1} = \frac{(k+x-1)(k+x-2)\ldots k}{x!} = (-1)^x \cdot {}^{-k}C_x$$

where,

$${}^{-k}C_x = \frac{(-k)(-k-1)\ \ldots(-k-x+1)}{x!}(-1)^x \cdot {}^{x+k-1}C_{k-1}$$

$p(x)$ can also be written as

$$p(x) = {}^{-k}C_x\, p^k(-q)^x,\ x = 0, 1, 2, \ldots \tag{2.6.6}$$

The probability distribution given by (2.6.6) is called *Pascal's distribution*. It has two parameters p and k. If $k = 1$ (2.6.6) reduces to the geometric distribution.

In the above distribution, k is an integer but the distribution remains meaningful even if k is not an integer but $k \geq 0$. In this case the distribution is called the *negative binomial distribution*. If, we put

$$k = \frac{1}{\beta},\ p = \frac{1}{1+\beta},\ q = \frac{\lambda}{1+\beta}$$

The distribution so obtained with two parameters β and λ is called Polya's distribution.

If we let $p \to 0$ and $k \to \infty$ in such a way that $\lim_{k\to\infty} kp = \lambda$ (a finite constant), then (2.6.6) takes the form

$$p(x) = \lim_{\substack{k\to\infty \\ p\to 0}} (1+p)^{-k} \frac{(k+x-1)\ldots k}{x!}\left(\frac{p}{1+p}\right)^x$$

$$= \lim_{\substack{k\to\infty \\ p\to 0}} \left(1+\frac{\lambda}{k}\right)^{-k} \frac{1}{x!}\left(1+\frac{x-1}{k}\right) \ldots 1.\ \lambda^x \left(1+\frac{\lambda}{k}\right)^{-x}$$

$$= \frac{e^{-\lambda}\lambda^x}{x!}$$

which is the expression for Poisson distribution.

(f) ***Hypergeometric distribution:*** The distribution having

$$p(x) = \frac{{}^mC_x\ {}^nC_{r-x}}{{}^{m+n}C_r},\ x = 0, 1, 2, \ldots, r;\ r \leq m,\ r \leq n \tag{2.6.7}$$

as the probability law is known as Hypergeometric distribution.

(g) ***Multinomial distribution:*** The binomial distribution can easily be generalized to n repeated independent trials when each trial may result in one of the several outcomes say $E_1, E_2, \ldots, E_k$ with respective probabilities $p_1, p_2, \ldots, p_k$ in each trial where $p_1 + p_2 + \ldots + p_k = 1$.

The probability that n trials will result in E_1 occurring n_1 times, E_2 occurring n_2 times, ..., E_k occurring n_k times in a fixed definite order is

$$p_1^{n_1} p_2^{n_2} \ldots p_k^{n_k}; \sum_{i=1}^{k} n_i = n.$$

But we are interested in events occurring in any order. The number of mutually exclusive ways in which this can happen is $\dfrac{n!}{n_1, n_2 \ldots n_k}$

Hence, the required probability is

$$p(n_1, n_2, \ldots, n_k) = \frac{n!}{n_1, n_2 \ldots n_k} \cdot p_1^{n_1} p_2^{n_2} \ldots p_k^{n_k};$$

$$\sum_{i=1}^{k} n_i = n \qquad (2.6.8)$$

This distribution is called multinomial probability distribution as the expression is the general term of the multinomial expansion of

$$(p_1 + p_2 + \ldots + p_k)^n.$$

This is not univariate distribution but is a multivariate distribution involving k variates $n_1, n_2, \ldots,$ n_{k-1}, n_k but only $(k - 1)$ variants are independent since $\sum_{i=1}^{k} n_i = n$.

(h) ***Normal distribution:*** Normal distribution is one of the most important continuous distributions in the field of statistics. Its graph, called the normal curve, is a bell-shaped curve (fig. below). The mathematical equation of this curve was developed by De Moivre in 1733. This distribution is also referred to as Gaussian distribution after the name of Gauss who also derived its equation for a study of errors in repeated measurements of the same quantity. A random variable x having the bell-shaped distribution is called a normal random variable.

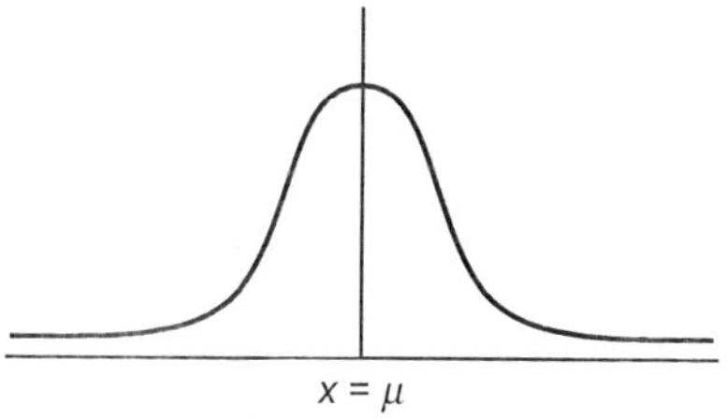

$x = \mu$

Normal curve

The mathematical equation for the probability distribution of the continuous normal variable depends on two parameters μ and σ, its mean and standard deviation. This distribution can be regarded as the limiting form of the Binomial distribution when n, the number of trials, is very large and neither p nor q is very small. The density function of a normal variable X, generally denoted by $N(x; \mu, \sigma)$, is given by

$$N(x; \mu, \sigma) = f(x) = \frac{1}{\sigma\sqrt{2\pi}} e^{-\frac{1}{2}\left(\frac{x-\mu}{\sigma}\right)^2}, \quad -\infty < x < \infty \tag{2.6.9}$$

The curve $y = f(x)$ is called the Normal Probability Curve.

The mathematical derivation of (2.6.9) involves a number of assumptions and mathematical calculations and therefore will not be attempted here. The mean, mode, median, etc. are given in Appendix.

As special features of the normal curve, we have

(i) The normal probability curve is symmetrical about the ordinate at $x = \mu$.

(ii) The ordinate decreases rapidly as x increases numerically.

(iii) The curve extends to infinity on either side of the mean.

(iv) The maximum ordinate is at $x = \mu$ and is given by $y_{max} = \dfrac{1}{\sigma\sqrt{2\pi}}$.

(v) The points of inflexion of the normal curve are equidistant from the mean.

(i) *Gamma distribution:* The continuous random variable X having probability density

$$f(x) = \frac{1}{\beta^{\alpha}\,\overline{|\alpha}}\, x^{\alpha-1} e^{-x/\beta}, \; x \geq 0, \; \alpha \geq 0 \tag{2.6.10}$$

is called a Gamma variate with parameters α and β and its distribution is called the Gamma distribution. The function $f(x)$ given by (2.6.10) represents a probability density since its integral over the range $(0, \infty)$ is unity.

As a special property of Gamma variate, we mention here that the sum of two independent Gamma variates with parameters α and β is a Gamma variate with parameter $\alpha + \beta$.

(j) *Beta distribution:* The continuous variable X, having probability density

$$\begin{aligned} f_1(x) &= \begin{cases} \dfrac{1}{B(\alpha, \beta)} x^{\alpha-1}(1-x)^{\beta-1}, 0 \leq x \leq 1; \alpha, \beta > 0 \\ \end{cases} \\ &= \begin{cases} 0, \text{otherwise} \end{cases} \end{aligned} \tag{2.6.11}$$

is called a Beta variate of first kind with parameters α and β.

The continuous variable X, having probability density

$$\begin{aligned} f_2(x) &= \begin{cases} \dfrac{1}{B(\alpha, \beta)} x^{\alpha-1}(1+x)^{-\alpha-\beta} \; x \geq 0; \alpha, \beta > 0 \\ \end{cases} \\ &= \begin{cases} 0, x < 0 \end{cases} \end{aligned} \tag{2.6.12}$$

is called a Beta variate of second kind with parameters α and β.

(k) *Cauchy distribution:* The continuous variable X, having probability density

$$f(x) = \frac{1}{\pi} \cdot \frac{1}{1 + (x - \lambda)^2}, \; -\infty < x < \infty \tag{2.6.13}$$

is called a Cauchy variate with parameter λ.

(l) ***Exponential distribution:*** A continuous random variable X is said to be exponentially distributed if it has the probability density function.

$$f(x) = \begin{cases} \dfrac{1}{\beta} \cdot e^{-x/\beta}, x > 0, \beta > 0 \\ 0, \text{elsewhere.} \end{cases} \tag{2.6.14}$$

with β as the parameter.

This distribution plays a key role in the theory of queues and reliability.

(m) ***Erlang distribution:*** A two-parameter generalisation of the exponential distribution is given by the Erlang distribution with density function

$$f(x) = \frac{\lambda k(\lambda k x)^{k-1}}{(k-1)!}, \; e^{-k\lambda x}, \; 0 \le x \le \infty \tag{2.6.15}$$

k is a positive integer. For $k = 1$, we get the exponential distribution.

This distribution also plays an important role in the theory of queues and reliability.

(n) ***Weibull distribution:*** The continuous random variable X has a Weibull distribution with parameters α and β if its density function is given by

$$f(x) = \begin{cases} \alpha\beta\, x^{\beta-1} e^{-\alpha x^{\beta}}, x > 0; \alpha, \beta > 0 \\ 0, x < 0. \end{cases} \tag{2.6.16}$$

This distribution was used by Weibull in 1951 to describe experimentally observed variation in the fatigue resistance of steel, its elastic limits, etc. This distribution is widely used in reliability theory.

(o) ***Chi-Squared distribution:*** The continuous random variable X has a chi-squared distribution with v degrees of freedom if its density function is given by

$$f(x) = \begin{cases} \dfrac{1}{2^{v/2}\sqrt{v/2}}\, x^{\frac{v}{2}-1}\, e^{-\frac{x}{2}}, x > 0 \\ 0, \text{elsewhere} \end{cases} \tag{2.6.17}$$

v is a positive integer. This distribution can be obtained from Gamma distribution putting

$$\alpha = \frac{v}{2}, \; \beta = 2.$$

Before closing the section, we mention here some more results about probability distributions in the form of Appendix.

APPENDIX

(a) The rth moment about any point x' of the discrete probability distribution with probability mass function p_i is given by

$$\mu'_r = E[x_i - x')^r] = \sum_{i=1}^{n} (x_i - x')^r \, p_i \tag{2.6.18}$$

If $x' = \bar{x}$ (mean) the moment is called the rth moment about mean and is denoted by μ_r. When $x' = 0$ and $r = 1$, the resulting expression from (2.6.18) is called the *expected* value or *expectation* or *mean value* of respective distribution. Also, when $x' = \bar{x}$, $r = 2$, we get from (2.6.18)

$$\mu_2 = E[(x - \bar{x})^2] \tag{2.6.19}$$

and is called *variance* of the respective distribution.

It may be proved that the expectation of the sum of two stochastic variates is equal to the sum of their expectations.

(b) The moment generating function (m.g.f.) about any point x' of the discrete probability distribution with probability mass function p_i and of the continuous probability distribution with probability density function $f(x)$ is given by the following:

$$M_x(t) = E[e^{t(x_i - x')}] = \sum_{i=1}^{n} e^{t(x_i - x')} \, p_i \tag{2.6.20}$$

$$M_x(t) = E[e^{t(x - x')}] = \int_{-\infty}^{\infty} e^{t(x - x')} f(x) dx, \tag{2.6.21}$$

(c) The characteristic function (c.f.) of the discrete probability distribution with probability mass function p_k and of the continuous probability distribution with probability density function $f(x)$ is defined as follows:

$$\phi(t) = E[e^{it} x_k] = \sum_{k=1}^{n} e^{it} x_k \, p_k, \tag{2.6.22}$$

$$\phi(t) = E[e^{itx}] = \int_{-\infty}^{\infty} e^{itx} f(x) dx, \tag{2.6.23}$$

Using the above formulae (2.6.18)—(2.6.23), the expected value, m.g.f. and c.f. may be evaluated for a distribution. These values are tabulated here for some of the distributions in simplified form.

2.7 A STOCHASTIC PROCESS IS A PROCESS WHICH HAS SOME RANDOM OR STOCHASTIC ELEMENT INVOLVED IN ITS STRUCTURE

Mathematically, a stochastic process is a set of random variables $\{x_t\}$ or $\{x(t)\}$ depending on some real parameter like the time t. A queueing system, turbulent fluid flow, random walk model, epidemic process and communication process are examples of stochastic process.

A stochastic process may be classified according as:

(i) the random variable is discrete or continuous

(ii) the parameter is discrete or continuous. If the joint distribution of x_{t_1} and x_{t_2} depends on only $t_1 - t_2$, the process is called stationary, otherwise it may be termed evaluationary. A stochastic process is called Gaussian or normal if the joint distribution of any number of variables is a normal distribution.

We now discuss some processes which are useful in the theory of queues and reliability.

(a) ***Markov process:*** If the probability of the system being in a given state at the next trial depends just on its state at present and not upon the states it may have been in earlier times, we call the property of a process as the Markov property. Mathematically, we have

$$P[A_{k+1}|A_k, A_{k-1}, \ldots, A_1] = P[A_{k+1}|A_k]. \tag{2.7.1}$$

This states that at the kth trial the conditional probability of any event A_{k+1}, dependent on the next trial, will not depend on what has happened in past trials but only on what is happening at the present. Sometimes, it is said that the trials have no memory. A process having Markov property is called Markov Process.

Let us observe at n times the state of a system which has r possible states. We number the states 1, 2, …, r (or 0, 1, 2, …, r – 1) and let $A_k^{(j)}$ be the event that the system is in state j at time k. If

$$P[A_k^{(j_k)} \mid A_{k-1}^{(j_{k-1})}, \ldots, A_1^{(j_1)}] = P[A_k^{(j_k)} \mid A_{k-1}^{(j_{k-1})}] \tag{2.7.2}$$

holds, we say that the system is a *Markov chain* with r possible states. In other words, it states that at any time the conditional probability of transition from one's present state to any other state does not depend on how one arrived in one's present state.

If the conditional probability is independent of time, the Markov chain is said to be *homogeneous* (or time homogeneous). A state is said to be *absorbing state* if once the system reaches to it and stays there. In such a case the chain is said to be an *absorbing chain.* A chain in which every two states can communicate with each other and in which the system cannot be left is called an *ergodic chain*. A Markov chain with infinite number of states is said to be denumerable or denumerably infinite.

Transition Matrix: Let P_{jk} be the transition probability (probability of transition from state j at nth trial to the state k at $(n + 1)$st trial) satisfying

$$P_{jk} \geq 0, \sum_k p_{jk} = 1 \text{ for all } j.$$

The probabilities may be written in the matrix form

$$P = \begin{bmatrix} p_{11} & p_{12} & \cdot & \cdot & \cdot & \cdot & p_{1k} \\ p_{21} & p_{22} & \cdot & \cdot & \cdot & \cdot & p_{2k} \\ \cdot & \cdot & \cdot & \cdot & \cdot & \cdot & \cdot \\ \cdot & \cdot & \cdot & \cdot & \cdot & \cdot & \cdot \\ p_{k1} & p_{k2} & \cdot & \cdot & \cdot & \cdot & p_{kk} \end{bmatrix}$$

This is called the transition matrix or matrix of transition probabilities of the Markov chain. P is a stochastic or Markov matrix, i.e., a square matrix with non-negative elements and unit row sums.

This is an important concept in the study of queueing and reliability problems. We illustrate it by an example:

EXAMPLE: A particle performs a random walk between two absorbing barriers, say at 0 and λ. Whenever it is at any position $r(0 < r < \lambda)$, it moves to $r + 1$ with probability p or to $(r - 1)$ with probability q, $p + q = 1$. But as soon as it reaches 0 or λ it remains there itself. Let X_n be the position of the particle after n moves. $\{X_n\}$ is a Markov chain with transition probabilities

$$\begin{aligned} P_r\{X_n = r + 1 \mid X_{n-1} = r\} &= p \\ P_r\{X_n = r - 1 \mid X_{n-1} = r\} &= q \\ P_r\{X_n = 0 \mid X_{n-1} = 0\} &= 1,\ 0 < r < \lambda \\ P_r\{X_n = \lambda \mid X_{n-1} = \lambda\} &= 1 \end{aligned}$$

The transition matrix is given by

$$\begin{bmatrix} 1 & 0 & 0 & 0 & \cdots & 0 \\ q & 0 & p & 0 & \cdots & 0 \\ 0 & q & 0 & p & \cdots & 0 \\ \cdot & \cdot & \cdot & \cdot & \cdots & \cdot \\ \cdot & \cdot & \cdot & \cdot & \cdots & \cdot \\ 0 & 0 & 0 & 0 & \cdots & 1 \end{bmatrix}$$

Similarly, if we have a particle performing a random walk between two reflecting barriers, the transition matrix will be as follows:

$$\begin{bmatrix} 0 & 1 & 0 & 0 & \cdots & 0 \\ q & 0 & p & 0 & \cdots & 0 \\ 0 & q & 0 & p & \cdots & 0 \\ \cdot & \cdot & \cdot & \cdot & \cdots & \cdot \\ \cdot & \cdot & \cdot & \cdot & \cdots & \cdot \\ 0 & 0 & 0 & 0 & \cdot\,\cdot\ 1 & 0 \end{bmatrix}$$

(b) ***Birth and Death Process:*** Consider a population and let X_t be the size of the population at a given time t. The probability law of X_t is specified by the following:

$$p_n(t) = P\{X_t = n\},\ n = 0, 1, 2, \ldots \tag{2.7.3}$$

To derive a differential equation for the probability mass function of X_t, we proceed as follows: Let $r_0(h)$, $r_1(h)$ and $r_2(h)$ be the functions defined for $h > 0$ with the property that

$$\underset{h\to 0}{\text{Lim}} \frac{r_0(h)}{h} = \underset{h\to 0}{\text{Lim}} \frac{r_1(h)}{h} = \underset{h\to 0}{\text{Lim}} \frac{r_2(h)}{h} = 0$$

Assume that $r_2(h)$ is the probability that in the time interval $(t, t + h)$ the population size will change by two or more. For $n \geq 1$, the event that $X_{t+h} = n$ (n members in population at time $t + h$) can then essentially happen in any one of following three mutually exclusive ways:

(i) the population size at time t is n and undergoes no change in the time interval $(t, t + h)$;
(ii) the population size at time t is $n - 1$ and increases by one in the time interval $(t, t + h)$;
(iii) the population size at time t is $n + 1$ and decreases by one in the time interval $(t, t + h)$.

Now let us define,

$\lambda_n h + r_1 h$ is the conditional probability that the population size will increase by one in the time interval $(t, t + h)$ for $h > 0$, and $\mu_n h + r_o(h)$ is the conditional probability that the population size will decrease by one in the time interval $(t, t + h)$ for $h > 0$, given that the population had size n at time t.

For $n = 0$, the event that $X_{t+h} = 0$ can happen only in ways (i) and (iii).

The events (i), (ii) and (iii) have probabilities

$p_n(t)(1 - \lambda_n h - \mu_n h)$, $p_{n-1}(t)\, \lambda_{n-1} h$ and $P_{n+1}(t)\, \mu_{n+1} h$ respectively.

Consequently, for $n \geq 1$, we get

$$p_n(t + h) = p_n(t)\,(1 - \lambda_n h - \mu_n h) + P_{n-1}(t)\,\lambda_{n-1} h + p_{n+1}(t)\,\mu_{n+1} h \tag{2.7.4}$$

For $n = 0$, we obtain

$$P_o(t + h) = p_o(t)(1 - \lambda_o h) + p_1(t)\,\mu_1 h. \tag{2.7.5}$$

Rearranging (2.7.4) and (2.7.5) and taking limits as h tends to zero, we obtain

$$\frac{d\,p_n(t)}{dt} = -(\lambda_n + \mu_n)p_n(t) + \lambda_{n-1}\,p_{n-1}(t) + \mu_{n+1}\,p_{n+1}(t) \tag{2.7.6}$$

$$\frac{d\,p_o(t)}{dt} = -\lambda_o p_o(t) + \mu_1 p_1(t) \tag{2.7.7}$$

If there is a maximum possible population size N then (2.7.4) holds only for $1 \leq n \leq N - 1$, whereas for $n = N$, we have

$$p_N(t + h) = p_N(t)(1 - \lambda_N h) + p_{N-1}(t)\,\lambda_{N-1} h. \tag{2.7.8}$$

Equations (2.7.6) and (2.7.7) can be solved by existing methods.

Due to its increasing and decreasing character, this process is called birth and death process. This process generally occurs in queueing and reliability problems.

Table

Probability law	Parameters	Probability mass/density function	Mean	Variance	m.g.f.	c.f.
Binomial	$n = 0, 1, 2, \ldots$ $0 \le p \le 1$	$p(x) = \binom{n}{x} p^x q^{n-x}, \quad x = 0, 1, 2, \ldots, n$ $= 0 \quad$ otherwise	np	npq	$(pe^t + q)^n$	$(pe^{it} + q)^n$
Geometric	$0 \le p \le 1$	$p(x) = pq^{x-1}, \quad x = 1, 2, \ldots$ $= 0 \quad$ otherwise	$\frac{1}{p}$	$\frac{q}{p^2}$	$\frac{pe^t}{1-qe^t}$	$\frac{pe^{it}}{1-qe^{it}}$
Hypergeometric	$n = 1, 2, \ldots$ $n = 1, 2, \ldots, N$ $p = 0, \frac{1}{N}, \frac{2}{N}, \ldots, 1$	$p(x) = \frac{{}^mC_x \, {}^nC_{r-x}}{{}^{m+n}C_r}, \quad x = 0, 1, \ldots, n$ $= 0 \quad$ otherwise		np	$npq\left(\frac{N-n}{N-1}\right)$	* *
Poisson	$\lambda > 0$	$p(x) = \frac{e^{-\lambda}\lambda^x}{x!}, \quad x = 0, 1, 2, \ldots$	λ	λ	$e^{(e^t-1)}$	$e^{(e^{it}-1)}$
Negative Binomial	$r > 0$ $0 \le p \le 1$	$p(x) = \binom{r+x-1}{x} p^r q^x$ $= \binom{-r}{x} p^r (-q)^x, \quad x = 0, 1, 2, \ldots$ $= 0 \quad$ otherwise	$\frac{rq}{p}$	$\frac{rq}{p^2}$	$\left(\frac{p}{1-qe^t}\right)^r$	$\left(\frac{p}{1-qe^{it}}\right)^r$
Normal	$-\infty < x < \infty$ $\sigma > 0$	$f(x) = \frac{1}{\sigma\sqrt{2\pi}} e^{-\frac{1}{2}\left(\frac{x-m}{\sigma}\right)^2}$	μ	σ^2	$e^{t + \frac{1}{2}t^2\sigma^2}$	$e^{it - \frac{1}{2}t^2\sigma^2}$
Exponential	$\lambda > 0$	$f(x) = \lambda e^{-\lambda x}, \; x > 0$ $= 0 \quad$ otherwise	$\frac{1}{\lambda}$	$\frac{1}{\lambda^2}$	$\left(1-\frac{t}{\lambda}\right)^{-1}$	$\left(1-\frac{it}{\lambda}\right)^{-1}$
Gamma	$r > 0$	$f(x) = \frac{\lambda}{\Gamma r} (\lambda x)^{r-1} e^{-\lambda x}, \; x > 0$ $= 0 \quad$ otherwise	$\frac{r}{\lambda}$	$\frac{r}{\lambda^2}$	$\left(1-\frac{t}{\lambda}\right)^{-r}$	$\left(1-\frac{it}{\lambda}\right)^{-r}$

EXERCISE

1. What is probability? Explain fundamental laws of probability.
2. What is random variable?
3. State and prove the Baye's theorem.
4. What is conditional probability?
5. State the difference between uniform distribution and binomial distribution.
6. Draw and explain normal distribution.
7. Let N be a random variable with a binomial distribution with parameters (n, p). Find E(N) and var(N).
8. Explain with examples, probability density function and probability distribution function.
9. The time to wear for a cutting tool is distributed normally with a mean of 2.8 hour and standard deviation of 0.6 hour. Find:
   (a) The probability that the tool will wear out in 1.5 hours
   (b) Find out the reliability for 1.5 hours
   (c) How often the cutting edge of the tool must be replaced in order to keep the failure less than 10 percentage?
10. A component with time to failure T has constant failure rate

$$z(t) = \lambda = 2.5 \text{ (hours)}^{-1}$$

   (a) Determine the probability that the component survives a period of 2 months without failure.
   (b) Find the MTTF of the component.
   (c) Find the probability that the component survives its MTTF.
11. Write short notes on:
   (a) Poisson distribution
   (b) Beta distribution
   (c) Cauchy distribution

3

Chapter

Important Concepts in Reliability Study

In the study of engineering systems it is required to evaluate certain reliability measures of the system under study, such as mean time to system failure (MTSF) or mean time to failure (MTTF), mean time to repair (MTTR), etc. In this chapter, we refer all important parameters mathematically along with the reliability expressions for some important systems.

3.1 MEAN TIME TO SYSTEM FAILURE (MTSF)

Mean time to system failure or the system mean time before failure or mean time to first failure is the most important concept generally required in system engineering. Let us suppose the reliability function for a system is given by $R(t) = 1 - F(t)$, where $F(t)$ is the failure time distribution function with probability density function $f(t)$.

The mean time to system failure is given by

$$\begin{aligned}\text{MTSF} &= \int_0^{\infty} t\, f(t)\, dt \\ &= \int_0^{\infty} t\left(\frac{dR(t)}{dt}\right) dt \\ &= -tR(t)\int_0^{+\infty} + \int_0^{+\infty} R(t)\,dt \\ &= \int_0^{+\infty} R(t)\,dt \qquad (3.1.1)\end{aligned}$$

[N.B. In eqn. (1.3.2) as dt approaches zero, we obtain the instantaneous probability, i.e., the probability density function $dR(t)/dt = -f(t)$]

Note that the MTSF denotes the expected or average time to failure and is neither the failure time which could be expected 50% of the time, nor it is the most probable time of system failure. The former interpretation only applies if the failure time density function is symmetric about the MTSF.

EXAMPLES:

(a) Exponential failure distribution

$$f(t) = \alpha\, e^{-at},\ \text{MTSF} = 1/\alpha$$

(b) Gamma failure distribution

$$f(t) = \frac{\alpha}{(r-1)!}\,(\beta t)^{r-1}\, e^{-at},\ \text{MTSF} = r/\alpha$$

(c) Weibull failure distribution

$$f(t) = \alpha\beta t^{\beta-1}\, e^{-at^{\alpha}},\ \text{MTSF} = \alpha^{-1/\beta}\,\Gamma\left(\frac{\beta+1}{\beta}\right)$$

where, $\Gamma(x)$ is the gamma function defined by

$$\Gamma(x) = \int_0^{+\infty} e^{-t}\, t^{x-1}\, dt$$

3.2 MEAN TIME TO REPAIR (MTTR)

The mean time to repair or mean downtime is defined by

$$\text{MTTR} = \int_0^{\infty} t\, f(t)dt \tag{3.2.1}$$

where, $f(t)$ is the repair time density function.

The total downtime, considered a random variable, is usually referred to as the repair time. If the total downtime t has probability density function $f(t)$, the probability of the failed system returns to service by time T is given by

$$p(t \le T) = \int_0^{T} f(t)\, dt. \tag{3.2.2}$$

This equation is called the maintainability equation.

EXAMPLES:

(a) If $f(t) = \alpha e^{-at}$,
then MTTR $= 1/\alpha$.

(b) If $f(t) = \alpha e^{-at}$, $\alpha > 0$
then $P(t \le T) = 1 - e^{-aT}$.

This is the exponential form of the maintainability equation.

3.3 MEAN TIME BETWEEN FAILURES (MTBF)

Another important concept in maintenance studies is the mean time between failures used with repairable equipment or systems. Generally, MTBF is used for both non-repairable components and repairable equipment and systems.

It is obvious that MTSF is the mean time to first failure while MTBF is the mean time between two successive component failures. These are not necessarily failures of identical components. The MTBF is normally conceived as being the mean time between the nth and the $(n + 1)$th failure in a system when n is relatively large. The relationship between the MTBF of a system and the MTSF of the individual components is

$$\frac{1}{\text{MTBF}} = \sum_{i=1}^{n} \frac{1}{T_i} \tag{3.3.1}$$

where the system has n components, all of different ages, each of which is replaced immediately on failure, and T_i is the MTSF of the ith component. Obviously, MTBF is a function of time.

When a system is first operated with all new components, the MTBF and the MTSF are identical. After that, the MTBF will fluctuate until, after many failures and replacements, it will stabilize at the value given by (3.3.1).

3.4 SYSTEM AVAILABILITY

As defined in Chapter 1, availability A of a device is defined as the probability that the device is operating satisfactorily at any point in time where the total time includes operating time, active repair time, logistics time and administrative time, i.e.,

$$A = \text{operable time/mission time} \tag{3.4.1}$$

$$A = \frac{\text{MTBF}}{\text{MTBF} + \text{MTTR}} \tag{3.4.2}$$

Availability $A(t)$ expresses the probability of being completely operable at any given time. The interval availability $A_I(t)$ for the interval $[0, T]$ is defined by

$$A_I(t) = \frac{1}{T}\int_0^T A(t)\,dt \tag{3.4.3}$$

The inherent availability of the system is defined by

$$A_I(\infty) = \lim_{T \to +\infty} A_I(t) \tag{3.4.4}$$

or more commonly called the up time ratio (UTR) given by

$$\text{UTR} = \frac{\text{MTSF}}{\text{MTSF} + \text{MTTR}} = \lim_{t \to \infty} p_0(t) \tag{3.4.5}$$

$$= \lim_{s \to 0} s p_0(s), \text{ if Laplace transform is taken.}$$

where, $p_0(t)$ is the point availability of the system and s the parameter of L.T. The system UTR is used to find the preventive maintenance scheduling. Another parameter is down time ratio (DTR) defined as:

$$\text{DTR} = 1 - \text{UTR} = \frac{\text{MTTR}}{\text{MTTR} + \text{MTSF}} \tag{3.4.6}$$

The down time ratio thus represents the probability—the system is down at any point in time, for sufficiently large times.

3.5 MISSION RELIABILITY

Mission reliability may be defined as the probability that the system will operate in the mode for which it was designed for the duration of a mission, given that it was operating in this mode at the beginning of the mission [4, 16].

The system is supposed to be in failed state if it does not become operable in a time less than τ, the constant time, measured from an instant at which failure occurred. We shall call the probability that the system failure does not occur during the time interval $(0, t)$, the mission reliability $R(t;\ \tau)$. If $\tau = 0$, then $R(t;\ \tau)$ becomes the usual reliability [4, 16].

Let $p_f(t)$ be the probability that at time t, the system is in failed state. Mission reliability function $R(t;\ \tau)$ for the system under consideration is defined as [ref. 4, 16]:

$$R(t;\ \tau) = 1 - p_f(t) \tag{3.5.1}$$

3.6 SYSTEM RELIABILITY

In this section, we mention, in short, the reliability formulae for some important systems:

(a) ***Series Systems:*** As mentioned in Chapter 1, a series system fails when any one of its independent components fails.

If $F_i(t)$ denotes the failure time distribution function of the ith component, then the failure time distribution function of the system is given by

$$F(t) = 1 - \prod_{i=1}^{n} (1 - F_i(t)) \tag{3.6.1}$$

with system failure time density

$$f(t) = \sum_{i=1}^{n} f_i(t) \prod_{\substack{j=1 \\ j \neq i}}^{n} (1 - F_j(t)) \tag{3.6.2}$$

where, $f_i(t) = dF_i(t)/dt$ for $1 \leq i \leq n$.

Let $R_i(t)$ denote the reliability of ith component for $i = 1, \ldots, n$. Since survival of the system, having components connected in series, requires all the components to be reliable, the reliability is given by

$$R(t) = \prod_{i=1}^{n} R_i(t) \tag{3.6.3}$$

(note that $R(t) = 1 - F(t)$).

Note that the reliability of a series system is less than or equal to that of the component with lowest reliability.

(b) ***Parallel Systems:*** In case of parallel systems all components of a system, having components connected in parallel, are required to operate simultaneously.

If $F_i(t)$, $f_i(t)$, $F(t)$, $f(t)$ and $R_i(t)$ stand for similar quantities as in (a) above, we have

$$F(t) = \prod_{i=1}^{n} F_i(t) \tag{3.6.4}$$

$$f(t) = \sum_{i=1}^{n} f_i(t) \prod_{\substack{j=1 \\ j \neq i}}^{n} F_j(t) \tag{3.6.5}$$

and the system reliability

$$R(t) = 1 - \prod_{i=1}^{n} (1 - R_i(t)) \tag{3.6.6}$$

Note that the reliability of a parallel system is greater than the reliability of the most reliable unit in the system.

(c) *Standby Systems:* In standby systems only one equipment is on-line at a time. When it fails, a standby equipment is immediately switched on-line and the failed equipment taken off-line. The system is declared failed when all equipment come to failure.

In such systems, the time to system failure is given by the sum of n independent random variables. The probability density function of the system failure time is equal to the n-fold convolution of the density function for each time, t_i say. If $y_i(t)$ denotes the density function for t_i and $f(t)$ is the corresponding system failure time density function, then

$$f(t) = y_1(t) * y_2(t) * \dots * y_n(t) \tag{3.6.7}$$

where * denotes the convolution. Equation (3.6.7) is the n-fold convolution of $y_1(t)$, $y_2(t)$, ..., $y_n(t)$. In particular for $n = 2$, we have

$$f(t) = \int_0^t y_1(x)\, y_2(t - x)\, dx. \tag{3.6.8}$$

The system reliability is given by

$$R(t) = 1 - \int_0^t (y_1 * \dots * y_n)(x)\, dx. \tag{3.6.9}$$

(d) *Parallel-Series Systems:* Parallel-series systems have been explained in Chapter 1.

If $R_i(t)$ equals the reliability of the ith type component (components are independent) for $i = 1, \dots, n$, the system reliability is given by

$$R(t) = 1 - \left(1 - \prod_{i=1}^{n} R_i(t)\right)^m \tag{3.6.10}$$

where n is the order of m identical series systems arranged in parallel.

(e) *Series-Parallel Systems:* Series-Parallel Systems have been explained in Chapter 1.

If $R_i(t)$ equals the reliability of the ith type-component (independent components) for $i = 1, \dots, n$, the system reliability is given by

$$R(t) = \left(1 - \prod_{i=1}^{n} \{1 - R_i(t)\}\right)^m \tag{3.6.11}$$

where, n is the order of m identical parallel systems arranged in series.

(f) *(k, n) Systems:* These systems (explained in Chapter 1) are also called 'k out of n' systems. If the n components have the same failure time distribution function $G(t)$ say, then the system reliability is equal to

$$R(t) = \sum_{i=k}^{n} \binom{n}{i} (1 - G(t))^i (G(t))^{n-i} \tag{3.6.12}$$

with system failure time distribution function $F(t)$ and system failure time density function $f(t)$ as follows:

$$F(t) = \sum_{i=0}^{k-1} \binom{n}{i} (1 - G(t))^i (G(t))^{n-i} \tag{3.6.13}$$

$$f(t) = \frac{n!}{(n-k)!(k-1)!} (G(t))^{n-k} \{1 - G(t)\}^{k-1} g(t) \tag{3.6.14}$$

where, $g(t) = dG(t)/dt$.

(g) *Systems with Arbitrary Repair:* When the repair density function is arbitrary, we call the system with arbitrary repair. A system consisting of only two states—operating and failed; is called a system of order 2.

Let us define the following for a simple system having only two components:

State 0 : both components are operable but only one is operating

State 1 : one component has failed and the other is operating

State 2 : both components have failed

$p_i(t)$: probability that at time t, the system is in state i. (i = 0, 1, 2).

The system reliability is given by

$$R(t) = 1 - p_2(t) = p_0(t) + p_1(t) \tag{3.6.15}$$

For detailed analysis of such systems see reference [4].

(h) *Systems with Parallel Repair:* Systems having more than one repairman or repair facility are called systems with parallel repair. In such situations more than one failed components can be repaired simultaneously (in parallel) and consequently there would be an improvement in system MTSF.

For an (m, n) system, if i denotes the system state in which exactly $i(i = 0, 1, \ldots, n - m)$ components have failed and are either receiving repair or are waiting for repair, then the system reliability is given by

$$R(t) = 1 - \sum_{j=1}^{n-m+1} C_j(e^{s_j t} - 1) \tag{3.6.16}$$

where, $$C_j = \frac{n}{(m-1)}\left(\prod_{\substack{i=1\\ i\neq j}}^{n-m+1}(s_j - s_i)\right)^{-1} \tag{3.6.17}$$

and s_i is ith root of the polynomial $\det A = \prod_{i=1}^{n-m+1}(s - s_i)$.

$A = (a_{ij})$ is an $(n - m + 1) \times (n - m + 1)$ matrix.

In case of standby systems if i denotes the system state in which exactly $i(i = 0, 1, ..., n)$ components have failed and are either receiving repair or are waiting for repair, then the system reliability is given by

$$R(t) = 1 - \sum_{j=1}^{n} C_j(e^{s_j t} - 1) \tag{3.6.18}$$

where, $$C_j = \left(s_j \prod_{\substack{i=1\\ i\neq j}}^{n}(s_j - s_i)\right)^{-1} \quad \text{for} \quad 1 \le j \le n. \tag{3.6.19}$$

and, $$\det A = \prod_{i=1}^{n}(s - s_i) \tag{3.6.20}$$

For detailed analysis see ref. [4], where the analysis is conducted by taking constant failure and repair rates.

3.7 RELIABILITY ALLOCATION

Reliability allocation is concerned with the allocation of individual component reliability so as to attain a prescribed level of system reliability subject to constraints, if any, on cost, weight and/or volume. The designer is required to choose the component (s) so as to yield an acceptable system reliability with minimum cost involved.

For example, consider a series system with n components in which the ith component has reliability $R_i(1 \le i \le n)$. The system reliability is given by

$$R = \prod_{i=1}^{n} R_i \tag{3.7.1}$$

and so $$\frac{R}{R_i} = \prod_{\substack{j=1\\ j\neq i}}^{n} R_j = \frac{R}{R_i} \tag{3.7.2}$$

It implies that R is most sensitive to increase in component i_0, if

$$\frac{R}{R_{i_0}} = \max_{1\le i\le n} \frac{R}{R_i} \tag{3.7.3}$$

or, equivalently,
$$R_{i_0} = \min_{1\le i\le n} R_i \tag{3.7.4}$$

which means that to increase the reliability of the system by increasing the reliability of one component, one should increase the reliability of the component for which R_i is the smallest.

Moreover, if cost is also taken into consideration, then let us suppose the cost per unit of reliability for the ith component is C_i. The total cost to increase R_i to $R_i + \Delta_i$ is equal to $C_i\Delta_i$. The designer is intended to increase R to R^0 by increasing the reliability of one component in such a way that the cost involved is minimum. Assume that $R^0 < \prod_{\substack{i=1\\ i\ne j}}^{n} R_i$ for each j which implies that it is feasible to increase the reliability of each component to a value less than 1 to obtain a total system reliability R^0. Now, if the ith component reliability is increased by an amount Δ_i, then we have

$$R^0 = \left(\prod_{\substack{j=1\\ j\ne i}}^{n} R_j\right)(R_i + \Delta_i) = R + \Delta_i \prod_{\substack{j=1\\ j\ne i}}^{n} R_j \tag{3.7.5}$$

giving

$$\Delta_i = (R^0 - R)/\prod_{\substack{j=1\\ j\ne i}}^{n} R_j \tag{3.7.6}$$

If R^0 can be attained by increasing R_i to $R_i + \Delta_i$ or R_j to $R_j + \Delta_j$, then

$$\left(\prod_{\substack{k=1\\ k\ne i}}^{n} R_k\right)(R_i + \Delta_i) = \left(\prod_{\substack{k=1\\ k\ne j}}^{n} R_k\right)(R_j + \Delta_j) \tag{3.7.7}$$

or,
$$R_j(R_i + \Delta_i) = R_i(R_j + \Delta_j) \tag{3.7.8}$$

or,
$$\Delta_i = \frac{R_i}{R_j}\Delta_j. \tag{3.7.9}$$

Therefore,
$$C_i\Delta_i = \frac{C_i R_i}{R_j}\Delta_j = \left(\frac{C_i R_i}{C_j R_j}\right) C_j\Delta_j \tag{3.7.10}$$

It is obvious from (3.7.10) that $C_i\Delta_i \le C_j\Delta_j$ if $C_iR_i \le C_jR_j$ and $C_i\Delta_i > C_j\Delta_j$ if $C_iR_i > C_jR_j$. This result gives an inference that the optimal component is i_0 where

$$C_{i_0}R_{i_0} = \min_{1\le i\le n} C_iR_i \tag{3.7.11}$$

Similarly, the case of a parallel system may be treated by taking the system reliability

$$R = 1 - \prod_{i=1}^{n} (1 - R_i) \tag{3.7.12}$$

and the corresponding results are obtained by simply replacing R_i by $1 - R_i$ in preceding analysis. Furthermore, component failure rate, weight and volume considerations are of extreme importance and must be reviewed in allocation process. (see ref. 4 for failure rate consideration).

3.8 PREVENTIVE MAINTENANCE

Adding redundancy to the system is a method of improving the reliability of the system. Another method to increase the system reliability is to perform preventive maintenance on the system according to some schedule. The maintenance action can be performed while the system is in operation without degrading its operation or by stopping the system briefly without any disastrous happening. In a periodic policy for preventive maintenance the maintenance is scheduled after some specified period of operation. The time preventive maintenance is a random variable with a scheduling density.

Now, suppose $R(t)$ denotes the system reliability for a period of length t without any preventive maintenance and let $R_1(t)$ denote the system reliability for a period of length t with preventive maintenance allowed if the schedule time falls within this interval. Let us suppose, $g(t)dt$ is the probability that preventive maintenance is performed in the interval $(t, t + dt)$ since the last failure. The system can be reliable at time t, either it does not fail prior to time t and no preventive maintenance was scheduled prior to t or it has preventive maintenances performed for the first time at some time $x < t$, is reliable up to time x and is reliable in the remaining length of time $t - x$. Hence, we have

$$R_1(t) = R(t) \int_t^{+\infty} g(x)dx + \int_0^t R_1(t - x)\ R(x)\ g(x)dx. \qquad (3.8.1)$$

The MTSF under preventive maintenance policy is given by

$$\text{MTSF} = \int_0^{+\infty} R_1(t)dt \qquad (3.8.2)$$

3.9 REDUNDANCIES

As defined in Chapter 1, redundancy is the existence of more than one means to perform a given task. Several types of redundancies have been defined in Chapter 1 and Chapter 2. In this section, we will define some more terms about redundancy.

Standby redundancy is one of the basic methods to increase reliability. Depending upon the state of the redundant units, the redundancy of unit falls into one of the three types: cold, hot and warm. *Cold standby* redundancy exists where the failure rate of a standby redundant unit is equal to zero, i.e., the standby unit cannot fail when not in use. In case of *warm standby* redundancy, the units in standby may fail according to some failure distribution, i.e., the failure rate of standby unit is not equal to zero. *Hot standby* redundancy is the special case of warm standby redundancy where the failure rate of unit in operation is equal to the failure rate of unit in standby.

3.10 DATA COLLECTION

Data collection is the way towards social occasion and measuring information on factors of enthusiasm, in a built-up efficient design that enables one to answer stated research questions, test theories, and assess results. The information assortment segment of research is common to all fields of study including physical and sociologies, humanities, business, and so on. While strategies shift by discipline, the emphasis on guaranteeing exact and genuine assortment continues as before.

3.11 DATA RECOVERY

In computing, data recovery is a procedure of rescuing (retrieving) inaccessible, lost, adulterated, harmed or organized information from optional stockpiling, removable media or records, when the information put away in them can't be gotten to in a typical manner. The information is regularly rescued from capacity media, for example, interior or outside hard disk drives (HDDs), solid-state drives (SSDs), USB streak drives, attractive tapes, CDs, DVDs, RAID subsystems, and other electronic gadgets. Recuperation might be required because of physical damage to the capacity gadgets or sensible harm to the file system that keeps it from being mounted by the host operating system (OS).

The most common data recovery recuperation situation includes a working system failure, breakdown of a capacity gadget, logical failure of capacity gadgets, accidental damage or erasure, and so on (commonly, on a single-drive, single-partition, single-OS system), in which case a definitive objective is basically to duplicate terrifically significant documents from the damaged media to another new drive. This can be effectively practised utilizing a live CD or DVD by booting straightforwardly from a ROM rather than the debased drive being referred to. Many live CDs or DVDs give a way to mount the system drive and reinforcement drives or removable media and to move the records from the system drive to the reinforcement media with a document chief or optical disc authoring software. Such cases can regularly be relieved by circle dividing and reliability putting away important information documents (or copies of them) on a different partition from the replaceable OS system files.

3.12 DATA ANALYSIS PROCEDURES

Data analysis is the procedure of efficiently applying statistical and/additionally consistent systems to describe and illustrate, consolidate and recap, and assess information. As per Shamoo and Resnik (2003) different investigative techniques "give a method for drawing inductive inferences from data and recognizing the signal (the wonder of enthusiasm) from the noise (measurable changes) present in the information". While information examination in subjective research can incorporate measurable strategies, commonly investigation turns into a progressing iterative procedure where information is constantly gathered and dissected at the same time. To be sure, specialists for the most part investigate for designs in perceptions through the whole information assortment stage (Savenye, Robinson, 2004). The type of the investigation is dictated by the particular subjective methodology taken (field study, ethnography content examination, oral history, life story, subtle research) and the type of the information (field notes, records, audiotape, tape).

A fundamental segment of guaranteeing information respectability is the precise and fitting investigation of research discoveries. Ill-advised measurable examinations mutilate logical discoveries, mislead causual readers (Shepard, 2002), and may adversely impact the open impression of research. Uprightness issues are similarly as pertinent to investigation of non-statistical data as well.

Considerations/issues in data analysis: There are various issues that researchers ought to be cognizant of concerning information investigation. These include:

1. Concurrently choosing information assortment strategies and fitting examination
2. Inappropriate subgroup analysis
3. Reliability and validity
4. Partitioning 'content' while investigating subjective information

5. Environmental/relevant issues
6. Drawing impartial deduction
7. Following worthy standards for disciplines
8. Lack of plainly characterized and target result estimations
9. Having the key aptitudes to separate
10. Data chronicle technique
11. Providing legit and exact examination
12. Extent of investigation
13. Inappropriate subgroup examination
14. Manner of displaying information
15. Determining measurable noteworthiness
16. Training of staff directing examinations

3.13 FAULT TREE ANALYSIS METHOD

The fault tree analysis (FTA) techniques are utilized to perform unwavering quality and security investigations of engineering systems. It was initially created to analyze the Minuteman Launch Control System in the early 1960s at Bell Laboratories. It can likewise be utilized to perform analysis of human blunder in support. The following example demonstrates the application of FTA in maintenance with respect to human error.

Example 1. Assume that the engineering system can flop because of an upkeep blunder brought about by components, for example, poor gear structure, insufficient preparation, poor workplace, utilization of lacking support manuals, or deficient work apparatuses. Two central points for poor gear configuration are oversight or no proper thought of the event of upkeep mistake. The "no conventional thought to the event of support blunder" factor can be brought about by either no necessity in plan particulars or lacking apportioned assets. Two significant variables for the poor workplace are deficient lighting and detachment. So also, two primary drivers for the utilization of inadequate upkeep manuals is inaccessibility of good support manuals or ineffectively composed support methodology. Build up a flaw tree for the top occasion "Designing framework flopped because of support blunder." The below figure shows a shortcoming tree for the model.

Example 2. Assume that the probability of event of occasions $E_1, E_2, \ldots, E_9$ appeared in figure below is 0.02. For free occasions, ascertain the likelihood of event of the top occasion T, i.e., engineering system failed due to maintenance error.

The probability of occurrence of event I_1 is

$$\begin{aligned} P(I_1) &= P(E_1) + P(E_2) - P(E_1)\,P(E_2) \\ &= 0.02 + 0.02 - (0.02 \times 0.02) \\ &= 0.0396 \end{aligned}$$

The probability of occurrence of event I_2 is given by

$$\begin{aligned} P(I_2) &= P(I_1) + P(E_7) - P(I_1)\,P(E_7) \\ &= 0.0396 + 0.02 - (0.0396 \times 0.02) \\ &= 0.0588 \end{aligned}$$

The probability of occurrence of event I_3 is

$$P(I_3) = P(E_6) + P(E_5) - P(E_6)\,P(E_5)$$
$$= 0.02 + 0.02 - (0.02 \times 0.02)$$
$$= 0.0396$$

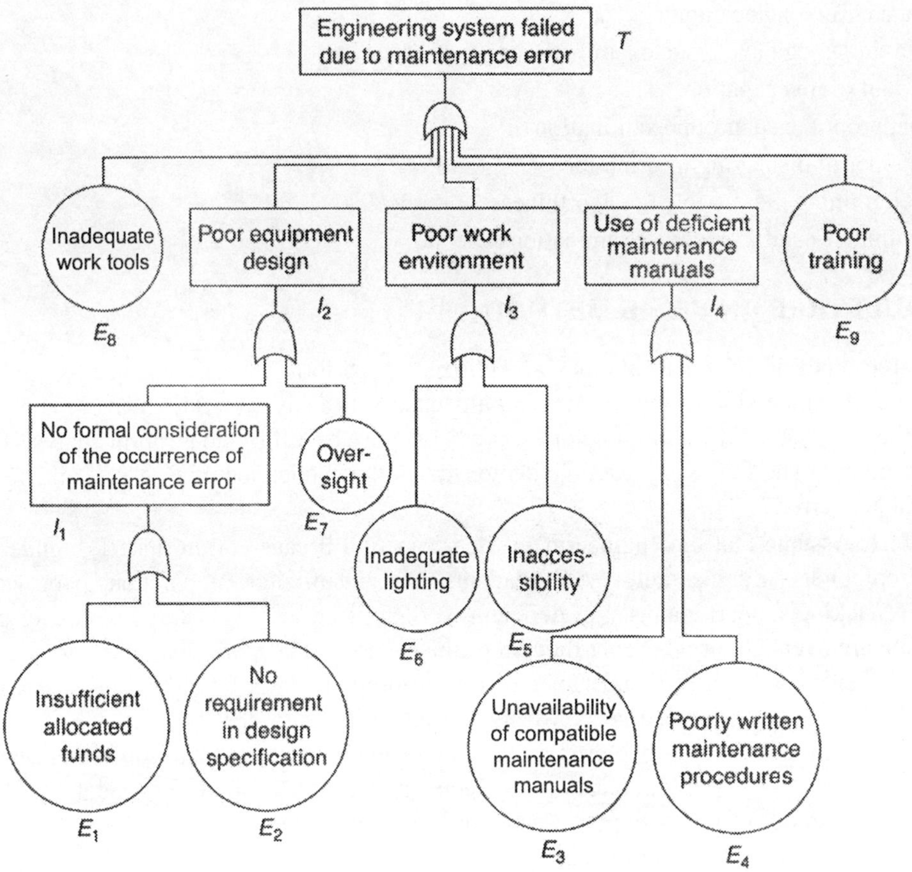

The occurrence probability of event I_4 is given by

$$P(I_4) = P(E_3) + P(E_4) - P(E_3)\,P(E_4)$$
$$= 0.02 + 0.02 - (0.02 \times 0.02)$$
$$= 0.0396$$

The top event T probability of occurrence is

$$P(T) = 1 - \{1 - P(E_8)\}\{1 - P(E_9)\}\{1 - P(I_2)\}\{1 - P(E_3)\}\{1 - P(I_4)\}$$
$$= 1 - (1 - 0.02)(1 - 0.02)(1 - 0.0588)(1 - 0.0396)(1 - 0.0396)$$
$$= 0.1663$$

There is approximately 17% chance the engineering system will fail due to maintenance error.

In some processing industries basically perception isn't sufficient; at that point fault tree analysis is finished. Fault tree analysis of a system is all around arranged method for distinguishing every single imaginable issue that may prompt system fail danger failures. The fault tree analysis gives a succinct and precise description of different and every single imaginable mix of occurrence within the system. It helps in recognizing and assessing the basic parts and flaw focuses and every conceivable blunder. In a fault tree analysis system will undoubtedly carry on as a perfect. This kind of investigation relies on its comprising parts. Coming up next are three principal segments of fault tree analysis of any system given as following:

(a) Gates that represent the outcomes
(b) Events that represent the input to the gates
(c) Out-of-tolerance failure of a system element failure due to excessive operational or environmental stress.

The following diagram shows the flowchart symbols that are used in fault free analysis of system:

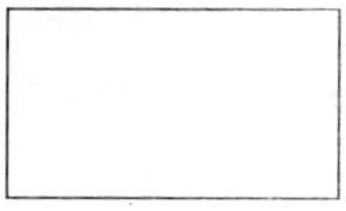

Rectangle is used for a fault or undesired event caused by other causes through logical gates.

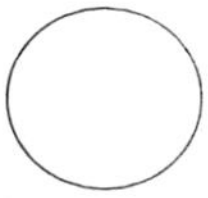

Circle is used for a primary failure or basic fault that does not require further development.

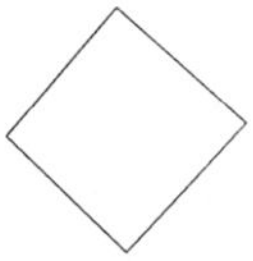

Diamond represents a secondary failure or undesired event.

OR gate signifies that a failure if any input fails.

Triangle represents the transfer event.

3.14 RELIABILITY TESTING

1. **Life Testing:** Reliability or life testing includes evaluating the expected durability after some time of a thing. This might be a whole system, an item, or an individual segment. We may likewise concentrate on a component of a segment, such has a material property.
2. **Requirements:** Reliability is a significant non-useful prerequisite for most software products so software requirements specification (SRSs) ought to contain a dependability necessity, and most do. In any case, one of our markers of the nature of a 'good' necessity is that it is testable, so it is sensible to ask whether the reliability requirements prerequisites in an SRS are testable as composed.

   Tests for useful prerequisites are usually binary. The item either supports the necessity or it doesn't and in this manner either passes or fails the assessment.

Testing for reliability isn't so direct. It might be hard to state, in a twofold way, that the item does or doesn't meet the reliability necessities.

For the clients of a system, it is the dependability of the system, in general, that is important however for experts and analysers it is critical to isolate the product prerequisites from the equipment necessities as there are some significant differences. There is no doubt that any bit of equipment will in the long run wear out and flop so equipment of item unwavering quality can be expressed and estimated as far as time to disappointment or time between disappointments. The arbitrary segment of the unwavering quality conjecture originates from the way that indistinguishable bits of equipment will work for different time spans before disappointment.

The software reliability can be an increasingly troublesome idea to get a handle on. A product item will flop under specific conditions, with specific information sources, and given similar data sources and conditions will bomb each time until the reason for the disappointment is rectified. Along these lines, the reliability of a product item is increasingly about the irregular disclosure of flaws coming about because of different contributions with the framework in different states. In spite of the fact that for little and straightforward frameworks it might be hypothetically conceivable to test each blend of state and contribution, for an arrangement of any size and intricacy but this isn't practical. The irregular idea of the deficiency disclosure process implies that we should utilize probabilities when we allude to programming reliability requirements and testing.

Reliability test results ought to be expressed regarding estimations. Estimations are taken during testing when we are gathering and analyzing information about the presentation of the software. As it were, we are following the event of failure during testing. In any case, a reliability prerequisite is an expectation or estimate of the presentation of the item later on. This means while we can gauge the number of failures every hour or per exchange in a framework test condition we can just give a gauge of the genuine presentation of the system in a future creation condition. Reliability is generally characterized as the likelihood that an item will work without disappointment for a predefined number of employments (exchanges) or for a predetermined timeframe. To be genuinely testable a prerequisite for software reliability ought to be expressed as an estimate and the test outcomes ought to show the certainty level related to the figure that the item will meet the necessities.

3. **Methods:** Here are the four most common ways of measuring reliability for any empirical method or metric: inter-rater reliability, test-retest reliability, parallel forms reliability, internal consistency reliability.

   Inter-Rater Reliability: The extent to which raters or observers react a similar path to a given marvel is one proportion of reliability. Where there's judgment there's disagreement.

   Indeed, even exceptionally prepared specialists differ among themselves while watching a similar marvel. Kappa and the relationship coefficient are two regular proportions of inter-rater reliability. A few models include:

   - Evaluators identifying interface problems
   - Experts rating the severity of a problem

   Test-Retest Reliability: Do clients give a similar arrangement of reactions when nothing about their experience or their frame of mind has changed? You don't need your estimation framework to change when every single other thing is static.

Parallel Forms Reliability: Getting the equivalent or fundamentally the same as results from slight minor departure from the inquiry or assessment strategy likewise builds up unwavering quality. One approach to accomplish this is to have, say, 20 things that measure one develop (fulfilment, dedication, ease of use) and to regulate 10 of the things to one gathering and the other 10 to another gathering, and afterward relate the outcomes. You're searching for high connections and no precise distinction in scores between the gatherings.

Internal Consistency Reliability: This is by far the most commonly utilized proportion of reliability in applied settings. It's well known on the grounds that it's the least demanding to process utilizing software—it requires just one example of information to gauge the interior consistency reliability. This proportion of reliability is depicted regularly utilizing Cronbach's alpha (at times called coefficient alpha).

Test-Planning: A reliability test plan is the elevated level arrangement that gets out the entirety of the reliability testing that will be performed on an item, including Design Verification Tests (DVTs), Highly Accelerated Life Tests (HALTs), Reliability Demonstration Tests (RDTs), Accelerated Life Tests (ALTs), and On-Going Reliability Tests (ORTs).

4. **Data Reporting System:** A quality data reporting system is a system to collect, investigate, and correct all imperfections and failures occurring during production and testing of a things, just as to assess and feedback the comparing and reliability data. The framework is generally computer aided.

3.15 RISK ANALYSIS AND MANAGEMENT

The primary reason for a risk management programme is to settle on the right choices at the key stages of the programme. It applies to all new product development and administration advancement. Common risk regions include execution, planning, generation, and resources. In the event that we have field encounters, at that point hazard, the board program is exceptionally simple and powerful. Dangers are for the most part characterized into two : one is transformative and the other one is progressive. Since dangers identified with parts and subsystems are amplified at the framework level, it is significant to be mindful of issues right off the bat in the programme and might be applied to determine these dangers.

Out of these techniques preventive and corrective upkeep increment, the existence time of machine parts through the standby system or redundancy in units increment the accessibility of the entire framework and system parameters. Working on any system may cease if any of the segments stop working because of software or equipment breakdown. The failure of an untrustworthy framework can prompt a catastrophe and overwhelming misfortunes. To reduce the hazards because of disappointments that can be halted by just routine inspection and preventive maintenance. We have a number of assets so we should use the resource efficiently and proficiently. It may be accomplished with dependable assets and appropriate structuring. We have a number of assets so we should use the resource efficiently and proficiently. It may be accomplished with dependable assets and appropriate structuring.

3.16 STRUCTURAL REDUNDANCY

Redundancy is a strategy to accomplish higher reliability. The presence or accessibility of more than one method for playing out the given undertaking is called redundancy and the methods are called excess units or redundants. There are two sorts of redundancies:

(a) **Active (or parallel) Redundancy:** It is wherein all the redundants are working at the same time as opposed to being turned on when required. If a unit is exceptionally basic, at that point beyond what one unit can be placed in active redundancy, to repay the unreliability of the basic unit, e.g., electric supply in I.C.U.

(b) **Stand-by Redundancy:** It is wherein the other elective methods (called the auxiliary units), stay inactive until the essential unit fizzles. Contingent on the condition of the redundant units, the excess is isolated into the accompanying three classes:

  (a) **Warm Redundancy:** It exists where the failure rate of the stand-by redundant unit is not equal to zero and it fails according to some failure distribution.

  (b) **Hot Redundancy:** It is a special and extreme case of a warm stand-by redundancy. It exists where the failure rate of the unit in hot stand-by mode is equal to that of the primary unit. Hot Redundancy is also called an 'Idling Standby' redundancy.

  (c) **Cold Redundancy:** It exists where the disappointment rate of the remain by the excess unit is zero, for example, the standby unit doesn't bomb when it isn't being used. Cold redundancy is likewise called an 'under no pressure redundancy', e.g., hard products and extra pieces of vehicles.

In many cases sliding stand-by, sliding stand-by with automatic fault locator are used to make the use of redundant easy and fault tree.

3.17 EXTREMAL DISTRIBUTIONS

Let X be the largest of n independent random variables, $Y_1, Y_2, \ldots, Y_n$. Then

$$F_x(x) = P\{Y_i \leq x\} = P\{Y_1 \leq x\}\, P\{Y_2 \leq x\} \ldots P\{Y_n \leq x\} = F_{y_1}(x)\, F_{y_2}(x) \ldots F_{y_n}(x) \tag{1}$$

If Y_i's are identically distributed, it follows that

$$F_x(x) = F_y(x))^n, \tag{2}$$

and the corresponding probability density function is given as

$$f_x(x) = n(F_y(x))^{n-1} f_y(x). \tag{3}$$

Gumbel proposed several asymptotic distributions for the extreme values of a random variable. These are described in this section.

(i) **Type 1: Extremal (Largest) Distribution (Gumbel Distribution)**

The distribution of X, the largest of many independent random variables Y (with an exponential upper tail distribution, $F_Y(y) = 1 - \exp\{-h(y)\}$) is given as

Probability Density

$$f_x(x) = \alpha \exp\{-\alpha(x-u) - e^{-\alpha(x-u)}\} - \infty < x < +\infty. \tag{4}$$

Probability Distribution

$$F_x(x) = \exp\{-e^{-\alpha(x-u)}\} \tag{5}$$

Parameters u and α are related to mean and variance of X. These are

$$\mu_x = u + 0.5772/\alpha, \tag{6}$$

$$\sigma^2_x = \pi^2/6\alpha^2 \tag{7}$$

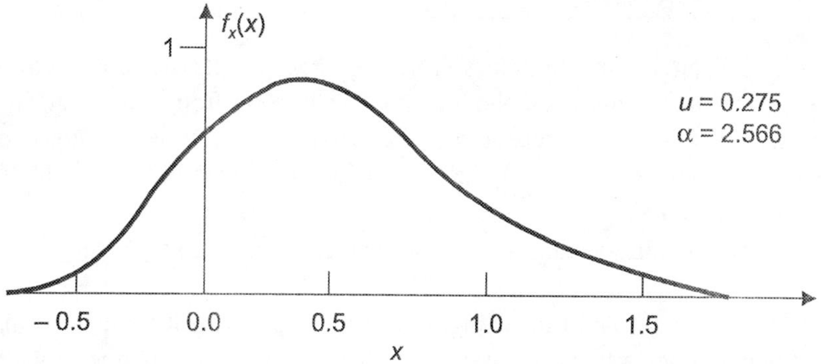

(ii) Type 2: Extremal (Smallest) Distribution (Gumbel Distribution)

The density and distribution functions of z, the smallest of many independent variables (with an exponential type lower tail distribution) are given as

Probability Density

$$f_z(z) = \alpha \exp\{\alpha(z-u) - e^{\alpha(z-u)}\} \quad -\infty < z < +\infty. \tag{8}$$

Probability Distribution

$$F_z(z) = 1 - \exp\{-e^{\alpha(y-u)}\} \tag{9}$$

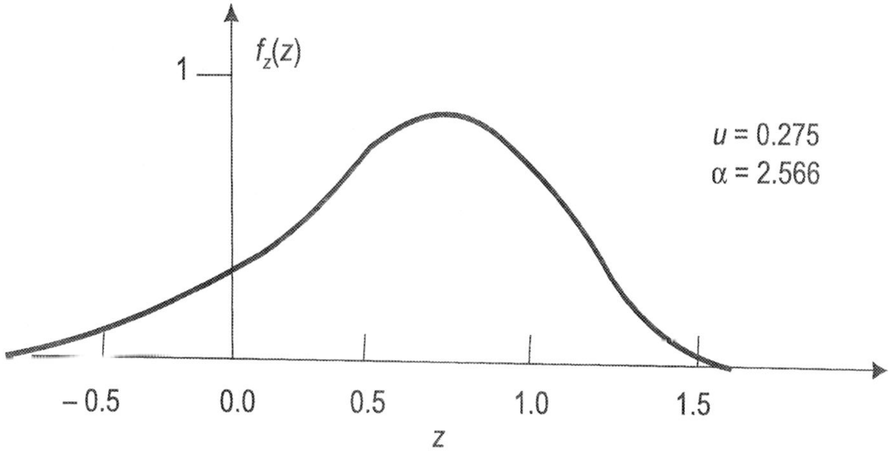

Parameters u and α are related to mean and variance of X. These are

$$\mu_z = u - 0.5772/\alpha, \tag{10}$$

$$\sigma^2_z = \pi^2/6\alpha^2 \tag{11}$$

3.18 ACCELERATED TESTS

There are three basic types of accelerated tests:

- accelerated life tests
- reliability enhancement tests, and
- environmental stress screening.

As shown in Fig. 3.1, environmental stress screening (ESS) is planned for uncovering new-born child mortality failures which would somehow or another happen from the get-go in the life of the item. Reliability enhancement testing leads to discover early disappointments identified with the item configuration, but on the other hand, it is utilized to decide the vigour of the item as for the arbitrary failures of the helpful life period.

The purpose of accelerated life testing is to find out how, when, and why wear out failures occur in the product.

Figure 3.1 shows how accelerated life testing, reliability improvement testing, and natural pressure screening fit into the item plan, advancement, and assembling cycle. The online forms have appear as square shapes. Online forms are pieces of the item structure and generation cycle. They are directed with tests of the genuine item. Disconnected procedures are those which are not part of the structure and creation cycle. They are typically not directed on real item tests, yet on nonexclusive examples speaking to the materials, parts, or procedures used to make the item. Two out of the three sorts of quickened tests (reliability enhancement testing and ESS) appeared above our online forms and the third (accelerated life testing) is an offline process.

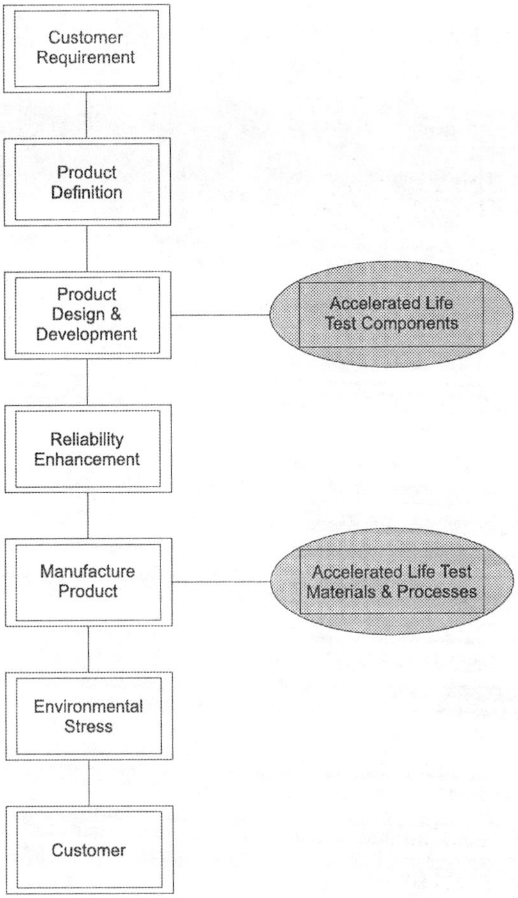

Fig. 3.1 A take from paper on "Accelerated testing for product reliability assurance" by Lioyd W. Coundra

3.19 RELIABILITY ENHANCEMENT TESTING (RET)

The reason for RET is to decide the sorts and levels of environment stresses which cause failures in an item, given that there are no imperfections in the materials and segments utilized in assembling it. In this sense, RET is a kind of assessment test for the item configuration, forms. Since RET is not coordinated towards discovering baby deserts, the example size can be extremely little. The perfect time to direct RET is towards the finish of the structure cycle, when the normal plan, materials, segments, and assembling forms are accessible; and the generation has not yet started. RET is not a capability test since its motivation is to discover shaky areas and right them before creation starts.

RET is typically led by applying the expected environmental and working burdens (either separately, consecutively, or at the same time) at first at low levels, and afterwards expanding them in ventures until one of the three things happen:

1. all sample fail
2. stress levels are reached which are well over those normal in administration, or
3. insignificant disappointments happen.

An important benefit of RET is to overview and decide the item's upper and lower district limits. This is valuable for deciding the heartiness of the item structure. The ESS levels ought not to be changed to reflect theories confined as it could pointless to excess the item. Test, Analyze and Fix (TAAF) is critical to an all-around created RET.

Accelerated Life Tests: Accelerated life tests (ALTs) are directed on segments, materials, and assembling forms (e.g., formation of a plated through hole) to decide their valuable life in the necessary item application. Their motivation is not to uncover abandons, however, to recognize and measure the disappointments and disappointment instruments that cause items to wear out towards the finish of their helpful life. Along these lines, quickened life tests should keep going long enough to make the examples under test flop by wear out. The test time may regularly shift from half a month to a couple of months.

Also, separate quickened life tests are led for every potential destroy system, since the anxieties which produce disappointments are diverse for every component. Customary quickened life test techniques include the use of single worries (for instance, just sine vibration or just temperature cycle). Be that, as it may, is progressively felt numerous potential disappointment instruments result from, or are quickened by, blends of natural conditions (for example, irregular vibration + high temperature).Accelerated life tests are generally called capability tests, since they are utilized to qualify a segment, materials, or procedures for all applications. A progressively sensible methodology is to gather quickened life test information for a given arrangements of test conditions, and afterwards to extrapolate these outcomes for every application.

Accelerated life tests normally take too long to even think about being directed online, as a feature of any item advancement cycle. In this way, they should be conducted off-line, well before the segments, materials, or procedures are required for a given application. Thus, ALT is generally directed, utilizing nonexclusive examples that speak to the materials, parts, and procedures utilized for an assortment of items.

The advantages of ALT are:

1. The capacity to gauge valuable existence of the item.
2. It additionally permits fashioner/producer of the items to distinguish and improve and control the basic segments, materials, and procedures so that the last item is hearty and adult.

Potential disappointment systems must be known; and the pressure condition of the item should be comprehended. Explicit quickening models must be accessible for every disappointment instrument; and the outcomes must be deciphered appropriately.

Environmental Stress Screening: Environmental stress screening (ESS) is helpful in limiting the early failures of the fabricated items by screening inert deformities. ESS is one of the most utilized reliability tests. Its motivation is to hasten dormant deformities, which are discernible just with the use of pressure. The imperfections preferably are those brought into the item during manufacturing, since design-related deformities ought to have been recognized and wiped out by reliability upgraded testing during the structure stages. ESS is effective only for a product with an infant mortality region, which is indicated by decreasing initial failure rate in Fig. 3.1. ESS ought to be founded on a comprehension of the potential kinds of idle deformities in the item, the failures of instruments, and the anxieties that cause them. ESS conditions are set up to encourage those imperfections and the information is utilized to decide their causes and taking preventive activities.

3.20 RELIABILITY AND ITS EQUIREMENT

Reliability is the capacity of a thing to play out a necessary capacity under-expressed conditions for an expressed timeframe. The following factors make it practically basic that an item experiences a reliability improvement programme.

1. Rapid advancement of new materials, segments, and procedures.
2. The complexity of the item being produced.
3. Need to fulfil client needs, for example, accessibility, security, and guarantee.
4. To meet wanted MTBF, item costs.
5. Compliance with the governmental guidelines.

3.21 RELIABILITY VS QUALITY CONTROL

The inspectors' view of reliability is that the item is surveyed against a particular or set of characteristics and having passed is conveyed to the client. Be that as it may, this methodology gives no proportion of value over some undefined timeframe. In this way, the requirement for a period based idea of value is created. The inspectors' concept is not time-subordinate. The item either breezes through a given assessment or it falls flat. Then again, unwavering quality is normally worried about failure in the time-space. This qualification denotes the contrast between conventional quality control and the way to deal with reliability. Quality control (QC) of assembling forms clearly makes a basic commitment to the unwavering quality of an item. QC can be considered a necessary piece of a general dependability programme. Reliability is commonly worried about failures during the life of an item.

EXERCISE

1. What is MTBF?
2. Define reliability, availability and maintainability.
3. Define maintainability and availability and compare with reliability.
4. What is reliability? Explain in detail the different techniques employed for improving the reliability.

5. Define MTTF and MTBF.
6. State the differences between reliability and quality.
7. Compare unit vs component redundancy with sketches.
8. Explain the fault tree analysis method.
9. How effective is preventive maintenance in product life cycle?
10. An electronic system has a MTBF of 1000 hours and a MTTR of 40 hours. Determine its availability.
11. Justify the use of exponential distribution in reliability evaluation of non-reparable system.
12. Explain inherent availability.
13. Explain the term MTTF. Also derive it with respect to reliability and CDF.
14. What is mixed redundancy?
15. Define reliability. Explain the reliability function.
16. What is the difference between quality assurance, quality control, and testing?
17. Obtain state space diagram for a two component system with condition that simultaneous failures do not occur.
18. Which operational phases are to be included in the analysis (e.g., start-up, steady state, maintenance, disposal)?
19. What are the objectives of the quality control?
20. Define the following terms:
    (a) Reliability
    (b) Accelerated test
    (c) Failure density
    (d) Extrenal distributions
    (e) Mean time to system failure (MTSF)
    (f) Mean time to repair (MTTR)
    (g) Mean time between failure (MTBF)

4

Chapter

System Modeling and Solution Techniques

In our daily life, we face several problems connected with production, administration, social development, employment, etc. If we have the clear-cut idea about the problem and the effective conditions, a solution to the problem is possible using reliability technology. In this chapter, we discuss the main points necessary to have a solution of the problem under consideration.

4.1 SYSTEM DESCRIPTION

Complete description of a system is necessary to prepare a mathematical model of the system. The working, dependence or independence of various units of the system (mode of failure and repair), common cause if any and human effect on the working of the system should be clearly explained or in other words all knowledge about the system structure and its working should be completely known. The initial and boundary conditions must be known. The next important part of system description is to prepare a transition diagram of the system. The transition diagram, if prepared correctly, gives us good knowledge about the system and helps in preparing mathematical model of the system.

For example, consider a system having two similar units with the assumptions: If one unit fails, the system works in reduced capacity while failure of both units causes complete failure of the system. Also both units may fail simultaneously. If there is only one repairman and the system is repairable, then the transition diagram is as given in Fig. 4.1.

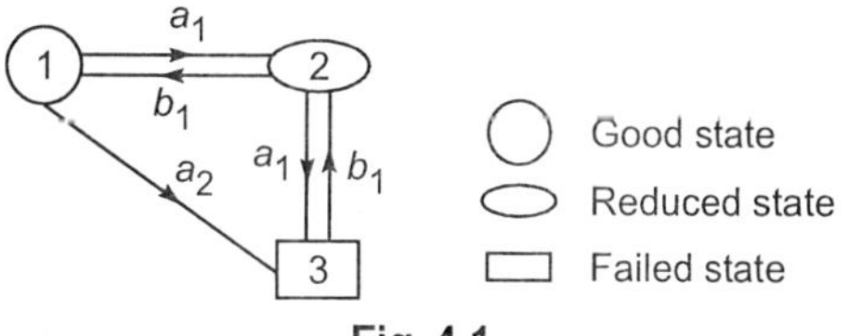

Fig. 4.1

where, a_1, a_2 are the failure rates and b_1 is the repair rate (i.e., transition rates). Initially, (i.e., at $t = 0$) the system is fully operable and is in state 1. Failures and repairs are independent and repaired system behaves like new system.

4.2 MATHEMATICAL MODELING

In Section 4.1 above, the system is described completely under certain assumptions about failure and repair modes. Following the assumptions, transition diagram Fig. 4.1 is drawn making use of probability laws and considering all possibilities the state equations are formed. Also using *mnemonic* rule to the transition diagram, differential equations governing the system may be written. According to mnemonic rule *derivative of probability at some state is equal to the sum of all probability flows which come from other states to the given state minus the sum of all the probability flows which go out from the given state to other states.*

These state equations are solved to obtain state probabilities using the initial and boundary conditions. The differential equations are solved with the help of Laplace transforms or by direct integration or numerically. For steady state behaviour of the system the derivatives of state probabilities are taken equal to zero as time (t) tends to infinity and the probabilities are independent of time. The resulting equations may be solved recursively to obtain state probabilities and hence the reliability of the system. Formation and solution of equations will be discussed in Sections 4.11 and 4.12. In the following sections, some techniques which are generally used to evaluate the reliability of complex systems, are discussed.

4.3 DECOMPOSITION METHOD

In this method a keystone component A is chosen. Using total probability law, the reliability R may be expressed in terms of A as

$$R = P(\text{System good}/A)\ P(A) + P(\text{System good}/\overline{A})\ P(\overline{A}). \tag{4.3.1}$$

In this expression first term stands for the probability that the system is good given that A is functioning and the second term stands for the probability that the system is functioning given that A is not functioning ($\overline{A}$).

For example, consider a complex system shown in Fig. 4.2 (Elsayed)

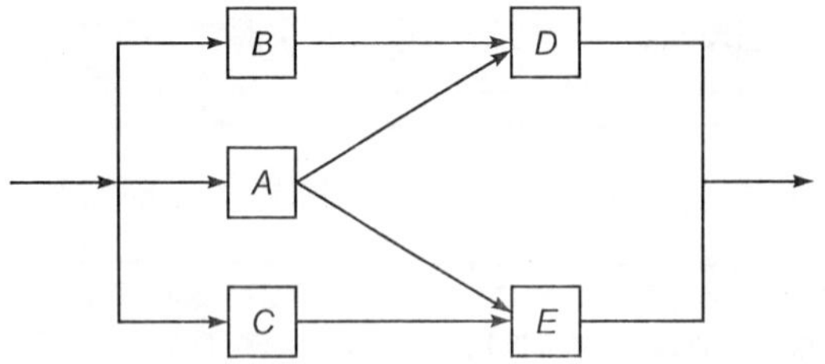

Fig. 4.2

Taking A as Keystone functioning component,

$$P(\text{System good}/A) = P(D) + P(E) - P(D)P(E) \tag{4.3.2}$$

$$P(\text{System good}/\overline{A}) = P(B)P(D) + P(C)P(E) - P(B)P(D)P(C)P(E) \tag{4.3.3}$$

Equation (4.3.2) is for the path starting from A and passing through D and E components, i.e., A

is functioning. Equation (4.3.3) is for the path when A is not functioning but the system is functioning through B, D, and C, E working in parallel as shown in the block diagrams.

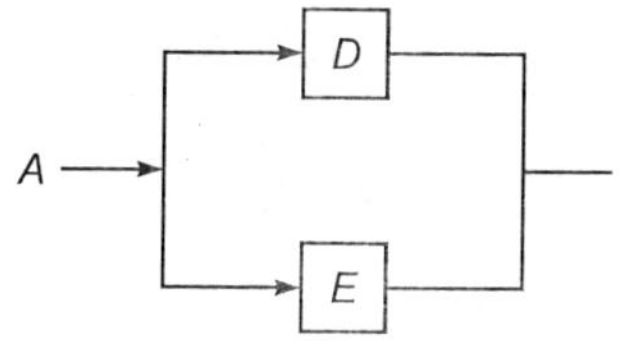

A is functioning

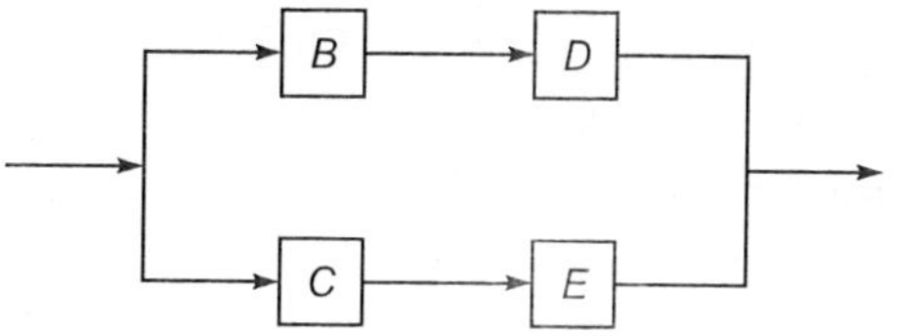

A is failed

Using (4.3.2) and (4.3.3) in (4.3.1), we have

$$R = [P(D) + P(E) - P(D)P(E)]\,P(A) + [P(B)P(D) + P(C)P(E) - P(B)P(D)\;P(C)P(E)]P(\overline{A})$$

where, $P(\overline{A}) = 1 - P(A)$.

4.4 CUT-SET METHOD

A cut-set is a set of components that interrupts all connections between the input and the output ends when removed from the reliability block diagram. A cut-set is called minimum if it contains no other cut-set (*s*) within it.

For example, consider the system shown in Fig. 4.3 (Elsayed).

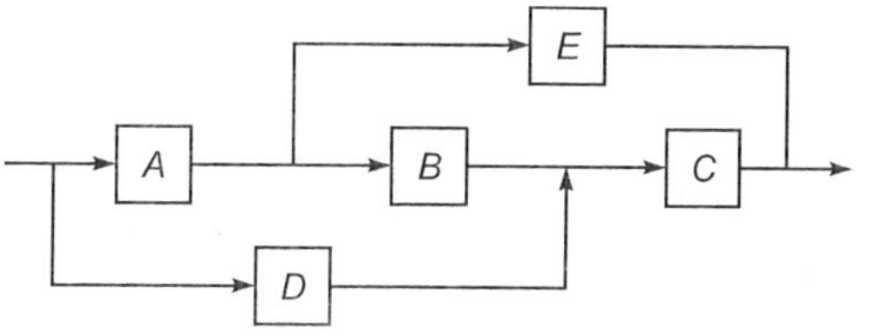

Fig. 4.3

The minimum cut-sets for Fig. 4.3 are $\overline{A}\,\overline{D}$, $\overline{E}\,\overline{C}$, $\overline{A}\,\overline{C}$ and $\overline{B}\,\overline{E}\,\overline{D}$. The reliability of the system is $R = 1 - P\{\overline{A}\,\overline{D} \cup \overline{E}\,\overline{C} \cup \overline{A}\,\overline{C} \cup \overline{B}\,\overline{E}\,\overline{D}\}$

where, $P\{\ldots\}$ is the probability of union of $\overline{A}\,\overline{D}$, $\overline{E}\,\overline{C}$, $\overline{A}\,\overline{C}$, $\overline{B}\,\overline{E}\,\overline{D}$ cut-sets. Assuming independence of probabilities,

$$\begin{aligned} R = 1 - [&P(\overline{A}\,\overline{D}) + P(\overline{E}\,\overline{C}) + P(AC) + P(\overline{E}\,\overline{B}\,\overline{D}) - P(\overline{A}\,\overline{D}\,\overline{E}\,\overline{C}) \\ &- P(\overline{A}\,\overline{D}\,\overline{C}) - P(\overline{A}\,\overline{D}\,\overline{B}\,\overline{E}) - P(\overline{E}\,\overline{C}\,\overline{A}) - P(\overline{E}\,\overline{C}\,\overline{B}\,\overline{D}) - P(\overline{A}\,\overline{C}\,\overline{B}\,\overline{E}\,\overline{D}) \\ &+ P(\overline{A}\,\overline{D}\,\overline{E}\,\overline{C}) + P(\overline{A}\,\overline{D}\,\overline{C}\,\overline{B}\,\overline{E}) + P(\overline{A}\,\overline{D}\,\overline{C}\,\overline{B}\,\overline{E}) + P(\overline{E}\,\overline{C}\,\overline{A}\,\overline{B}\,\overline{D}) \\ &- P(\overline{A}\,\overline{B}\,\overline{C}\,\overline{D}\,\overline{E}))] \end{aligned}$$

4.5 TIE-SET METHOD

A tie-set is a complete path through the reliability block diagram. A tie-set is called minimum if it contains no other tie-set (*s*) within it.

As an example, consider the block diagram of the system shown in Fig. 4.3. The minimum tie-sets of the system are *AE*, *DC* and *ABC*. The reliability of the system is given by

$$
\begin{aligned}
R = P\{AE \cup DC \cup ABC\} &= P(AE) + P(DC) + P(ABC) \\
&- P(AEDC) - P(AEBC) - P(DCAB) + P(AEDCB). \\
&= P(A)P(E) + P(D)P(C) + P(A)P(B)P(C) \\
&- P(A)P(E)P(D)P(C) - P(A)P(E)P(B)P(C) - P(D)P(C)P(A)P(B) \\
&+ P(A)P(E)P(D)P(C)P(B) \text{ if events are independent.}
\end{aligned}
$$

4.6 EVENT SPACE METHOD

The event space method is based on listing all the possible occurrences of the events. The reliability of the system is the probability of the union of the occurred events. As an example, again consider the system shown in Fig. 4.3. The possible events occurred are:

(i) No failure: $ABCDE = e_1$

(ii) One failure: $\bar{A}BCDE = e_2$, $A\bar{B}CDE = e_3$, $AB\bar{C}DE = e_4$, $ABC\bar{D}E = e_5$, $ABCD\bar{E} = e_6$.

(iii) Two failures: $\bar{A}\bar{B}CDE = e_7$, $\bar{A}B\bar{C}DE = e_8$, $\bar{A}BC\bar{D}E = e_9$ $\bar{A}BCD\bar{E} = e_{10}$, $A\bar{B}\bar{C}DE = e_{11}$, $A\bar{B}C\bar{D}E = e_{12}$ $A\bar{B}CD\bar{E} = e_{13}$, $AB\bar{C}\bar{D}E = e_{14}$, $AB\bar{C}D\bar{E} = e_{15}$ $ABC\bar{D}\bar{E} = e_{16}$.

(iv) Three failures: $AB\bar{C}\bar{D}\bar{E} = e_{17}$, $A\bar{B}C\bar{D}\bar{E} = e_{18}$, $A\bar{B}\bar{C}D\bar{E} = e_{19}$, $A\bar{B}\bar{C}\bar{D}E = e_{20}$, $\bar{A}BC\bar{D}\bar{E} = e_{21}$, $\bar{A}B\bar{C}D\bar{E} = e_{22}$ $\bar{A}B\bar{C}\bar{D}E = e_{23}$, $\bar{A}\bar{B}CD\bar{E} = e_{24}$, $\bar{A}\bar{B}C\bar{D}E = e_{25}$ $\bar{A}\bar{B}\bar{C}DE = e_{26}$.

(v) Four failures: $A\bar{B}\bar{C}\bar{D}\bar{E} = e_{27}$, $\bar{A}B\bar{C}\bar{D}\bar{E} = e_{28}$, $\bar{A}\bar{B}C\bar{D}\bar{E} = e_{29}$ $\bar{A}\bar{B}\bar{C}D\bar{E} = e_{30}$, $\bar{A}\bar{B}\bar{C}\bar{D}E = e_{31}$

(vi) Five failures: $\bar{A}\bar{B}\bar{C}\bar{D}\bar{E} = e_{32}$.

In above occurrences, events e_8, e_9, e_{15}, e_{17}, e_{18}, e_{19}, e_{21}, e_{22}, e_{23}, e_{25}, e_{26}, e_{27}, e_{28}, e_{29}, e_{30}, e_{31} and e_{32} are the failed events of the system. Therefore, reliability of the system is

$$
\begin{aligned}
R &= P(e_1 + e_2 + e_3 + e_4 + e_5 + e_6 + e_7 + e_{10} + e_{11} + e_{12} + e_{13} + e_{14} + e_{16} + e_{20} + e_{24}). \\
&= \sum_{i=1}^{24} P(e_i),\ i \neq 8, 9, 15, 17, 18, 19, 21, 22, 23 \text{ if events are independent.}
\end{aligned}
$$

4.7 PATH-TRACING METHOD

While using path-tracing method, we find the successful paths. The reliability of the system is the probability of the union of the successful paths. Applying the method to the system shown in Fig. 4.3, we observe that the successful paths are *AE*, *DC* and *ABC*. Thus, the reliability of the system is:

$$R = P\{AE \cup DC \cup ABC\}$$
$$= P(AE) + P(DC) + P(ABC) - P(AEDC) - P(AEBC) - P(DCAB) + P(ABCDE).$$

4.8 BOOLEAN TABLE METHOD

In this method, a Boolean truth table is constructed for the system. The reliability of the system is equal to the sum of all the probabilities of functioning states. Referring to Fig. 4.3, we have the following Boolean truth table:

A	B	C	D	E	System State	Probability (Assuming independence of events)
1	1	1	1	1	1	$P(A)P(B)P(C)P(D)P(E) = p_1$
1	1	1	1	0	1	$P(A)P(B)P(C)P(D)P(\bar{E}) = p_2$
1	1	1	0	1	1	$P(A)P(B)P(C)P(\bar{D})P(E) = p_3$
1	1	0	1	1	1	$P(A)P(B)P(\bar{C})P(D)P(E) = p_4$
1	1	1	0	0	1	$P(A)P(B)P(C)P(\bar{D})P(\bar{E}) = p_5$
1	1	0	1	0	0	$P(A)P(B)P(\bar{C})P(D)P(\bar{E}) = p_6$
1	1	0	0	1	1	$P(A)P(B)P(\bar{C})P(\bar{D})P(E) = p_7$
1	1	0	0	0	0	$P(A)P(B)P(\bar{C})P(\bar{D})P(\bar{E}) = p_8$
1	0	1	1	1	1	$P(A)P(\bar{B})P(C)P(D)P(E) = p_9$
1	0	1	1	0	1	$P(A)P(\bar{B})P(C)P(D)P(\bar{E}) = p_{10}$
1	0	1	0	1	1	$P(A)P(\bar{B})P(C)P(\bar{D})P(E) = p_{11}$
1	0	1	0	0	0	$P(A)P(\bar{B})P(C)P(\bar{D})P(\bar{E}) = p_{12}$
1	0	0	1	1	1	$P(A)P(\bar{B})P(\bar{C})P(D)P(E) = p_{13}$
1	0	0	1	0	0	$P(A)P(\bar{B})P(\bar{C})P(D)P(\bar{E}) = p_{14}$
0	1	1	1	1	1	$P(\bar{A})P(B)P(C)P(D)P(E) = p_{15}$
0	1	1	1	0	1	$P(\bar{A})P(B)P(C)P(D)P(\bar{E}) = p_{16}$
0	1	1	0	1	0	$P(\bar{A})P(B)P(C)P(\bar{D})P(E) = p_{17}$
0	1	1	0	0	0	$P(\bar{A})P(B)P(C)P(\bar{D})P(\bar{E}) = p_{18}$
0	1	0	1	1	0	$P(\bar{A})P(B)P(\bar{C})P(D)P(E) = p_{19}$
0	1	0	1	0	0	$P(\bar{A})P(B)P(\bar{C})P(D)P(\bar{E}) = p_{20}$
0	1	0	0	1	0	$P(\bar{A})P(B)P(\bar{C})P(\bar{D})P(E) = p_{21}$
0	1	0	0	0	0	$P(\bar{A})P(B)P(\bar{C})P(\bar{D})P(\bar{E}) = p_{22}$
0	0	1	1	1	1	$P(\bar{A})P(\bar{B})P(C)P(D)P(E) = p_{23}$
0	0	1	1	0	1	$P(\bar{A})P(\bar{B})P(C)P(D)P(\bar{E}) = p_{24}$
0	0	1	0	1	0	$P(\bar{A})P(\bar{B})P(C)P(\bar{D})P(E) = p_{25}$
0	0	1	0	0	0	$P(\bar{A})P(\bar{B})P(C)P(\bar{D})P(\bar{E}) = p_{26}$
0	0	0	1	1	0	$P(\bar{A})P(\bar{B})P(\bar{C})P(D)P(E) = p_{27}$
0	0	0	1	0	0	$P(\bar{A})P(\bar{B})P(\bar{C})P(D)P(\bar{E}) = p_{28}$
0	0	0	0	1	0	$P(\bar{A})P(\bar{B})P(\bar{C})P(\bar{D})P(E) = p_{29}$
0	0	0	0	0	0	$P(\bar{A})P(\bar{B})P(\bar{C})P(\bar{D})P(\bar{E}) = p_{30}$

Assuming independent probabilities, the reliability is given by

$$R = \sum_{i=1}^{5} p_i + p_7 + p_9 + p_{10} + p_{11} + p_{13} + p_{15} + p_{16} + p_{23} + p_{24}$$

4.9 REDUCTION METHOD

The method involves the construction of truth table first. Then rows resulting in the successful functioning of the system are listed in first column. Product terms are formed by comparison for those terms in first column that differ by an inverse letter. Once a term is used in a comparison, it is eliminated from all further comparisons ensuring that all remaining terms are still mutually exclusive. It is repeated until no further comparisons are possible. The union of all the states that cannot be further compared gives the reliability of the system. From Section 4.8 above, the successful states are listed below in column first. Then other columns are prepared as follows:

Reduction Table

Column First	Column Second	Column Third
$ABCDE$	$ABCD$	ACD (from $ABCD$, $A\bar{B}CD$)
$ABCD\bar{E}$		
$ABC\bar{D}E$	ABE	AE (from ABE, $A\bar{B}E$)
$AB\bar{C}DE$		
$ABC\bar{D}\bar{E}$	$AB\bar{D}$	
$AB\bar{C}\bar{D}E$		
$A\bar{B}CDE$	$A\bar{B}CD$	
$A\bar{B}CD\bar{E}$		
$A\bar{B}C\bar{D}E$	$A\bar{B}E$	
$A\bar{B}\bar{C}DE$		
$\bar{A}BCDE$	$\bar{A}BCD$	$\bar{A}CD$ (from $\bar{A}BCD$, $\bar{A}\bar{B}CD$)
$\bar{A}BCD\bar{E}$		
$\bar{A}\bar{B}CDE$	$\bar{A}\bar{B}CD$	
$\bar{A}\bar{B}CD\bar{E}$		

The reliability of the system is given by

$$\begin{aligned} R &= P\{ACD \cup AE \cup AB\overline{D} \cup \overline{A}CD\} \\ &= P(ACD) + P(AE) + P(AB\overline{D}) + P(\overline{A}CD)\text{s} \end{aligned}$$

4.10 RENEWAL THEORETIC APPROACH

The system life cycle can be described by a sequence of up and down states, i.e., the system is operating, fails and is repaired, returns to an operating or operable state, fails again after some random time of operation and is repaired, etc. Such a cycle of up and down states constitutes what is called a renewal process. A renewal process is a sequence of independent and non-negative random variables $\{V_1, V_2, \ldots, V_n, \ldots\}$, which are not all zero with probability one. Let x_1 denotes the time from initial system operation to first failure and y_1 denotes the time from first failure to next system operation, $V_1 = x_1 + y_1$ denotes the time of first renewal. Thus, V_1 is the sum of an up time and a down time. If the system returns to operation after repair immediately without delay, then y_1 denotes the actual repair time of the system. Similarly, if x_2 denotes the time from second operation to second failure and y_2 denotes the time from second failure to third operation, then taking $V_2 = x_2 + y_2$, $V_1 + V_2$ gives the time of second renewal. Thus, $V_i = x_i + y_i$ for $i = 1, 2, \ldots, n$ where x_i denotes the time from $(i-1)$th return to operation until ith failure occurs and y_i denotes the time from ith failure until $(i+1)$th return to operation, forms a renewal process. These V_i's, x_i's and y_i's are mutually independent.

If the system starts initially in a failed state and a renewal corresponds to return to operation, then $V_1 = y_1$ is the time of first renewal, $V_2 = x_2 + y_2$ is the time of second renewal, and so on. In general, $V_i = x_i + y_i$, $i \geq 2$ corresponds to ith renewal. If the system starts initially in an operating state, then taking $V_1 = x_1$, $V_2 = y_1 + x_2$, $V_i = y_{i-1} + x_i$ for $i \geq 2$ is a renewal corresponding to a system failure. Thus, by defining what constitutes a renewal, a large class of different problems can be conveniently treated as renewal process by defining properly a sequence of V_i's.

There are two important random variables for every renewal process

$$N(t) = \text{the number of renewals occurred by time } t,$$
$$T(n) = \text{the time of } n\text{th renewal},$$

where $N(t)$ is a discrete random variable having values 0, 1, 2, ... and $T(n)$ is a continuous random variable with values all positive real numbers. Since n is an integer, it is obvious that event $\{N(t) \leq n\}$ is the same as event $\{N(t) < n + 1\}$ for fixed t. Also for fixed t, the event $\{N(t) < n\}$ is same as event $\{T(n) > t\}$. Taking probability of these events, we have

$$P\{N(t) < n\} = P\{T(n) > t\}. \tag{4.10.1}$$

Since $T(n) = V_1 + V_2 + V_3 + \ldots + V_n$, expected time until nth renewal is given by

$$E\{T(n)\} = \sum_{i=1}^{n} E(V_i) \tag{4.10.2}$$

If each $V_i = x_i + y_i$ for $i = 1, 2, \ldots$ where $E(x_i) = X$ and $E(y_i) = Y$ for all i, then

$$E\{T(n)\} - n(X + Y).$$

If x_i denotes an operating time, y_i denotes a down time and the system initially starts in an operating state, then X = MTBF (mean time before failure) and Y = MTTR (mean time to repair), hence in this case

$$E\{T(n)\} = n(\text{MTBF} + \text{MTTR})$$

the expected time to return to operating state for nth time. Expected number of renewals $E\{N(t)\}$ in the time interval $[0, t]$ is called the *renewal function.* By definition of expectation,

$$E\{N(t)\} = \sum_{n=0}^{\infty} n p_{N(t)}(n) = M(t), \text{ say.}$$

$$= \sum_{n=0}^{\infty} n[P\{N(t) < n + 1\} - P\{N(t) < n\}] \tag{4.10.3}$$

where, $p_{N(t)}(n)$ is the probability of exactly n renewals in time t, i.e., $p_{N(t)}(n) = P\{N(t) = n\}$.

If $h(t)$ denotes failure time density and $g(t)$ the repair time density, then the density of $V_i = x_i + y_i$ is given by convolution $(h * g)(t)$. Assuming that the system is initially operating, in order for the system to be up at time t either it has never failed prior to t or it has failed at least once, a repair occurs in the interval $(x, x + dx)$ with probability $m(x)dx$ and then it does not fail in the remaining time $t - x$ with probability $R(t - x)$. Thus, the availability of the system is given by

$$A(t) = R(t) + \int_0^t R(t-x)\, m(x)dx \tag{4.10.4}$$

with initial condition $A(0) = R(0) = 1$. Note that if no repair is performed, then $R(t) = A(t)$ and

$$R(t) = 1 - \int_0^t h(x)dx.$$

Again $$P\{N(t) < n\} = P\{T(n) > t\} = 1 - P\{T(n) \le t\} = 1 - F_n(t)$$
$$P\{N(t) > n\} = F_{n+1}(t).$$

But $P\{N(t) < n\} + P\{N(t) = n\} + P\{N(t) > n\} = 1$

$\therefore$ $$P\{N(t) = n\} = 1 - P\{N(t) < n\} - P\{N(t) > n\} = F_n(t) - F_{n+1}(t).$$

and, $$E\{N(t)\} = \sum_{n=0}^{\infty} n[P\{N(t) = n\}] = \sum_{n=0}^{\infty} n[F_n(t) - F_{n+1}(t)]$$

or, $$M(t) = \sum_{n=1}^{\infty} F_n(t) = F(t) + \sum_{n=1}^{\infty} F_{n+1}(t) \tag{4.10.5}$$

where, $F_{n+1}(t)$ is the convolution of $F_n(t)$ and F. Let f be the probability density function of F, then

$$F_{n+1}(t) = \int_0^t F_n(t-x)\, f(x)dx$$

and the renewal function

$$M(t) = F(t) + \sum_{n=1}^{\infty} \int_0^t F_n(t-x)\, f(x)dx$$

$$= F(t) + \int_0^t \left\{ \sum_{n=1}^{\infty} F_n(t-x) \right\} f(x)dx$$

or, $$M(t) = F(t) + \int_0^t M(t-x)\, f(x)dx. \tag{4.10.6}$$

Equation (4.10.5) is called *fundamental renewal equation*. The *renewal density* $m(t)$ is given by

$$m(t) = \frac{dM(t)}{dt} = \frac{dF(t)}{dt} + \frac{d}{dt}\int_0^t M(t-x)\, f(x)dx$$

or, $$m(t) = f(t) + \int_0^t m(t-x)\, f(x)dx \tag{4.10.7}$$

Equation (4.10.7) is known as *renewal density equation*. Equations (4.10.4), (4.10.5) and (4.10.6) can be solved using Laplace transforms or using usual methods for solving integral equations.

From eqn. (4.10.6), whenever $dM(t)/dt$ exists,

$$m(t) = \sum_{n=1}^{\infty} f^{(n)}(t) \tag{4.10.8}$$

where $f^{(n)}(t)$ is the n fold convolution of $f(t)$. $f^{(n)}(t)dt$ denotes the probability that the nth renewal occurs in the interval $(t, t+dt)$ and $\sum_{n=1}^{\infty} f^{(n)}(t)dt$ denotes the probability that at least one renewal occurs in the interval $(t, t+dt)$. Hence, the probability of at least one renewal in the interval $(t, t+dt)$ is given by $m(t)dt$. It is for this reason that $m(t)$ is referred to as the renewal density. Although $m(t)$ is not a probability density function, since $\int_0^{\infty} m(t)dt \neq 1$. It is better to call $m(t)$ as the renewal rate or expected instantaneous renewal rate. If each x_i has the same density $h(t)$ and each y_i has the same density $g(t)$, then the density of each V_i is

$$f(t) = \int_0^t h(t-x)\, g(x)dx = \int_0^t g(t-x)\, h(x)dx = (h * g)(t) \tag{4.10.9}$$

Also note that $$M(t) = \int_0^t m(x)dx.$$

Taking Laplace transforms of eqns. (4.10.4), (4.10.6), (4.10.7) and (4.10.9), we obtain

$$\overset{*}{A}(s) = \overset{*}{R}(s) + \overset{*}{R}(s)\,\overset{*}{m}(s);\ \overset{*}{M}(s) = \overset{*}{F}(s) + \overset{*}{M}(s)\,\overset{*}{f}(s);$$
$$\overset{*}{m}(s) = \overset{*}{f}(s) + \overset{*}{m}(s)\,\overset{*}{f}(s);\ \overset{*}{f}(s) = \overset{*}{h}(s)\,\overset{*}{g}(s).$$

Solving these equations for $\overset{*}{m}(s)$, $\overset{*}{M}(s)$, and noting that $\overset{*}{R}(s) = \frac{1}{s} - \frac{\overset{*}{h}(s)}{s}$, we get

$$\overset{*}{m}(s) = \frac{\overset{*}{h}(s)\overset{*}{g}(s)}{1-\overset{*}{h}(s)\overset{*}{g}(s)};\ \overset{*}{M}(s) = \frac{\overset{*}{h}(s)\overset{*}{g}(s)}{s[1-\overset{*}{h}(s)\overset{*}{g}(s)]};$$

$$\overset{*}{A}(s) = \overset{*}{R}(s)[1 + \overset{*}{m}(s)] = \frac{1-\overset{*}{h}(s)}{s[1-\overset{*}{h}(s)\overset{*}{g}(s)]}.$$

4.11 MARKOV METHOD

This method can be used to estimate the time-dependent reliability of the system. The method is valid when both the failure and repair rates are constant. When these rates are time-dependent, the Markov process breaks down, except in some special cases. If the probability law of its future state of existence depends only upon the state it is in and not on how the system arrived in that state, then the system state behaviour is described by a process called Markov process, i.e., in a Markov process given the 'present' of the process, the 'future' is independent of its 'past'.

Consider a repairable system consisting of two units each with constant failure rate λ and constant repair rate μ. Assume that initially both units are operable and when one unit operates the other unit is in cold standby (i.e., inactive). As soon as the working unit fails, the standby unit is immediately switched on by a perfect switch and sensing device. The failed unit is repaired by a repairman immediately and the repaired unit goes in standby when its repair is completed. If a failure occurs when there is no standby unit, then both units are in failed state. At any time, the system is in any of the three states: operable state 0 when both units are good with one operating and the other in standby; operable state 1 when one unit is under repair and the other is operating, state 2 when both the units are failed. Let $p_i(t)$, $i = 0, 1, 2$ be the probability that the system is in ith state at time t, then the probability $p_i(t + \Delta t)$, $i = 0, 1, 2$ that the system is in ith state at time $t + \Delta t$ is as follows:

$$p_0(t + \Delta t) = p_0(t)\,(1 - \lambda\Delta t) + p_1(t)\,\mu\Delta t + 0(\Delta t) \tag{4.11.1}$$

i.e., the probability $p_0(t + \Delta t)$ that both the units are good at time $t + \Delta t$ is equal to the probability that the units are good at time t (i.e., $p_0(t)$) and there is no failure during time Δt (i.e., in the interval $(t, t + \Delta t)$ with probability $(1 - \lambda(t))$ or the probability that at time t one unit is failed (i.e., $p_1(t)$) while the other unit is operating and the failed unit is repaired during time Δt with probability $\mu\Delta t$. Applying product law of probability, measure both the terms on R.H.S. of (4.11.1) are obtained. Again using the addition law of probability, we get R.H.S. of eqn. (4.11.1). The term $0(\Delta t)$ is very small. The eqn. (4.11.1) is written assuming that during the small interval $(t, t + \Delta t)$ not more than one failure or repair occurs. Similarly, we have

$$p_1(t + \Delta t) = p_1(t)\,(1 - \lambda\Delta t)\,(1 - \mu\Delta t) + p_0(t)\,\lambda\Delta t + p_2(t)\,\mu\Delta t + 0(\Delta t) \tag{4.11.2}$$

$$p_2(t + \Delta t) = p_2(t)\,(1 - \mu\Delta t) + p_1(t)\,\lambda\Delta t + 0(\Delta t) \tag{4.11.3}$$

First term on R.H.S. of eqn. (4.11.2) is the probability that at time t one unit is failed while other unit is operating and during the interval $(t, t + \Delta t)$ there is no failure (probability $1 - \lambda\Delta t$) and no repair (probability $1 - \mu\Delta t$) in the system. Second term of (4.11.2) is the probability that both the units are good at time t and during the interval $(t, t + \Delta t)$ one unit is failed. The third term is the probability that both units are in failed state at time t and during the interval $(t, t + \Delta t)$ one unit is repaired. Similar arguments may be given for eqn. (4.11.3).

Equations (4.11.1)–(4.11.3) are (neglecting terms of higher than one power in Δt):

$$\frac{p_0(t + \Delta t) - p_0(t)}{\Delta t} = -\lambda p_0(t) + \mu p_1(t) + \frac{0(\Delta t)}{\Delta t},$$

$$\frac{p_1(t + \Delta t) - p_1(t)}{\Delta t} = -(\lambda + \mu)\,p_1(t) + \lambda p_0(t) + \mu p_2(t) + \frac{0(\Delta t)}{\Delta t},$$

$$\frac{p_2(t + \Delta t) - p_2(t)}{\Delta t} = -\mu p_2(t) + \lambda p_1(t) + \frac{0(\Delta t)}{\Delta t},$$

Taking the limit $\Delta t \to 0$, the equations become

$$p_0'(t) = -\lambda p_0(t) + \mu p_1(t) \tag{4.11.4}$$

$$p_1'(t) = -(\lambda + \mu)\, p_1(t) + \lambda p_0(t) + \mu p_2(t) \tag{4.11.5}$$
$$p_2'(t) = -\mu p_2(t) + \lambda p_1(t) \tag{4.11.6}$$

The eqns. (4.11.4)–(4.11.6) are to be solved using the initial condition $p_0(0) = 1$ otherwise zero. These equations may be solved either using Laplace transforms or by other methods like matrix method, numerical methods for first order ordinary differential equation, etc.

For steady state, we take that as $t \to \infty$, $\dfrac{dp_i(t)}{dt} \to 0$ and $p_i(t) \to p_i$. The resulting equations are solved recursively and using the normalizing condition $\sum_{i=0}^{2} p_i = 1$ to find the probabilities p_0, p_1, p_2.

Reliability of the system is given by

$$R(t) = 1 - p_2(t) = p_0(t) + p_1(t).$$

The process considered in this section is Markov one and is known as birth-death process.

4.12 SUPPLEMENTARY VARIABLES TECHNIQUE

When the repair rate or failure rate or both are time-dependent, the system loses its Markov character. In this situation, the future event will not depend on the present only (like Markov events) but will depend on the past also. These events are known as non-Markovian events. By introducing a new variable, called supplementary variable, the non-Markovian nature of the events is changed to Markovian.

As an example, consider the system consisting of two units, discussed in Section 4.11 with constant failure rate λ and, arbitrary repair rate $\mu(x)$ with probability density function

$$s(x) = \mu(x) \exp\left\{-\int_0^x \mu(x)dx\right\}$$

Now the system is of non-Markovian character. As in 4.11, let $p_i(t)$, $i = 0, 1, 2$, define $\mu(x)\,\Delta x =$ the probability that a failed unit is repaired in the interval $(x, x + \Delta x)$ given that the elapsed repair time is x. Define $p_i(x, t)\, dx =$ the probability that the system is in state i at time t and the system stays there between x and $x + \Delta x$ time units, $i = 1, 2$. Introducing this supplementary variable x, the character of the system becomes Markovian. Arguing as in Section 4.11, we have the following state equations:

$$p_0(t + \Delta t) = p_0(t)\,(1 - \lambda\Delta t) + \left(\int_0^t p_1(x, t)\, \mu(x)dx\right) \Delta t + 0(\Delta t) \tag{4.12.1}$$

$$p_1(x + \Delta x, t + \Delta t) = p_1(x, t)\,(1 - \mu(x)\,\Delta x)\,(1 - \lambda\Delta t) + p_0(t)\,\lambda\Delta t + 0(\Delta t, \Delta x) \tag{4.12.2}$$
$$p_2(x + \Delta x, t + \Delta t) = p_2(x, t)\,(1 - \mu(x)\,\Delta x) + p_1(x, t)\,\lambda\Delta t + 0(\Delta t, \Delta x) \tag{4.12.3}$$

or,
$$\frac{p_0(t+\Delta t) - p_0(t)}{\Delta t} = -\lambda p_0(t) + \int_0^t p_1(x, t)\, \mu(x)dx + \frac{0(\Delta t)}{\Delta t}$$

$$p_1(x + \Delta x, t + \Delta t) - p_1(x, t) = -(\mu(x)\,\Delta x + \lambda\Delta t - \lambda\mu(x)\,\Delta x\Delta t)\, p_1(x, t) + p_0(t)\,\lambda\Delta t + 0(\Delta t, \Delta x)$$
$$p_2(x + \Delta x, t + \Delta t) - p_2(x, t) = -\mu(x)\, p_2(x, t)\,\Delta x + p_1(x, t)\,\lambda\Delta t + 0\,(\Delta t, \Delta x)$$

Taking the limits $\Delta t \to 0$, $\Delta x \to 0$ and neglecting higher order (more than first order) terms in Δx and Δt, we obtain

$$p_0'(t) = -\lambda p_0(t) + \int_0^t p_1(x, t)\, \mu(x) dx, \tag{4.12.4}$$

$$\frac{\partial p_1(x,t)}{\partial x} + \frac{\partial p_1(x,t)}{\partial t} = -(\mu(x) + \lambda)\, p_1(x, t) + \lambda p_0(t), \tag{4.12.5}$$

$$\frac{\partial p_2(x,t)}{\partial x} + \frac{\partial p_2(x,t)}{\partial t} = -\mu(x)\, p_2(x, t) + \lambda p_1(x, t) \tag{4.12.6}$$

Equation (4.12.4) is ordinary first order differential equation while eqns. (4.12.5) and (4.12.6) are first order partial differential equations (Lagrange's type). These equations are to be solved using the following initial and boundary conditions:

$$p_0(0) = 1,\ p_1(x, 0) = p_2(x, 0) = 0 \tag{4.12.7}$$

$$p_1(0, t) = \lambda p_0(t) \tag{4.12.8}$$

$$p_2(0, t) = \lambda \int_0^t p_1(x, t) dx \tag{4.12.9}$$

and,

$$p_i(t) = \int_0^t p_i(x, t) dx,\ i = 1, 2 \tag{4.12.10}$$

The governing equations (4.12.4) to (4.12.6) may be solved using Laplace transforms method given in several texts [4]. Here, another method used by Singh [46, 47] is given.

The subsidiary equations for (4.12.5) and (4.12.6) are

$$\frac{dx}{1} = \frac{dt}{1} = \frac{dp_1(x,t)}{-(\mu(x)+\lambda)p_1(x,t)+\lambda p_0(t)} \tag{4.12.11}$$

and,

$$\frac{dx}{1} = \frac{dt}{1} = \frac{dp_2(x,t)}{-\mu(x)p_2(x,t)+\lambda p_1(x,t)} \tag{4.12.12}$$

Solutions of (4.12.11) and (4.12.12) are

$$p_1(x, t) = [\phi_1(t - x) + \lambda \textstyle\int p_0(t) \exp (\int \mu(x)dx + \lambda x)dx] \exp \{-\lambda x - \int \mu(x)dx\} \tag{4.12.13}$$

$$p_2(x, t) = [\phi_2(t - x) + \lambda \textstyle\int p_1(x, t) \exp (\int \mu(x)dx)dx] \exp (-\int \mu(x)dx) \tag{4.12.14}$$

where, $\phi_1(t) = p_1(0, t) = \lambda p_0(t)$ and $\phi_2(t) = p_2(0, t) = \lambda \int_0^t p_1(x, t)dx$.

Solution of (4.12.4) is given by {using (4.12.7)}

$$p_0(t) = 1 + \int \left[\int_0^t p_1(x,t)\mu(x)dx \right] \exp (\lambda t) dt \tag{4.12.15}$$

where, $p_1(\cdot)$ is given by (4.12.13) in terms of $p_0(\cdot)$. Thus, the probabilities $p_1(\cdot)$ and $p_2(\cdot)$ are obtained in terms of $p_0(\cdot)$ while $p_0(\cdot)$ is given by (4.12.15).

For steady-state case, take all the derivatives with respect to 't' equal to zero and solve the equations. The steady state probabilities p_1 and p_2 are obtained in terms of p_0 which is obtained using the normalizing condition $\sum_{i=0}^{2} p_i = 1$.

In steady state, eqns. (4.12.5) and (4.12.6) become

(taking $\frac{\partial p_i(\cdot)}{\partial(t)} = 0$, $i = 1, 2$ and probabilities independent of t)

$$\frac{dp_1(x)}{dx} + (\mu(x) + \lambda)\, p_1(x) = \lambda p_0 \tag{4.12.16}$$

$$\frac{dp_2(x)}{dx} + \mu(x)\, p_2(x) = \lambda p_1(x) \tag{4.12.17}$$

with conditions
$$p_1(0) = \lambda p_0,\ p_2(0) = 0 \tag{4.12.18}$$

Solutions of (4.12.16) and (4.12.17) are given by (using (4.12.18)

$$p_1(x) = [\lambda p_0\{1 + \int \exp\,(\lambda x + \int \mu(x)dx)dx\}].\ \exp\,(-\lambda x - \int \mu(x)dx) \tag{4.12.19}$$

$$p_2(x) = [\lambda \int p_1(x)\ \exp(\int \mu(x)dx)\,dx]\ \exp\,(-\int \mu(x)dx) \tag{4.12.20}$$

and,
$$p_1 = \int p_1(x)dx,\ p_2 = \int p_2(x)dx.$$

Thus, $p_1(x)$ and $p_2(x)$ are obtained in terms of p_0, which may be evaluated using $\sum_{i=0}^{2} p_i = 1$. Note that although p_0 is given by (4.12.4) but if we use this equation taking $p_0'(t) = 0$, then we obtain $p_0 = 0$ which means all the probabilities vanish and hence meaningless analysis.

The reliability of the system is

$$R(t) = 1 - \int p_2(x, t)dx,$$

and long-run availability is

$$A = 1 - p_2 = p_0 + p_1.$$

4.13 DISCRETE TRANSFORM METHOD

Consider a system of difference-differential equations obtained using supplementary variable technique for some n unit system. Let the equations are

$$\left[\frac{\partial}{\partial t} + \frac{\partial}{\partial x} + \eta(x) + \lambda b_n + \gamma_n\right] p_n(x, t) = \lambda b_{n-1}\, p_{n-1}(x, t) + \gamma_{n+1} p_{n+1}(x, t)\ (n = 1, 2 \ldots, N) \tag{4.13.1}$$

$$\frac{d}{dt} p_0(t) = -\lambda p_0(t) = \int_0^\infty p_1(x, t)\ \eta(x)dx \tag{4.13.2}$$

with initial and boundary conditions

$$p_n(x, 0) = \delta(x)\ \delta_{ni};\ p_n(0, t) = \int_0^{\infty} p_{n+1}(x, t)\ \eta(x)dx + \lambda p_0(t)\delta_{n,1} \tag{4.13.3}$$

where, $$S(x) = \eta(x)\ \exp\left\{-\int_0^x \eta(u)\,du\right\}.$$

Equations (4.13.1) and (4.13.2) are to be solved using (4.13.3).

In matrix notation, the system of equations can be written as

$$(\theta\bar{I} - \bar{A})\ \bar{P}(x, t) = \bar{O} \tag{4.13.4}$$

$$\bar{P}(0, t) = \int_0^{\infty} \bar{D}\cdot\bar{P}(x, t)\ \eta(x)dx + \lambda p_0(t)\cdot\bar{d}_1, \tag{4.13.5}$$

where,

$$\bar{P}(x, t) = \begin{bmatrix} p_1(x,t) \\ \vdots \\ P_N(x,t) \end{bmatrix},\ \bar{d}_1 = \begin{bmatrix} 1 \\ 0 \\ 0 \\ \vdots \\ 0 \end{bmatrix}$$

$\bar{I}$ is the $N \times N$ identity matrix,

$$A = \begin{bmatrix} -(\lambda b_1 + r_1) & r_2 & 0 & \vdots & \vdots & 0 \\ \lambda b_1 & -(\lambda b_2 + r_2) & r_2 & \vdots & \vdots & \vdots \\ \vdots & \vdots & \vdots & \vdots & \vdots & \vdots \\ 0 & 0 & 0 & 0 & \lambda b_{N-1} & -(\lambda b_N + r_N) \end{bmatrix} \tag{4.13.6}$$

$\bar{D}$ is the $N \times N$ shift matrix

$$\bar{D} = \begin{bmatrix} 0 & 1 & 0 & \ldots & 0 \\ 0 & 0 & 1 & \ldots & 0 \\ \ldots & \ldots & \ldots & \ldots & \ldots \\ 0 & 0 & 0 & \ldots & 1 \\ 0 & 0 & 0 & \ldots & 0 \end{bmatrix} \tag{4.13.7}$$

and θ is the operator $\dfrac{\partial}{\partial t} + \dfrac{\partial}{\partial x} + \eta(x)$.

A linear transform $\bar{G}(x, t)$ of $\bar{P}(x, t)$ is called a discrete transform when such a transformation reduces (4.13.4) into a diagonal form in $\bar{G}(x, t)$.

Let $\bar{H}$ and $\bar{H}^{-1}$ be the left and right eigenmatrices of $\bar{A}$. It is well known that

$$\bar{A} = \bar{H}^{-1}\bar{K}\bar{H} \tag{4.13.8}$$

where, $$\overline{K} = \text{diag}\,(K_1, K_2, \ldots, K_N) \tag{4.13.9}$$

$K_1, K_2, \ldots, K_N$ being the eigenvalues of $\overline{A}$.

Premultiplying (4.13.4) by $\overline{H}$ and using (4.13.8), we get

$$(\theta\overline{I} - \overline{K})\,\overline{G}(x, t) = \overline{O} \tag{4.13.10}$$

where, $$\overline{G}(x, t) = \overline{H}\,\overline{P}(x, t). \tag{4.13.11}$$

$\overline{G}(x, t)$ is the required discrete transform of $\overline{P}(x, t)$ and $\overline{H}$ is the transformation matrix. Once $\overline{G}(x, t)$ is determined, $\overline{P}(x, t)$ can be obtained by the inverse transform

$$\overline{P}(x, t) = \overline{H}^{-1}\overline{G}(x, t) \tag{4.13.12}$$

Premultiplying (4.13.5) by $\overline{H}$, we obtain

$$\overline{G}(0, t) = \int_0^\infty \overline{H}\,\overline{D}\,\overline{H}^{-1}\,\overline{G}(x, t)\;\eta(x)dx + \overline{H}\,\overline{d}_1\,\lambda p_0(t) \tag{4.13.13}$$

Defining the Laplace transform $\overline{G}(x, s) = \int_0^\infty e^{-st}\,\overline{G}(x, t)dt$ and using in (4.13.10) and (4.13.13), we obtain

$$(\hat{\theta}\overline{I} - \overline{K})\;\overline{G}(x, s) = \overline{C}\,\delta(x) \tag{4.13.14}$$

$$\overline{G}(0, s) = \overline{H}\,\overline{D}\,\overline{H}^{-1}\int_0^\infty \overline{G}(x, s)\;\eta(x)dx + \overline{H}\,\overline{\alpha}_1\lambda\overline{p}_0(s) \tag{4.13.15}$$

where, $$\hat{\theta} = \frac{\partial}{\partial x} + \eta(x) + s \text{ and}$$

$$\overline{C} = \overline{G}(x, 0) = \overline{H}\,\overline{P}(x, 0) = \overline{H}\begin{bmatrix} 0 \\ 0 \\ \vdots \\ 1 \\ \vdots \\ 0 \end{bmatrix} = \overline{H}\,\overline{d}_i. \tag{4.13.16}$$

The solution of (4.13.14) is given by

$$\overline{G}(x, s) = e^{-N(x)}\,e^{-sx}\,e^{x\overline{K}}\,\hat{\overline{G}}(0, s) \tag{4.13.17}$$

where, $\hat{\overline{G}}(0, s) = \overline{G}(0, s) + \overline{C}$ and $N(x) = \int_0^x \eta(u)du$. Substituting in (4.13.15), we obtain

$$\overline{G}(0, s) = \overline{H}\,\overline{D}\,\overline{H}^{-1}\int_0^\infty e^{-N(x)}\,e^{-sx}\,e^{x\overline{K}}\,\hat{\overline{G}}(0, s)\;\eta(x)dx + \overline{H}\,\overline{d}_1\,\lambda\overline{p}_0(s) + \overline{C}. \tag{4.13.18}$$

Now, since $e^{x\overline{K}} = \text{diag}\,(e^{xK_1}, e^{xK_2}, \ldots, e^{xK_N})$, (4.13.18) gets simplified to

$$\hat{\overline{G}}\,(0, s) = \overline{H}\,\overline{D}\,\overline{H}^{-1}\overline{S}\,(s - K)\;\overline{G}\,(0, s) + \overline{H}\,\overline{d}_1\,\lambda\,\overline{p}_0(s) \tag{4.13.19}$$

where, $\overline{S}\,(s - K) = \text{diag}\,[\overline{S}\,(s - K_1), \overline{S}(s - K_2) \ldots, \overline{S}\,(s - K_N)]$; $\overline{p}_0(s) = Lp_0(t)$; $\overline{S}(s) = L\{S(x)\}$. Thus,

$$\overline{G}\,(0, s) = [\overline{I} - \overline{H}\,\overline{D}\,\overline{H}^{-1}\,\overline{S}(s - K)]^{-1}\;[\overline{H}\,\overline{d}_1\,\lambda\,\overline{p}_0(s) + \overline{C}\,] \tag{4.13.20}$$

From (4.13.3), we get

$$(s + \lambda)\;\overline{p}_0(s) = \int_0^{\infty} \overline{H}^{(1)}\,\overline{G}\,(x, s)\;\eta(x)dx + \delta_{i,0} \tag{4.13.21}$$

where,
$$\overline{H}^{-1} = \begin{pmatrix} \overline{H}^{(1)} \\ \overline{H}^{(2)} \\ \vdots \\ \overline{H}^{(N)} \end{pmatrix} \tag{4.13.22}$$

Using (4.13.17) and (4.13.20) in (4.13.21), we get finally

$$[(s + \lambda) - \lambda\,\overline{H}^{(1)}\,\overline{S}\,(s - K)\;\{\overline{I} - \overline{H}\,\overline{D}\,\overline{H}^{-1}\,\overline{S}(s - K)\}^{-1}\,\overline{H}\,\overline{d}_1\,]\;\overline{p}_0(s)$$
$$= \overline{H}^{(1)}\,\overline{S}\,(s - K)\;[\overline{I} - \overline{H}\,\overline{D}\,\overline{H}^{-1}\,\overline{S}(s - K)]^{-1}\,\overline{C} + \delta_{i,0} \tag{4.13.23}$$

This gives $\overline{p}_0(s)$ and hence all the probabilities. It is to mention here that explicit solution without much algebraic manipulation will only be possible if the matrix product $\overline{H}\,\overline{D}\,\overline{H}^{-1}$ reduces to a simple form [Ref. Rao and Jaiswal 1969]. A similar matrix method is used by Singh [1973] for a queueing system which is simple one. He used the method for solving differential equations by matrix method [Bellman]. Singh and Goyal [2006] developed a matrix method for solving set of differential equations (first order) where the evaluation of eigenvalues is not needed.

4.14 BOOLEAN FUNCTION TECHNIQUE

The reliability evaluation becomes more complicated when complexities increase in a system. Therefore, the derivation of symbolic reliability expression in a simplified and compact form for a general system is very helpful. In this section, a method is given which gives reliability in symbolic form.

Consider a system consisting of three subsystems A, B and C (e.g., power generators) arranged in parallel connected with six switching devices S_1, S_2, …, S_6. Such a system is considered by Gupta and Agarwal [1983] and analysed using this method (explained in this section). The assumptions and notations used to explain the system are:

(i) the reliabilities of all constituent components are known,

(ii) all components are always operating,

(iii) there is no standby or switching redundancy,

(iv) the states of all components are statistically independent,

(v) the system or the components are either good (operating) or bad (failed),

(vi) the failure of each component follows arbitrary distribution, and

(vii) there is no repair facility.

W_1, W_2, W_3	states of subsystems A, B, C
W_{i+p}	states of switching device S_i, $i = 1, \ldots, 6$; $p = 1, 2, 3$.
W'	negation of W.
W_i'	0 in good state and 1 in bad state, $i = 1, \ldots, 9$.

Following the assumptions and notations mentioned above, the system configuration is given below.

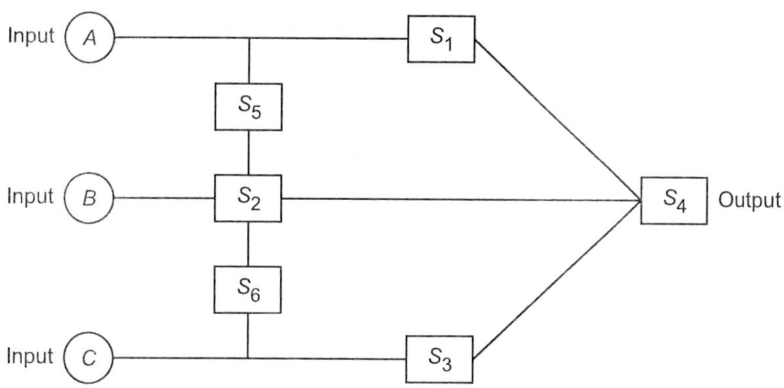

The power supply, generated by subsystems A, B, C, is given through switching device S_4. Using Boolean Function technique [Premo 1963, Christiaanse 1970, Nakzawa 1977], the conditions of capability for the successful operation of the system in terms of logical matrix, are expressed as

$$\sigma(W_1, W_2, \ldots, W_9) = \begin{vmatrix} W_1 & W_4 & W_7 & & & \\ W_1 & W_8 & W_5 & W_7 & & \\ W_1 & W_8 & W_5 & W_9 & W_6 & W_7 \\ W_2 & W_5 & W_7 & & & \\ W_2 & W_5 & W_8 & W_4 & W_7 & \\ W_2 & W_5 & W_9 & W_6 & W_7 & \\ W_3 & W_6 & W_7 & & & \\ W_3 & W_9 & W_5 & W_7 & & \\ W_3 & W_9 & W_5 & W_8 & W_4 & W_7 \end{vmatrix} \tag{4.14.1}$$

Using algebra of logic, (4.14.1) can be written as

$$\sigma(W_1, W_2, \ldots, W_9) = |W_7\ \phi(W_1, W_2, W_3, W_4, W_5, W_6, W_8, W_9)| \tag{4.14.2}$$

where,

$$\phi(W_1, W_2, W_3, W_4, W_5, W_6, W_8, W_9) = \begin{vmatrix} W_1 & W_4 & & & \\ W_2 & W_5 & & & \\ W_3 & W_6 & & & \\ W_1 & W_5 & W_8 & & \\ W_3 & W_5 & W_9 & & \\ W_2 & W_4 & W_5 & W_8 & \\ W_2 & W_5 & W_6 & W_9 & \\ W_1 & W_5 & W_6 & W_8 & W_9 \\ W_3 & W_4 & W_5 & W_8 & W_9 \end{vmatrix} \tag{4.14.3}$$

writing $W_1W_4 = M_1$, $W_2W_5 = M_2$, $W_3W_6 = M_3$, $W_1W_5W_8 = M_4$, $W_3W_5W_9 = M_5$, $W_2W_4W_5W_8 = M_6$, $W_2W_5W_6W_9 = M_7$, $W_1W_5W_6W_8W_9 = M_8$, $W_3W_4W_5W_8W_9 = M_9$ in (4.14.3), we obtain

$$\phi(\ldots) = \begin{vmatrix} M_1 & & & & & & & & \\ M_1' & M_2 & & & & & & & \\ M_1' & M_2' & M_3 & & & & & & \\ M_1' & M_2' & M_3' & M_4 & & & & & \\ M_1' & M_2' & M_3' & M_4' & M_5 & & & & \\ M_1' & M_2' & M_3' & M_4' & M_5' & M_6 & & & \\ M_1' & M_2' & M_3' & M_4' & M_5' & M_6' & M_7 & & \\ M_1' & M_2' & M_3' & M_4' & M_5' & M_6' & M_7' & M_8 & \\ M_1' & M_2' & M_3' & M_4' & M_5' & M_6' & M_7' & M_8' & M_9 \end{vmatrix} \tag{4.14.4}$$

using algebra of logic,

$$M'_1 M_2 = \begin{vmatrix} W_1' & W_2 & W_5 & \\ W_1 & W_2 & W_4' & W_5 \end{vmatrix} \tag{4.14.5}$$

$$M'_1 M'_2 M_3 = \begin{vmatrix} W_1' & W_2' & W_3 & W_6 & & \\ W_1' & W_2 & W_3 & W_5' & W_6 & \\ W_1 & W_2' & W_3 & W_4' & W_6 & \\ W_1 & W_2 & W_3 & W_4' & W_5' & W_6 \end{vmatrix} \tag{4.14.6}$$

$$M'_1 M'_2 M'_3 M_4 = \begin{vmatrix} W_1 & W_2' & W_3' & W_4' & W_5 & W_8 & \\ W_1 & W_2' & W_3 & W_4 & W_5 & W_6' & W_8 \end{vmatrix} \tag{4.14.7}$$

$$M'_1 M'_2 M'_3 M'_4 M_5 = \begin{vmatrix} W_1' & W_2' & W_3 & W_5 & W_6' & W_9 & & \\ W_1 & W_2' & W_3 & W_4' & W_5 & W_6' & W_8' & W_9 \end{vmatrix} \tag{4.14.8}$$

$$M'_1\, M'_2\, M'_3\, M'_4\, M'_5\, M_6 = 0 \tag{4.14.9}$$
$$M'_1\, M'_2\, M'_3\, M'_4\, M'_5\, M'_6\, M_7 = 0 \tag{4.14.10}$$
$$M'_1\, M'_2\, M'_3\, M'_4\, M'_5\, M'_6\, M'_7\, M_8 = 0 \tag{4.14.11}$$
$$M'_1\, M'_2\, M'_3\, M'_4\, M'_5\, M'_6\, M'_7\, M'_8\, M_9 = 0 \tag{4.14.12}$$

Putting (4.10.5) – (4.10.12) into (4.14.4), one gets

$$\phi(W_1, W_2, W_3, W_4, W_5, W_6, W_8, W_9) = \left|\begin{array}{llllllll} W_1 & W_4 & & & & & & \\ W'_1 & W_2 & W_5 & & & & & \\ W_1 & W_2 & W'_4 & W_5 & & & & \\ W'_1 & W'_2 & W_3 & W_6 & & & & \\ W'_1 & W_2 & W_3 & W'_5 & W_6 & & & \\ W_1 & W'_2 & W_3 & W'_4 & W_6 & & & \\ W_1 & W_2 & W_3 & W'_4 & W'_5 & W_6 & & \\ W_1 & W'_2 & W'_3 & W'_4 & W_5 & W_8 & & \\ W_1 & W'_2 & W_3 & W'_4 & W_5 & W'_6 & W_8 & \\ W'_1 & W'_2 & W_3 & W_5 & W'_6 & W_9 & & \\ W_1 & W'_2 & W_3 & W'_4 & W_5 & W'_6 & W'_8 & W_9 \end{array}\right| \tag{4.14.13}$$

The probability of successful operation (i.e., reliability) of the system is given by

$$\begin{aligned} R_s &= P\{\sigma(W_1, W_2, \ldots, W_9) = 1\} \\ &= R_7\,[R_1 R_4 + Q_1 R_2 R_5 + R_1 R_2 Q_4 R_5 + Q_1 Q_2 R_3 R_6 + Q_1 R_2 R_3 Q_5 R_6 + \\ &\quad + R_1 Q_2 R_3 Q_4 R_6 + R_1 R_2 R_3 Q_4 Q_5 R_6 + R_1 Q_2 Q_3 Q_4 R_5 R_8 + \\ &\quad + R_1 Q_2 R_3 Q_4 R_5 Q_6 R_8 + Q_1 Q_2 R_3 R_5 Q_6 R_9 + \\ &\quad + R_1 Q_2 R_3 Q_4 R_5 Q_6 Q_8 R_9] \end{aligned} \tag{4.14.14}$$

where, $R_i (i = 1, 2, \ldots, 9)$ are reliabilities of subsystems A, B, C switching devices $S_i (i = 1, \ldots, 6)$ respectively and $Q_i (i = 1, 2, \ldots, 9)$ are corresponding unreliabilities.

In particular, if reliabilities of each component is R, then (4.14.14) gives

$$R_s = R^3[R^6 - 5R^5 + 8R^4 - R^3 - 7R^2 + 2R + 3].$$

when failure rates follow exponential distribution, the reliability of the complex system at an instant t is

$$R_s(t) = e^{-3\lambda t}[e^{-6\lambda t} - 5e^{-5\lambda t} + 8e^{-4\lambda t} - e^{-3\lambda t} - 7e^{-2\lambda t} + 2e^{-\lambda t} + 3]$$

If failure rates follows Weibull distribution with positive parameter α, the reliability of the system at an instant t is

$$R_s(t) = e^{-3\lambda t^\alpha}[e^{-6\lambda t^\alpha} - 5e^{-5\lambda t^\alpha} + 8e^{-4\lambda t^\alpha} - e^{-3\lambda t^\alpha} - 7e^{-2\lambda t^\alpha} + 2e^{-\lambda t^\alpha} + 3].$$

4.15 GUPTA-SINGH TECHNIQUE

When we deal with availability and other parameters such as MTSF, busy period of the server for inspection/instructions/repairs/replacements and number of server's visits doing different types of jobs such as repairs/replacements/inspection, etc. we get very complicated calculations. If Laplace transform technique is used, then for complex systems it is very difficult to find inverse Laplace transform when there is more than 16 degree term in denominator. Also, if differential equation technique is used, a lengthy calculation is to be faced. Similar is the case for regenerative point technique. This difficulty is overcome by V.K. Gupta and Jai Singh [2009] providing four formulae in closed form. This technique is simple and computer oriented. The technique provides direct results to reliability engineers even for fuziness in the system, if any. This technique may be called Gupta-Singh Technique.

Before discussing the technique, let us define some terms which are needed during the further discussion. The probability density function (p.d.f.) of the first passage time from a regenerative state i to a regenerative state j or to a failed state j, without visiting any other regenerative state in $(0, t]$ is defined as

$$q_{i,j}(t) = p_r\{X_{n+1} = j | X_n = i\}$$

where, X_n denotes the state of the system at nth epoch of time. The cumulative distribution function (c.d.f.) of first passage time, given that the system entered regenerative state i at $t = 0$ is

$$Q_{i,j}(t) = \int_0^t q_{i,j}(t)\,dt = p_r\{X_{n+1} = j,\ t_{n+1} - t_n \le t | X_n = i\}$$

$$= p_{i,j}(t)$$

Also

$$\lim_{t\to\infty} Q_{i,j}(t) = \lim_{t\to\infty} p_{i,j}(t) = p_{i,j}.$$

The steady-state transition probability factor of a state 'j' reachable from the state 'i' is defined as the sum of the steady-state path probabilities of transition from the state 'i' to the state 'j' along all the directed simple paths $(i \xrightarrow{sr} j)$ for different values of 'r' and subject to the condition that the system does not visit any circuit/cycle along any path $(i \xrightarrow{sr} j)$ under consideration for any fixed value of r. A simple path $(i \xrightarrow{sr} j)$ for a given value of r from the state 'i' to the state 'j' in the state-transition diagram may have the regenerative state (s) k at which k-cycle (s) are formed. The steady-state transition probability factor of state j reachable from state 'i' is

$$V_{i,j} = \sum_{sr} \left\{ \frac{p_r(i \xrightarrow{sr} j)}{\prod_k \{1 - \Sigma p_r(k\text{-cycle})\}} \right\} \tag{4.15.1}$$

where, k is a regenerative state belonging to the simple path $(i \xrightarrow{sr} j)$ where k-cycle (s) are formed. $\Sigma p_r(k\text{-cycle}) = \Sigma V_{k,k}$, the sum of probability factor of all the different k-simple circuit (s) formed at the given regenerative state k and for a given value of r. If no k-cycle is formed, then this sum is zero.

(a) *Mean Time to System Failure (MTSF):* The mean time to system failure is the statistical average time for which the system is operative before any failure of the system. MTSF is used when the system undergoes either preventive or corrective maintenance actions. MTSF under steady-state conditions, is given by $\Sigma V_{0,i} \cdot \mu_i$ where i is an unfailed regenerative state in the state-transition diagram of the system to which the system can transit before failure. On using (4.15.1)

$$\text{MTSF} = \left[\sum_{i,sr}\left\{\frac{\{p_r(o \xrightarrow{sr(sff)} i)\} \cdot \mu_i}{\prod_{k_1 \neq 0}\{1 - \Sigma p_r(k_1\text{-}\overline{\text{cycle}})\}}\right\}\right] \div \left[1 - \sum_{sr}\left\{\frac{\{p_r(o \xrightarrow{sr(sff)} o)\}}{\prod_{k_2 \neq 0}\{1 - \Sigma p_r(k_2\text{-}\overline{\text{cycle}})\}}\right\}\right] \quad (4.15.2)$$

i : a regenerative unfailed state to which the system can transit before entering any failed state while entering the initial 0-state at time $t = 0$.

k_1 : a regenerative state along the path $(o \xrightarrow{sr(sff)} i)$, at which $k_1\text{-}\overline{\text{cycle}}$ is formed through regenerative unfailed states.

k_2 : a regenerative state along the path $(o \xrightarrow{sr(sff)} o)$, at which a $k_2\text{-}\overline{\text{cycle}}$ is formed through unfailed states. In the numerator $p_r(o \xrightarrow{sr(sff)} o) = 1$ and in denominator $p_r(o \longrightarrow o) = p_{0,0}$ (provided there is a loop at the o-state).

(b) *Steady-state availability:* It is defined as the proportion of time that the system is operational when the time interval is very-very large and the corrective, preventive maintenance, down times and the waiting times are included

$$A_0 = \frac{\text{MTBM}}{\text{MTBM} + \text{MDT}}$$

where, MTBM $\rightarrow$ mean time between maintenance;

MDT $\rightarrow$ mean down time which is the statistical mean of the down times caused due to breakdowns, including supply down time, administrative down time.

The steady-state availability is given by

$$A_0 = \left[\sum_j V_{0,j} \cdot \mu_j\right] \div \left[\sum_i V_{0,i} \cdot \mu_i^1\right]$$

where, μ_i^1 is the total unconditional time spent before transiting to any other regenerative state (s), given that the system entered regenerative state i at $t = 0$. In case the system fails partially and is not fully available for its purpose, then the availability of the system is discounted according to the proportions to the fuzziness measure of the states that the system can visit and

$$A_0 = \left[\sum_j V_{0,j} \cdot \mu_j \cdot f_j\right] \div \left[\sum_i V_{0,i} \cdot \mu_i^1\right]$$

where, f_j is the fuzziness measure of the unfailed state j,

$$A_0 = \left[\sum_{j,sr}\left\{\frac{\{p_r(o \xrightarrow{sr} j)\} f_j \cdot \mu_j}{\prod_{k_1 \neq 0}\{1 - \Sigma p_r(k_1\text{-cycle})\}}\right\}\right] \div \left[\sum_{i,sr}\left\{\frac{\{p_r(o \xrightarrow{sr} i)\} \cdot \mu_i^1}{\prod_{k_2 \neq 0}\{1 - \Sigma p_r(k_2\text{-cycle})\}}\right\}\right] \tag{4.15.3}$$

j : a reachable state which may be down/reduced state and is an available state.

i : a regenerative state.

k_i ($\neq 0$) : a regenerative state (may be interior or terminal state along the path) at which k_i-cycle is formed (may be formed through non-regenerative/failed states). k_1 is a regenerative state visited along the path ($o \xrightarrow{sr} j$) and it can be equal to j. k_2 is regenerative state visited along the path ($o \xrightarrow{sr} i$) and it can be equal to i.

In case a down state of the system is treated as a failed state, then for availability purposes the said state is to be treated as unavailable state.

(c) *Busy period of the server:* It is under steady-state conditions defined as

$$B_0 = \frac{\text{MTTR}}{\text{MTBM} + \text{MDT}}$$

where, MTTR → mean time to repair and is $\propto \Sigma V_{0,j} \cdot \eta_j$; η_j is the expected time spent in the state j, given that the system entered regenerative state j at $t = 0$. MTBM + MDT $\propto \sum_i V_{0,i} \cdot \mu_i^1$, where μ_i^1 is the unconditional time spent before transiting to any other regenerative state, given that the system entered regenerative state i at $t = 0$.

$$B_0 = \left[\sum_j V_{0,j} \cdot \eta_j\right] \div \left[\sum_i V_{0,i} \cdot \mu_i^1\right]$$

$$= \left[\sum_{j,sr}\left\{\frac{\{p_r(o \xrightarrow{sr} j)\} \cdot \eta_j}{\prod_{k_1 \neq 0}\{1 - \Sigma p_r(k_1\text{-cycle})\}}\right\}\right] \div \left[\sum_{i,sr}\left\{\frac{\{p_r(o \xrightarrow{sr} i)\} \cdot \mu_i^1}{\prod_{k_2 \neq 0}\{1 - \Sigma p_r(k_2\text{-cycle})\}}\right\}\right] \tag{4.15.4}$$

j : a regenerative state (may be partially or totally failed state) where the server is busy doing any given job.

i : a regenerative state.

k_i ($\neq 0$) : a regenerative state along the path (interior or terminal state) at which a k_i-cycle is formed (cycle may be through non-regenerative/failed states). k_1 and k_2 as explained in (b) above.

(d) *Number of Server's Visits/Number of replacements:* It is defined by,

$$V_0 = \left[\sum_j V_{0,j} \cdot \delta_j\right] \div \left[\sum_i V_{0,i} \cdot \mu_i^1\right]$$

$$= \left[\sum_{j,sr} \left\{\frac{\{p_r(o \xrightarrow{sr} j)\}}{\prod_{k_1 \neq 0} \{1 - \Sigma p_r(k_1\text{-cycle})\}}\right\}\right] \div \left[\sum_{i,sr} \left\{\frac{\{p_r(o \longrightarrow i)\} \cdot \mu_i^1}{\prod_{k_2 \neq 0} \{1 - \Sigma p_r(k_2\text{-cycle})\}}\right\}\right] \quad (4.15.5)$$

where, $\delta_j = 1$, if the visit of the server for the job/replacement is fresh at the regenerative state j, otherwise $\delta_j = 0$.

j : a regenerative state where the server visits afresh (may be after staying, existing, absenting and then again entering the system). In the later case j appears as an interior state of the path $(o \xrightarrow{sr} j)$ or replacement/preventive maintenance/corrective maintenance action occur for the first time along the path $(o \xrightarrow{sr} j)$.

i : a regenerative state.

k_i $(\neq 0)$: a regenerative state visited along a path (may be interior or terminal point) at which a k_i-cycle is formed (may be through non-regenerative or failed states), k_1 is regenerative state visited along $(o \xrightarrow{sr} j)$ and k_2 is regenerative state visited along $(o \xrightarrow{sr} i)$, $k_1 \neq j$ but k_2 can be equal to i.

If state j occurs only at the end of the path $(o \xrightarrow{sr} j)$, then $k_1 \neq j$ and in case state j occurs as an interior state of the path then k_1 may be equal to j.

These four formulae, i.e., (4.15.2) to (4.15.5) constitute the whole technique. Using these formulae all the parameters for the system under consideration can be evaluated.

Now to understand the method, let us consider a system consisting of one main unit and two Programmable Logic Controllers (PLCs) used for controlling the operations. Initially, one of the PLCs is operative and the other is in hot standby mode 18, 2009. Inspection is carried out to detect the type of fault/failure of other PLCs. The system consists of two subsystems A and B. In the subsystem A there are two PLC units A_1 and A_2 composed of several components in series. One of these is on-line while the other is in hot standby mode. The subsystem B consists of several repairable components in series. Failure of any one causes the system failure. The block diagram is as shown below:

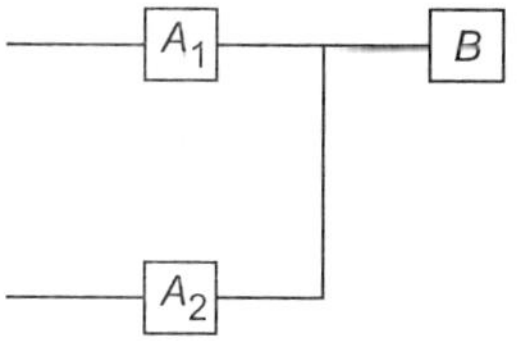

Fig. 4.4

The following assumptions and notations are taken:

1. Two types of failures are: (a) Type-I, the redundant unit on failure is found to be repairable after inspection or that it needs some part(s) to be replaced called RepI, (b) Type-II, the redundant unit on failure is found to be unrepairable owing to total burnt of the unit and needs full replacement called RepII (after inspection).
2. Nothing can further fail when system is in failed state.
3. If the main unit fails, the system is in failed state and the PLCs are put in the down state and the main unit is immediately put under repair.
4. The system is in down state in case both the redundant units are under repair/RepI/RepII.
5. On failure of a redundant unit, an inspection is carried out to detect the type of failure. It is assumed that the inspection for the type of failure finishes before any other failure of the main unit or of the standby unit.
6. The failure of a redundant unit may be of casual nature which needs no repair or replacement, etc. and the system starts functioning as such after some time.
7. It is assumed that the repair/RepI/RepII of a redundant unit finishes first with probability 1 before the repair of the main unit is completed.
8. The repairman is available on demand and the server who inspects the failed redundant unit remains available for the subsequent repair/replacement, if any.
9. The main unit is put under down state when either both the redundants fail or are undergoing RepI/RepII.
10. The failure times and replacement/repair times of the redundants are assumed to have exponential distributions whereas the repair time of the main unit and inspection time of the redundants may follow the general distributions.

p_r	: probability
$\underline{k}$	: state k is a non-regenerative state.
$p_{i,j}$	: steady-state transition probability from a regenerative state i to a regenerative state j without visiting any other state.
$A_i(t)$	: probability that the system is available in up-state at time t, given that the system entered regenerative state i at $t = 0$.
$B_i(t)$	: probability that the server is busy doing a particular job at epoch t, given that the system entered regenerative state i at $t = 0$.
$V_i(t)$	: the expected number of visits of the server for a given job in $(0, t]$, given that the system entered regenerative state i at $t = 0$.
$W_i(t)$	: waiting time probability that the server is busy doing a particular job at epoch t without transiting to any other regenerative state or returning to the state i through one or more non-regenerative states, given that the system entered the regenerative state i at $t = 0$.
η_j	: expected waiting time spent by the server doing a given job, given that the system entered regenerative state j at $t = 0$.

μ_i : the mean sojourn time spent in state i before visiting any other state (s), $\mu_i = \int_0^{\infty} R_i(t)dt$, where $R_i(t)$ is the reliability that the system remains in regenerative state i at time t given that the system entered the regenerative state i at $t = 0$.

μ_i^1 : the total unconditional time spent before transiting to any other regenerative state (s) given that the system entered regenerative state i at $t = 0$.

f_i : fuzziness measure of the i-state. $f_i = 0$, if i is failed state, $f_i = 1$ if i is a good state and $f_i \in (0, 1)$ if i is a partially failed state.

$B_i^{RI}(t)$: probability that the server is busy doing RepI of a redundant unit at epoch t, given that the system entered regenerative state i at $t = 0$.

$V_i^M(t)$: expected number of visits by the server to repair the main unit in $(0, t]$ given that the system entered regenerative state i at $t = 0$.

Cycle/circuit : a directed path in which the end states are the same.

Simple circuit : a circuit in which no state is repeated in the interior of the circuit.

Simple path : a directed path in which no state is repeated except with the possibility that end states of the path may be the same.

$\overline{\text{Cycle}}$: a circuit formed through unfailed states.

k-cycle : a circuit whose terminals are at the regenerative state k. The circuit may be formed through regenerative or non-regenerative states and failed states.

$k\text{-}\overline{\text{cycle}}$: a circuit whose terminals are at the regenerative state k and formed through only regenerative or non-regenerative states which are not the failed states.

$(i \xrightarrow{sr} j)$: rth-directed simple path from i state to j state. r takes positive integral values for different paths from the state i to j.

$(o \xrightarrow{(sff)} i)$: a directed simple failure free path from 0 state to i state.

$(i, j)/(i, j, k)$: $(i, j) = p_{i,j}$; $(i, j, k) = (i, j) \cdot (j, k) = p_{i,j} \cdot p_{j,k}$.

$(i, \underline{k}, j)/(i, j, \underline{k}, l)$: $(i, \underline{k}, j) = p_{i,j} \cdot k$; $(i, j, \underline{k}, l) = (i, j) \cdot (j, \underline{k}, l) = p_{i,j}\, p_{j,l}, k$

$p_r(i \to j) = \prod_{x,y} p_{x,y}$; x, y are any two consecutive regenerative states on the simple circuit free path $(i \to j)$.

$\Sigma pr(k\text{-cycle})$ = sum of probability factors of all k-cycles formed at the regenerative path point k

$\Sigma pr(k\text{-}\overline{\text{cycle}})$ = sum of probability factors of all $k\text{-}\overline{\text{cycles}}$ formed at the regenerative path point k.

M_o/M_d : main unit is in operative/down state.

C_o/C_s : redundant unit is in operative/hot standby.

F_{ur}/F_{UR} : main unit is under repair/repair of main unit is continuing from previous state.

$C_d/C_{ui}/C_{wi}$: redundant unit is in down state/under inspection/waiting for inspection.

C_r/C_R : redundant unit is under repair/repair of the unit is continuing from previous state.

$C_{\text{rep I}}/C_{\text{Rep I}}$: RepI is taking place/RepI continuing from previous state.

$C_{\text{rep II}}/C_{\text{Rep II}}$: RepII is taking place/RepII continuing from previous state.

λ : constant failure rate of the main unit.

α_1/α_2 : constant failure rate of the redundant unit (PLC) in operation/in hot standby mode, $\alpha_1 + \alpha_2 = \alpha$.

$p_1/p_2/p_3$: probability that a failed unit has failure of type-I/type-II/probability that a failed redundant unit is found O.K. after inspection; $p_1 + p_2 + p_3 = 1$.

$\beta/\beta_1/\beta_2$: constant rate of repair/RepI/RepII of a redundant unit.

$h(t)$: probability density function of the inspection time of a failed redundant unit.

$g(t)$: probability density function of the repair time of the main unit.

(X, Y, Z) : is the triplet where X denotes the status of main unit, Y denotes the status of PLC on-line and Z denotes the states of PLC in hot standby mode.

Using the above notations and assumptions, the transition diagram is as given in Fig. 4.5.

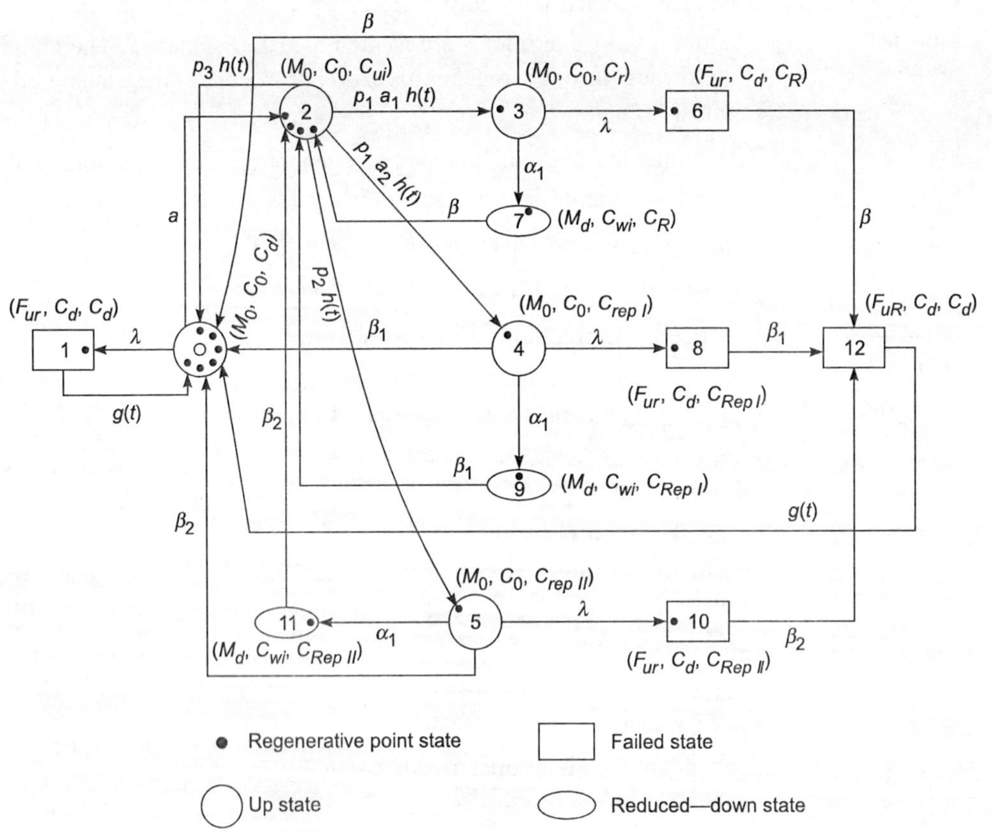

Fig. 4.5

System Parameters

(a) *MTSF:* The mean time to system failure is evaluated by the formula given by (4.15.2).

From Fig. 4.5, the regenerative unfailed states to which the system can transit before entering any failed state are i = 0, 2, 3, 4, 5, 7, 9 and 11; k_1 = 2, k_2 = 2. Using (4.15.2), we have

$$\text{MTSF} = \left[\frac{(1)\mu_0 + \left(\dfrac{1}{1-\bar{L}_2}\right)\begin{array}{l}\{(0,2)\mu_2 + (0,2,3)\mu_3 + (0,2,4)\mu_4 + \\ (0,2,5)\mu_5 + (0,2,3,7)\mu_7 + (0,2,4,9)\mu_9 + \\ (0,2,5,11)\mu_{11}\}\end{array}}{1 - \left(\dfrac{1}{1-\bar{L}_2}\right)\{(0,2,0) + (0,2,3,0) + (0,2,4,0) + (0,2,5,0)}\right]$$

$$= [(1-\bar{L}_2)\mu_0 + p_{0,2}(\mu_2 + p_{2,3}\cdot\mu_3 + p_{2,4}\cdot\mu_4 + p_{2,5}\cdot\mu_5 + p_{2,3}\cdot p_{3,7}\cdot\mu_7 + p_{2,4}$$
$$\cdot p_{4,9}\cdot\mu_9 + p_{2,5}\cdot p_{5,11}\cdot\mu_{11})] \div [(1-\bar{L}_2) - p_{0,2}(p_{2,0} + p_{2,3}\cdot p_{3,0}$$
$$+ p_{2,4}\cdot p_{4,0} + p_{2,5}\cdot p_{5,0})]$$

where, $\bar{L}_2 = (2, 3, 7, 2) + (2, 4, 9, 2) + (2, 5, 11, 2)$

$$= p_{2,3}\cdot p_{3,7}\cdot p_{7,2} + p_{2,4}\cdot p_{4,9}\cdot p_{9,5} + p_{2,5}\, p_{5,11}\cdot p_{11,2}$$

(b) *Steady-state availability*: Using the formula (4.15.3) availability can be evaluated. Again from Fig. 4.5, the regenerative states at which the system is in upstates are: j = 0, 2, 3, 4, 5; f_j = 1; the regenerative states are i = 0 to 11; k_1 = 2, k_2 = 2.

Using (4.15.3), the availability is

$$A_0 = \left[(0,\ 0)\mu_0 + \left(\frac{1}{1-L_2}\right)\{(0,\ 2)\mu_2 + (0,\ 2,\ 3)\mu_3 + (0,\ 2,\ 4)\mu_4 + (0,\ 2,\ 5)\mu_5\right]$$

$$\div \left[(0,\ 0,)\mu_0^1 + (0,\ 1)\mu_1^1 + \left(\frac{1}{1-L_2}\right)\{(0,\ 2)\mu_2^1 + (0,\ 2,\ 3)\mu_3^1 + (0,\ 2,\ 4)\mu_4^1\right.$$
$$+ (0,\ 2,\ 5)\mu_5^1 + (0,\ 2,\ 3,\ 7)\mu_7^1 + (0{,}2,\ 4,\ 9)\mu_9^1 + (0,\ 2,\ 5,\ 11)\mu_{11}^1$$
$$\left. + (0,\ 2,\ 3,\ 6)\mu_6^1 + (0,\ 2,\ 4,\ 8)\mu_8^1 + (0,\ 2,\ 5,\ 10)\mu_{10}^1\}\right]$$

$$= [(1-L_2)\mu_0 + (0,\ 2)\ \{\mu_2 + (2,\ 3)\mu_3 + (2,\ 4)\mu_4 + (2,\ 5)\mu_5\}] \div [(\mu_0^1 + (0,\ 1)\mu_1^1$$
$$(1-L_2) + (0,\ 2)\{\mu_2^1 + (2,\ 3)\mu_3^1 + (2{,}4)\mu_4^1 + (2,\ 5)\mu_5^1 + (2,\ 3,\ 7)\mu_7^1 +$$
$$(2,\ 4,\ 9)\mu_9^1 + (2,\ 5,\ 11)\mu_{11}^1 + (2,\ 3,\ 6)\mu_6^1 + (2,\ 4,\ 8)\mu_8^1 + (2,\ 5,\ 10)\mu_{10}^1\}]$$

$$= [(1-L_2)\mu_0 + p_{0,2}\ (\mu_2 + p_{2,3}\cdot\mu_3 + p_{2,4}\cdot\mu_4 + p_{2,5}\cdot\mu_5] \div D_1$$

where, $D_1 = [(\mu_0^1 + \mu_1^1\, p_{0,1})\,(1 - L_2) + p_{0,2}\,\{\mu_2^1 + p_{2,3}\cdot\mu_3^1 + p_{2,4}\cdot\mu_4^1 + p_{2,5}\cdot\mu_5^1 + p_{2,3}\cdot p_{3,7}\cdot\mu_7^1 + p_{2,4}\cdot p_{4,9}\cdot\mu_9^1 + p_{2,5}\cdot p_{5,11}\cdot\mu_{11}^1 + p_{2,3}\cdot p_{3,6}\,\mu_6^1 + p_{2,4}\cdot p_{4,8}\cdot\mu_8^1 + p_{2,5}\cdot p_{5,10}\cdot\mu_{10}^1\}]$

and L_2 is already specified.

(c) *Busy period of the server (for RepI):* From Fig. 4.5, the regenerative states where the server is busy doing RepI are $j = 4, 8, 9$; $k_1 = 2$, $k_2 = 2$. Using formula (4.15.4) the busy period of the server for doing the replacement of type-I is

$$B_0^{RI} = \left[\left(\frac{1}{1-L_2}\right)\{(0,\,2,\,4)\cdot\eta_4 + (0,\,2,\,4,\,8)\cdot\eta_8 + (0,\,2,\,4,\,9)\cdot\eta_9\}\right] \div [D_1/(1 - L_2)]$$
$$= (0,\,2)\{(2,\,4).\mu_4 + (2,\,4,\,8).\mu_9 + (2,\,4,\,9).\mu_9\} \div D_1$$
$$= [p_{0,2}\cdot p_{2,4}\,[\mu_4 + (p_{4,8} + p_{4,9})\mu_9\} \div D_1$$

where, D_1 is already specified.

(d) *Expected number of server's visits for doing the repairs of the main unit:* The expected number of server's visits can be calculated from the formula (4.15.5). From Fig. 4.5, the regenerative states where the server visits (afresh) for the repairs of the main unit are: $j = 1, 6, 8, 10$. The regenerative states are $i = 0$ to 11; $k_1 = 2$, $k_2 = 2$; $k_1 \neq j$. Thus, using the formula (4.15.5), we have

$$V_0^M = \left[(0,\,1) + \left(\frac{1}{1-L_2}\right)\{(0,\,2,\,3,\,6) + (0,\,2,\,4,\,8) + (0,\,2,\,5,\,10)\}\right] \div [D_1/(1 - L_2)]$$
$$= [(0,\,1)(1 - L_2) + (0,\,2)\{(2,\,3\,,\,6) + (2,\,4,\,8) + (2,\,5,\,10)\}] \div D_1$$
$$= [p_{0,1}\,(1 - L_2) + p_{0,2}\,(p_{2,3}\cdot p_{3,6} + p_{2,4}\cdot p_{4,8} + p_{2,5}\cdot p_{5,10})] \div D_1$$

D_1 and L_2 have already specified.

Thus, making use of this technique all the calculations are done and a lot of time and space may be saved.

(e) *Busy period of the server for RepII*: From Fig. 4.5, the regenerative states where the server is busy doing replacement of type-II are: $j = 5$, 10 and 11; $k_1 = 2$, $k_2 = 2$.
On using (4.15.4), the busy period of the server doing RepII is

$$B_0^{RII} = \left[\left\{\frac{1}{(1-L_2)}\right\}\{(0,2,5)\cdot\eta_5 + (0,2,5,10)\cdot\eta_{10} + (0,2,5,11)\cdot\eta_{11}\}\right] \div [D_1/(1 - L_2)]$$
$$= [p_{0,2}\cdot p_{2,5}\,\{\mu_5 + (p_{5,10} + p_{5,11})\cdot\mu_{11}\}] \div D_1.$$

(f) *Busy period of the server for inspection:* From Fig. 4.5, the regenerative state where the server is busy doing inspection is: $j = 2$; $k_1 = 2$, $k_2 = 2$. On using (4.15.4), the busy period of the server doing inspection is:

$$B_0^I = \left[\left\{\frac{(0,2)}{(1-L_2)}\right\}\eta_2\right] \div [D_1/(1 - L_2)] = p_{0,2}\cdot\mu_2 \div D_1$$

(g) *Busy period of the server for repairs of the main unit:* From Fig. 4.5, the regenerative states where the server is busy doing repairs of the main unit are: j = 1, 6, 8, 10; k_1 = 2, k_2 = 2; $\eta_1 = \eta_6 = \eta_8 = \eta_{10} = \mu_1$.

On using (4.15.4), the busy period of the server doing repairs of the main unit is:

$$B_0^M = \left[(0,1)\eta_1 + \left\{\frac{1}{(1-L_2)}\right\}\left\{\begin{matrix}(0,2,3,6)\eta_6 + (0,2,4,8)\eta_8 \\ +(0,2,5,10)\eta_{10}\end{matrix}\right\}\right] \div [D_1/(1-L_2)]$$

$$= [\{p_{0,1}\cdot(1-L_2)] + p_{0,2}\cdot(p_{2,3}\cdot p_{3,6} + p_{2,4}\cdot p_{4,8} + p_{2,5}\cdot p_{5,10})\}\mu_1] \div D_1.$$

(h) *Busy period of the server for repairs of redundants:* From Fig. 4.5, the regenerative states where the server is busy doing repairs of the redundant units are: j = 3, 6, 7; k_1 = 2, k_2 = 2.

On using (4.15.4), the busy period of the server doing repairs of the redundant units is:

$$B_0^R = \left[\left\{\frac{1}{(1-L_2)}\right\}\left\{\begin{matrix}(0,2,3)\cdot\eta_3 + (0,2,3,6)\cdot\eta_6 \\ +(0,2,3,7)\cdot\eta_7\end{matrix}\right\}\right] \div [D_1/(1-L_2)]$$

$$= [p_{0,2}\cdot p_{2,3}\{\mu_3 + (p_{3,6} + p_{3,7})\mu_7\}] \div D_1.$$

(i) *Expected number of RepI:* From Fig. 4.5, the regenerative state(s) where the RepI takes place (afresh) along different paths: j = 4; k_1 = 2, k_2 = 2; $k_1 \neq j$. Using (4.15.5), the expected number of RepI is:

$$R_0^I = \left[\left\{\frac{(0,2,4)}{1-L_2}\right\}\right] \div [D_1/(1-L_2)] = [p_{0,2}\cdot p_{2,4}] \div D_1.$$

(j) *Expected number of RepII:* From Fig. 4.5, the regenerative state(s) where the RepII takes place (afresh) along different paths: j = 5; k_1 = 2, k_2 = 2; $k_1 \neq j$. Using (4.15.5), the expected number of RepII is:

$$R_0^{II} = \left[\left\{\frac{(0,2,5)}{1-L_2}\right\}\right] \div [D_1/(1-L_2)] = [p_{0,2}\cdot p_{2,5}] \div D_1.$$

(k) *Expected number of server's visits for the inspection/repairs/replacements of the redundant units:* From Fig. 4.5, the regenerative state(s) where the server visits (afresh) for the inspection/repairs/replacements of the redundant units along different paths are: j = 2; k_1 = 2, k_2 = 2; $k_1 \neq j$. Using (4.15.5), the expected number of server's visits is:

$$V_0^R = [(0,2)] \div [D_1/(1-L_2)] = [p_{0,2}(1-L_2)] \div D_1$$

In this chapter, thirteen methods are given for availability evaluation of systems. These methods are useful to analyse a system for its behaviour in different situations/states.

4.16 THE SIGN TEST

The sign test is the least difficult of the non-parametric tests. Its name originates from the way that it depends on the heading (or signs for pluses or minuses) of a pair of observation and not on their numerical greatness. In any issue where sign test is utilized, we check:

Number of + signs

Number of – signs

Number of 0's (i.e., which can't be incorporated either as positive or negative).

We take H_0; $P = 0.5$ (Null theory)

On the off chance that the thing that matters is because of chance impacts the probability of a + sign for a specific pair is 1/2, similar to the probability of a – sign. In the event that R is the occasion the less incessant sign happens, at that point R has the binomial conveyance with $p = 1/2$.

The basic incentive for a two-sided elective at $\alpha = 0.05$ can be advantageously found by the articulation

$$M = (m - 1) / 2 - (0.98) \sqrt{m}$$

H_0 is rejected in $R \leq M$ for the sign test. The sign test can be of two kinds:

- The one-example sign test.
- The matched example sign test.

In a one-example test we test the invalid speculation $\mu = \mu_0$ against a proper option based on an irregular example of size m, we supplant then μ_0 with a short sign and dispose of test esteem precisely equivalent to (put 0). We at that point test the invalid theory that these in addition to and less sign are estimations of an irregular variable having the binomial dissemination with $p = 1/2$.

4.17 THE PAIRED-SAMPLE SIGN TEST

The sign test has significant application in issue including paired information, for example, information identifying with the accumulation of records receivable when another gathering strategy; reactions of mother and daughter towards perfect family measure, and so forth in these issue, each pair of test esteems can be supplanted with one or more signs if the principal esteem is more noteworthy than the second, a less sign if the primary esteem is littler than the second, or be disposed of if the two qualities are equivalent.

Example: A typing school guarantees that in a six-week concentrated course, it can train students to type, on the normal, in any event 60 words for every moment. An arbitrary example of 15 graduates is given a typing test and the middle number of words every moment composed by every one of these students is given beneath. Test the theory that the middle typing speed of graduates is at any rate 60 words for each moment.

Students	Words per minute	WPM-60 Di
A	87	+21
B	76	+16
C	53	–7
D	71	+11
E	66	+6
F	59	–1
G	88	+28
H	73	+13
I	80	+20
J	66	+6
K	58	–2
L	70	+10
M	60	0
N	56	–4
O	55	–5
		R = 5

Solution: Using the trial method, we have:

1. $H_0 : Md = 60$

   $H_1 : Md > 60$
2. Dimension of hugeness; 5 per cent
3. Criteria: Since the substitute speculation is one-sided, we acquire the probability of our exploration from the above table and contrast and 0.05. On the off chance that the probability is under 0.05, we dismiss H_0.

   Result:

   No. of + signs = 9

   No. of – signs = 5

   No. of 0's = 1

   All out = 15

   $$M = (m - 1)/2 - (0.98)\sqrt{m}$$

   $$= (14 - 1)/2 - (0.98)\sqrt{14} = 6.5 - 3.67 = 2.83$$

   Since $R > M$ the invalid theory is acknowledged.

The sign test is frequently utilized for perceptions that have been arbitrarily chosen in pairs, utilizing a paired difference experiment. The sign test is one of only a handful couple of tests that can be employed when the main data accessible is that and A perception surpasses a B perception (or the other way around). Be that as it may, the sign test can likewise be utilized to look at two populace conveyances where test of equivalent size have been haphazardly and freely chosen from the two populaces. At that point the sets are framed by arbitrarily coordinating every perception in test A with an observation in simple B. A few analysts article to the utilization of sign test for the independent random samples case in light of the fact that, in certain circumstances the dismissal or non-dismissal may prompt dismissal of the invalid speculation, another might show non-dismissal. This reality may bother a few clients, however the test is substantial and the probability 'an' of dismissing the invalid speculation when it is genuine will be actually the esteem set by the experiments.

4.18 A RANK SUM TEST: THE MANN-WHITNEY U TEST

The sign test comparing two populace circulations disregards the actual magnitude of the matched perceptions and along these lines disposes of data that would be valuable in recognizing a take-off from the invalid theory. A statistical test that somewhat goes around this misfortune by using the overall extents of the perceptions was proposed by Mann and Whitney and is proportional to a test proposed independent by Wilcoxon. Rank sum test is an entire group of tests. Here we will talk about just a solitary one of the sort, the Mann-Whiney U test. With this test we can test the invalid speculation $\mu_1 = \mu_2$ without accepting whether the populaces inspected have generally the state of ordinary distributions. This test causes us to decide if two irregular examples have originated from identical populations. On the off chance that the facts demonstrate that the examples have originated from similar populaces it is sensible to expect that the methods for the positions appointed to have the estimations of the two samples are pretty much the equivalent. The alternative hypothesis is that the methods for the populace are not rise to and if so, the vast majority of the littler position will be go to the estimations of one sample, while the greater part of the higher positions will go to those of the other sample.

The trial of the invalid theory that the two samples originate from indistinguishable populaces may either be founded on R_1, the total of the positions of the estimation of first example, or on R_2, the entirety of the positions of the estimations of the second sample. It might be seen that practically speaking it doesn't make a difference which test we call test 1 and which we call test 2. On the off chance that the sample estimates are m_1 and m_2 the whole of R_1 and R_2 is simply the sum of first $m_1 + m_2$ positive number, which is known to be

$$(m_1 + m_2)(m_1 + m_2 + 1) / 2$$

This equation empowers us to discover R_2 on the off chance that we know R_1 and the other way around.

At the point when the utilization of the rank entireties was first proposed as a non-parametric option in contrast to the two-sample t-test, the choice depended on R_1 or R_2, however at this point the choice is normally founded on both of the related measurements:

$$U_1 = m_1 m_2 + m_1 (m_1 + 1) / 2 - R_1$$

$$U_2 = m_1 m_2 + m_2 (m_2 + 1) / 2 - R_2$$

where, m_1 and m_2 are the measures of the samples and R_1 and R_2 are the rank aggregates of the relating tests. For samples, if both m_1 and m_2 are under 10 (a few analysts state 8) extraordinary tables must be used, and if U is smaller than the critical value, H_0 can be related to the standard normal curve by the statistics.

$$Z = U - (m_1 m_2)/2 / \sqrt{m_1 m_2 (m_1 + m_2)} / 12$$

Example: Twenty-four candidates for a position are met by three administrators and appraised on a size of 5 as to appropriateness for the position. Each candidate is given a "suitability" score which is the whole of the three numbers. In spite of the fact that college education is not a prerequisite for the position, a work force chief felt that it might have some bearing on suitability for the position. Raters made their rating based on individual had finished in any event multi year of college. Utilize the Mann-Whitney U test to decide if there was a distinction in the scores of the two groups. Utilize a 0.05 dimension of centrality group C had an educational background of under two years of college, while group D had finished in any event two years of college.

"Suitability" Scores

Group C	Group D
7	8
11	9
9	13
4	14
8	11
6	10
12	12
11	14
9	13
10	9
11	10
11	8

Solution: To rank the information, it is every now and again supportive to have a worksheet, especially when there is a vast mass of information. Such a worksheet could list the qualities from low unreasonably high for each gathering (may be utilizing a tally system), at that point list all ranks, computing the common marks. At last, the ranks for scores in each sample can be recorded. The finished sheet may take after the accompanying:

Scores C	Scores D	Ranks	Common Ranks	Rank C	Rank D
4		1	1	1	
6		2	2	2	
7		3	3	3	
8	8, 8	4, 5, 6	5	5	10
9, 9	9, 9				
		7, 8, 9, 10	8.5	17	17
10	10, 10	11, 12, 13	12	12	24
11,11,11,11	11	14, 15, 16, 17, 18	15.5	46.5	15.5
12	12	19, 20	18.5	18.5	18.5
	13, 13	21, 22	20.5		41
	14, 14	23, 24	22.5		45
			Total	105	171

A valuable check is to see that the whole of the two rank sums must be equivalent to 1/2 M (M + 1), where M is the all-out number of perceptions. Here, M = (23.5) (23) (24) = 276 = 105 + 0171. This serves check for inner consistency.

$$U_1 = m_1 m_2 + m_1 (m_1 + 1) / 2 - R_1$$
$$m_1 = 11,\ m_2 = 12,\ R_1 = 105$$
$$U_1 = 93$$
$$U_1 = m_1 m_2 + m_3 (m_2 + 1) / 2 - R_2$$
$$= 39$$

On the off chance that a table with basic estimations of U for m_1 = 11 and m_2 = 12 is accessible, we may utilize it. If not, we perform the transformation to Z.

$$Z = U - (m_1 m_2)/2 / \sqrt{m_1 m_2 (m_1 + m_2)} / 12$$
$$U = 39,\ m_1 = 11,\ m_2 = 12$$
$$Z = 1.74$$

Since the determined esteem is under 1.96, the invalid theory is acknowledged. Consequently, there is no significant difference in the scores of the two gatherings.

4.19 THE ONE SAMPLE RUNS TEST

The different tests examined above depend on the suspicion that we are dealing with random samples. In any case, there are numerous situations where it is hard to choose this assumption is justifiable. This is especially obvious when we have little or no control over the choice of the information. For instance, on the off chance that we need to anticipate the offers of a readymade piece of clothing store for a given month, we must choose the option to utilize deals information from earlier years and consider the financial conditions by and large. None of this data would constitute a random sample in the severe sense.

Various strategies have been created for making a decision about the arbitrariness of a sample on the basis of the order in which the perceptions are taken. It is conceivable to decide if the samples that look suspiciously non-arbitrary might be credited to risk. The method examined beneath depends on the hypothesis of runs. A run is the progression of indistinguishable letters (or different sorts of images) which is followed or preceded by various letter or no letters by any stretch of the imagination. To represent think about the accompanying plan of letters:

BBB AAAA CCCC DDDDDD EE FFF

Using underlines to combine the letters which constitute the runs. We find that first share is a run of three *B*'s, then a run of four *A*'s, then a run of four *C*'s, then a run of six *D*'s, then a run of two *E*'s and lastly a run of three *N*'s. In all, there are six runs of varying lengths.

It might be called to attention that the complete number of runs showing up in an understanding is frequently a decent sign of a conceivable absence of haphazardness.

On the off chance that there are too few runs, a definite grouping, bunching or pattern might be suspected. Then again, if there are such a large number of runs a type of continued exchanging example might be suspected. Along these lines it might be conceivable to demonstrate that too numerous or too few keeps running in a sample indicated something other than chance when the things were chosen. The quantity of runs, or r, is a measurement with its own special sampling distribution and its very own test. To derive the mean of the testing circulation of r measurement, the accompanying equation is utilized

$$\mu_1 = (2m_1m_2) / (m_1 + m_2) + 1$$

μ_1 = mean of the r statistic

The standard mistake of the r statistic is determined by the formula:

$$\sigma_1 = \frac{\sqrt{2m_1m_2(2m_1m_2 - m_1 - m_2)}}{(m_1 - m_2)^2 (m_1 + m_2 - 1)}$$

It might be noticed that the sampling distributions of r can be intently approximated by the typical appropriation if either m_1 or m_2 is bigger than 20.

$$Z = r - \mu_r / \alpha_r$$

EXERCISE

1. What is a tie-set method and cut-set method?
2. What do you mean by minimal tie-set method?
3. Draw and explain Boolean function technique.
4. Define the renewal theoretic approach. Explain the Markov method.
5. Explain the paired sample test method.
6. Define system description and decomposition method.
7. Compute Markov analysis of availability model for two component stand-by system.

5

Chapter

Reliability Analysis of Some Electrical-Electronics, Telecommunication and Mechanical Systems

In our daily life, the words 'reliable' and 'reliability' are frequently used indicating the property of a person, an animal, an equipment and a machine, etc. Thus, reliability is a part of our daily life. To analyze daily life problems, a number of theoretical and practical methods are used. Some of these methods and the basic things helpful in the analysis of the problems have been discussed in Chapters 1 to 4. In this chapter, analysis of various systems is given. Simple problems are discussed in most of the texts on reliability. In this chapter, some systems of complex nature are discussed from where simple systems may be extracted after some manipulation/adjustments. The analysis of these systems includes the use of techniques given in Chapter 4 and some more techniques which are not mentioned in Chapter 4 but are useful for the analysis purpose. Although the problems given in Chapters 6–8 are all of mechanical systems but in this chapter a complex mechanical system (container manufacturing plant) is discussed. Besides these, a number of good concepts generally occurring in practical problems, viz. priority repair, warm standby, waiting time, common cause failure, failure due to human error, role of imperfect switch, cost function, etc. are discussed. The objective of this chapter is to introduce the readers about the use of the various methods to our daily life problems.

5.1 k-OUT-OF-n SYSTEMS

In the literature these systems are mentioned in two ways—(i) k-out-of-n : G system, (ii) k-out-of-n : F system. k-out-of-n : G system is a configuration which works if and only if at least k of its n units are good (i.e., functioning) while k-out-of-n : F system is a system which fails as soon as k out of n units fail. Such systems are used in electrical-electronics and telecom systems. For example, consider a company which services communication satellites using n reusable orbiters. Orbiters are considered failed due to damage sustained on re-entry and perhaps other requirements. It is 1-out-of-n : G system since system is survived even with one satellite unit. As another example, consider a telecommunication system with n relay stations (either satellites or ground stations). Suppose a signal emitted from station 1 is received by stations 2 and 3 and a signal relayed by station 2 can be received by stations 3 and

4, and so on. Thus, when station 2 fails the telecommunication system is still able to transmit a signal from station 1 to station n. If both stations 2 and 3 fail, a signal cannot be transmitted from station 1 directly to station 4. Therefore, the system fails. Similarly, if any two consecutive stations fail, the system fails. This is called a consecutive 2-out-of-n : F system. In real world, there are many other systems, such as electric generators sharing an electric load in a plant, multiple regulate power supplies across a buss, bolts used to hold a machine member and pumps or valves in a hydraulic system. These are load sharing k-out-of-n : G systems. In this section, we consider such systems.

(a) 1-out-of-*n* : *G* system – An interesting system discussed by Linton [1989] is given under the following assumptions:

(i) The n units are identical and their failure rate is constant, λ.
(ii) An unit has three states – good, failed and waiting for repair, failed and in repair.
(iii) Repair is done on first come first served basis.
(iv) There is only one repair facility.
(v) Repair times follow a general distribution and repair times are statistically independent of failure times.
(vi) When all n units are failed, the system is failed.
(vii) Initially, all units are good.
(viii) Switch over device is perfect.

Let s be the Laplace transform parameter

$p_0(t)$: probability density that at time t there is no failed unit.

$p_i(t, z)$: probability density that at time t there are i failed units in the system and the elapsed repair time is z, $1 \le i \le n - 1$.

$q_k(t, z)$: probability density that at time t there are k failed units and the system is waiting for repair upto time z, $2 \le k \le n - 1$.

The state transition equations are given by

$$\frac{dp_0(t)}{dt} = -n\lambda\, p_0(t) + \int_0^t e^{-(n-1)\lambda z}\, g_1(t, z)\,dF(z) \tag{5.1.1}$$

$$\frac{\partial g_1}{\partial t} + \frac{\partial g_1}{\partial z} = 0 \tag{5.1.2}$$

$$\frac{\partial q_k}{\partial t} + \frac{\partial q_k}{\partial z} = \begin{cases} -(n-z)\lambda q_2(t,z) + (n-1)\,e^{-(n-1)\lambda z} g_1(t,z), & k=2 \quad (5.1.3) \\ -(n-k)\lambda q_k(t,z) + (n-k+1)\lambda q_{k-1}(t,z), & 3 \le k \le n-1 \quad (5.1.4) \end{cases}$$

with initial condition $p_0(0) = 1$ and boundary conditions

$$p_k(t,0) = \begin{cases} n\lambda p_0(t) + \int_0^t q_2(t,z)\,dF(z), & k=1 \quad (5.1.5) \\ \int_0^t q_{k+1}(t,z)\,dF(z), & 2 \le k \le n-2 \quad (5.1.6) \\ o, & k = n, n-1 \quad (5.1.7) \end{cases}$$

where, $p_1(t, z) \equiv g_1(t, z)\, e^{-(n-1)\lambda z}\, \bar{F}(z)$, $p_k(t, z) \equiv q_k(t, z)\, \bar{F}(z)$, $2 \le k \le n - 1$ and

$$\bar{F}(z) = 1 - F(z)$$

Equation (5.1.1) is ordinary differential equation of first order while eqns. (5.1.2)–(5.1.4) are partial differential equations of first order. These equations may be solved using initial and boundary conditions as discussed in Section 4.12. Linton solved this system of equations using a theorem—A solution of eqns. (5.1.3) and (5.1.4) in terms of arbitrary functions $[g_k(t-z)]_{k=2}^{n-1}$ is

$$q_k(t, z) = \sum_{i=0}^{k-1} (-1)^i \binom{n-k+i}{i} e^{-(n-k+i)\lambda z}\, g_{k-i}^{(t-z)}, \tag{5.1.8}$$

for $2 \le k \le n - 1$, $n \ge 4$.

Linton derived some results regarding waiting time, etc. Elsayed (1996) also discussed 1-out-of-n : F system.

(b) *k*-out-of-*n* : *G* system – A shared load k-out-of-n : G system is considered by Shao and Lamberson [1991]. The shared load redundant configuration is more complicated and more practical than simple k-out-of-n : G system. The discussion about shared load system can be reduced to simple system. The shared-load system is designed such that a controller monitors the system, detects and disconnects failed units and adjusts the operating parameters of the surviving units to keep the system working most efficiently and economically. The component failure rates can depend on the system states. The very practical and interesting system is discussed by Shao and Lamberson under the following assumptions:

(i) Each unit is either working or failed.

(ii) The failure rate of all functioning units in the system is the same and constant in reaching success state. When a unit fails, the system state changes and the failure rate for every functioning unit changes. For i units functioning, every functioning unit has the constant failure rate λ_i, $i = k, \ldots, n$. The unit failure rates for each success state obey

$$k\lambda_n < \lambda_{n-1} < \ldots\ldots < \lambda_k.$$

(iii) The failed unit harms the system and must be detected and disconnected by a controller. The probability of successful detection and switching of the controller is 'α', for each failure event. If the controller cannot detect and disconnect a failed unit or the controller itself has failed, the system fails. The controller failure rate is a constant λ_c.

(iv) At most r units can be in repair at one time, the constant repair rate is μ. The repair rate for j units failed is $\mu_j = \mu \min(j, r)$. A repaired unit is as good as new and is immediately reconnected to the system.

(v) The switch over time and the time for re-energizing a repaired unit are negligible. The controller is never repaired or replaced during a mission.

State j ($j = 0, 1, \ldots, n - k$): j units have failed and have been disconnected from the network. $n - j$ units and the controller are functioning.

State $n - k + 1$: the system fails because $k - 1$ units are functioning.

However, the system can return to the working state $(n - k)$ at a repair rate μ_{n-k+1}.

State f: the system fails because the controller cannot detect and disconnect a failed unit. The transition diagram for the system is as follows (Fig. 5.1):

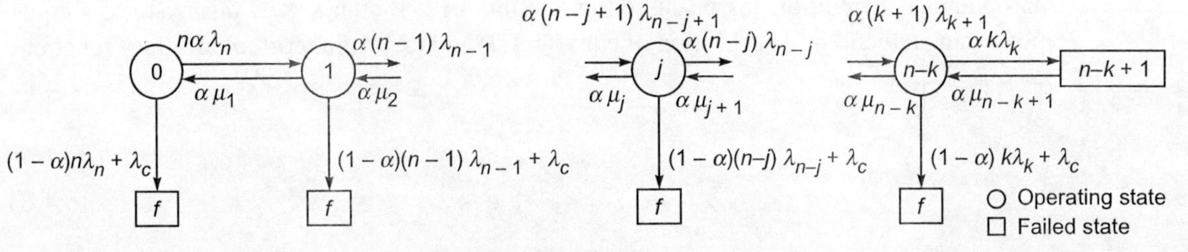

Fig. 5.1

The differential-difference equations associated with the above transition diagram are

$$\frac{dp_0(t)}{dt} = -(n\lambda_n + \lambda_c)\, p_0(t) + \alpha\mu_1\, p_1(t), \tag{5.1.9}$$

$$\frac{dp_j(t)}{dt} = -[(n-j)\lambda_{n-j} + \lambda_c + \alpha\mu_j]\, p_j(t) + \alpha(n-j+1)\lambda_{n-j+1}\, p_{j-1}(t) + \alpha\mu_{j+1}p_{j+1}(t), \text{ for } j = 1, 2, \ldots, n-k \tag{5.1.10}$$

$$\frac{dp_{n-k+1}(t)}{dt} = -\alpha\mu_{n-k+1}p_{n-k+1}(t) + \alpha k\lambda_k\, p_{n-k}(t) \tag{5.1.11}$$

with initial conditions $p_i(0) = 1$ for $i = 0$ and zero otherwise.

Equations (5.1.9) to (5.1.11) are ordinary differential equations of first order which may be solved using either Laplace transforms or by other methods like matrix, etc. Shao and Lamberson solved the equations with the aid of Laplace transforms. System availability $A(t)$ of the system is

$$A(t) = \sum_{j=0}^{n-k} p_j(t).$$

Availability can be considered the generalized reliability when the system is repairable or component-replaceable. System reliability is the availability of the same system without repair and replacement. Let $\mu_j = 0$ when $j = 0, 1, \ldots, n-k+1$, the reliability function of the shared load system considered is

$$R(t) = \sum_{j=0}^{n-k} p_j(t).$$

For long-run availability $A(\infty)$, taking all derivatives equal to zero as $t \to \infty$ and solving recursively all the probabilities are obtained in terms of p_0 which is evaluated using the normalizing condition.

(c) *K*-out-of-($N + M$) : *G* System In 5.1(a) and 5.1(b), we have discussed systems concerned with telecom and power control. In 5.1(a) controller is perfect while in 5.1(b), controller is not perfect and the working of controller is considered assigning a probability 'α' for successful working. In this subsection, a new different system is given where the controller is perfect. There are N working units while M units are in warm standby. Moreover, there may be common cause failure in the system. In many industrial establishments and electrical/electronics systems warm standby and common cause failure situations occur. Singh (1989) considered such a system and quoted a number of practical examples of such situations. Making the discussion more realistic (taking general repair time instead of constant) and departing from the usual Laplace transform technique used by reliability research workers for solving the governing system equations, he solved the equations by direct integration. Following are the assumptions governing the system:

(i) The system consists of $N + M$ identical units and 'r' repair facilities. N units are active while M units are in warm standby. Initially, i.e., at time $t = 0$, the system is perfectly good.

(ii) Switching device is perfect and switch-over time is negligible.

(iii) Failures and repairs are statistically independent. The service discipline is first come first served. A repaired unit is as good as new.

(iv) Upon failure, if all repair facilities are busy, the failed unit joins the end of the queue of non-operating units. Intermediate ordinary failures are attended by repairmen on first come first served basis.

(v) Failure rates are constant and failure rate of an active unit is different from the standby unit while repair times are arbitrarily distributed.

(vi) Common cause failure can only occur with more than one operable unit in the system.

(vii) System comes in failed state if less than K units remain good or a common cause failure occurs.

Let b, d : respective constant failure rate of an active unit and standby unit,

c : failure due to common cause,

i : number of good units, $i = k, k + 1, k + 2, \ldots, N + M$,

$p_{N+M}(t)$: probability density that the system is in good state at time t,

$p_i(x, t)$: probability density that at time t, the system is in ith state and has an elapsed repair time x,

$p_i(y, t)$: probability density that at time t, the system is in ith failed state and has an elapsed repair time y,

$\mu_i(x)$: repair rate of a unit when the system is in ith failed state and has an elapsed repair time x, with $\mu_i(x) = min(N + M - i, r)\ \mu(x)$,

α : constant common cause failure rate of the system. $\alpha = 0$ if $k = 1$ and $\alpha = \alpha_0$ for $k > 1$,

$\beta_j(y)$: repair rate when the system is in jth failed state and has an elapsed repair time y,

r : number or repair facilities.

$$p_i(t) = \int_0^\infty p_i(x, t)dx, \; p_j(t) = \int_0^\infty p_j(y, t)dy.$$

Using these assumptions and symbols, the transition diagram is given in Fig. 5.2.

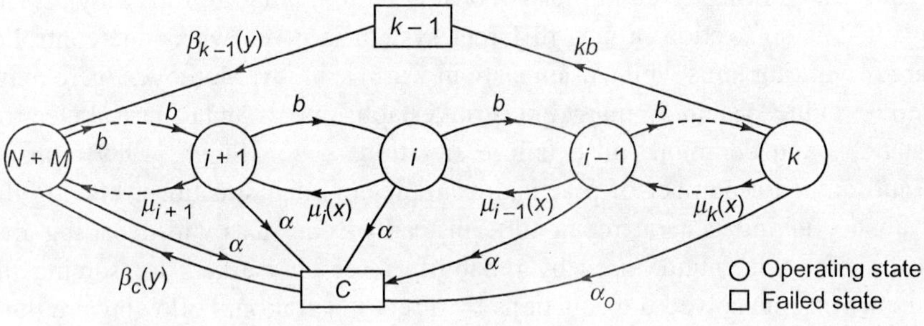

Fig. 5.2

The differential-difference state equations associated with the transition diagram (Fig. 5.2) are as follows:

$$\left(\frac{d}{dt} + Nb + Md + \alpha\right) \quad p_{N+M}(t) = \int p_{N+M-1}(x, t)\; \mu_1(x)dx + \sum_{j=c,\,k-1} \int p_j(y, t)\; \beta_j(y)\; dy \tag{5.1.12}$$

$$\left(\frac{\partial}{\partial x} + \frac{\partial}{\partial t} + A_i + \mu_i(x) + \alpha\right) p_i(x, t) = B_i, \; k \le i \le M + N - 1 \tag{5.1.13}$$

$$\left(\frac{\partial}{\partial y} + \frac{\partial}{\partial t} + \beta_j(y)\right) \; p_j(y, t) = 0, \, j = c, \, k - 1 \tag{5.1.14}$$

with boundary and initial conditions

$$p_i(0, t) = a \int p_{i+1}(x, t)dx \tag{5.1.15}$$

$$\left.\begin{array}{l} p_j(y, 0) = p_i(x, 0) = 0 \text{ for all } i, j \\ p_{N+M}(0) = 1, \; p_i(0) = 0 \;\text{ for } i \ne N + M \end{array}\right\} \tag{5.1.16}$$

where,

$$A_i = \begin{cases} kb & \text{if } i = k \\ Nb + (M - i)d & \text{if } k + 1 \le i \le M + N - 1 \end{cases}$$

$$B_i = \begin{cases} (k+1)b\, p_{k+1}(x,t), \text{ if } i = k \\ A_i\, p_{i-1}(x,t) + \int p_{i+1}(x,t)\, \mu_{i+1}(x)dx, \text{ if } k+1 \le i \le M+N-1 \end{cases}$$

$$a = \begin{cases} (k+1)b & \text{if } i = k \\ b_{i+1} & \text{if } k+1 \le i \le M+N-1, b_{i+1} = b(i+1) \end{cases}$$

$$p_j(0,\ t) = \begin{cases} \alpha \sum_{i=k}^{N+M} \int p_i(x,t)\, dx & \text{if } j = c \\ kb \int p_k(x,t)\, dx & \text{if } j = k-1 \end{cases}$$

and limits of integration are 0 to ∞.

Using initial and boundary conditions, solutions of eqns. (5.1.12)—(5.1.14) are given by (direct integration method given by Singh [1977], Kumar et al., [1988])

$$p_j(y,\ t) = D_j(t-y)\ e^{-\int \beta_j(y)dy},\ j = c,\ k-1 \tag{5.1.17}$$

$$p_k(x,\ t) = [a \int p_{k+1}(x,\ t) A_k(x)dx + a p_{k+1}(t-x)] A_k^{-1}(x) \tag{5.1.18}$$

$$p_i(x,\ t) = [b_{i+1} \int p_{i+1}(x,\ t) A_i(x)dx + \int \{\int p_{i-1}(x,\ t) U_{i-1}(x)dx\} A_i(x)dx + b_{i+1}\, p_{i+1}\ (t-x)] A_i^{-1}(x)$$

$$(k+1 \le i \le M+N-1), \tag{5.1.19}$$

$$p_{N+M}(t) = \left[1 + \int \left\{ \int p_{N+M-1}(x,t)\, \mu_1(x)\, dx + \sum_{j=c,\,k-1} \int D_j(t-y) Q_j(y)\, dy \right\} A(t)\, dt \right] A^{-1}(t) \tag{5.1.20}$$

where, $A_i(x) = \exp\{\int (b_i + \mu_i(x) + \alpha) dx\}$, $(b_k = kb)$

$$A(t) = (Nb + Md + \alpha)t\ ;\ B_j(y) = \beta_j(y) e^{-\int \beta_j(y)dy}$$

$$Q_j(t-y) = kb \int p_{k-1}(x,\ t-y)dx, \text{ if } j = k-1$$

$$Q_j(t-y) = \alpha \sum_{j=k}^{N+M} \int p_i(x,\ t-y)dx, \text{ if } j = c.$$

The system availability $A(t) = \sum_{i=k}^{N+M} \int p_i(x,\ t)dt.$

Taking $\frac{d}{dt} \to 0$, $\frac{\partial}{\partial t} \to 0$, as $t \to \infty$ and the probabilities independent of t, steady-state probabilities can be evaluated in terms of p_{N+M} which, in turn, is evaluated using the normalizing condition, i.e., sum of all the probabilities is equal to one. For $d = 0$, results for cold standby redundant system may be obtained. If $d = b$, the results for hot standby redundant system are obtained. Also putting $\alpha = \alpha_0 = 0$ and $\beta_c(y) = 0$ the results are obtained for the system without common cause failure. A particular case of 7-out-of-14 : G is analyzed by Gupta et al., [2005] in detail.

(d) 1-out-of-*N* : *G* (SPECIAL) – In 5.1(a), 5.1(b) and 5.1(c) different systems are discussed. In 5.1(a), 1-out-of-*n* : *G* system is considered with perfect switch-over device and waiting time while in 5.1(b), *k*-out-of-*n* : *G* system is given having imperfect switch and state dependent failure and repair rates. In 5.1(c), an entirely different problem with perfect switching device, state dependent failure (in Fig. 5.2 although these are shown by b simply but in the analysis it is clearly written by b_{i+1}) and repair rates and the practical concept of common cause failure is discussed. Also the idea of warm standby redundancy is discussed. A common cause failure may occur as a result of design deficiencies and abnormal environmental conditions, etc. Failures may occur because of human errors caused by maintenance errors, misinterpretation of instruments or wrong actions. In this subsection, we consider the system consisting of N identical units and 'r' repairmen where failures occur due to common cause failures and critical human errors. For example, in train lighting system both types of failure occur. On complete failure, the system is sent to the workshop for repair, whereas intermediate failures are attended to by repairmen. Singh and Dayal [1991] analysed the system taking the assumption that completely failed system is not repaired. Other assumptions are similar as given in 5.1(c) with some changes.

Let us assume the following:

H: failure caused by critical human error,

h_i: constant failure rate from ith state caused by critical human error,

α_i: constant failure rate from ith state caused by common cause failure,

a_i: constant failure rate from ith state,

$b_i(x)$: repair rate when the system is in ith state and has an elapsed repair time x,

$p_0(t)$: probability density that at time t the system is good,

$p_i(x, t)$: probability density that at time t the system is in ith state and the elapsed repair time is x,

$p_H(t)$: probability density that at time t the system is in failed state due to critical human error,

$p_c(t)$: probability density that at time t the system is in failed state because of common cause failure.

Using these notations and the assumptions stated in 5.1(c), the transition diagram is given in Fig. 5.3.

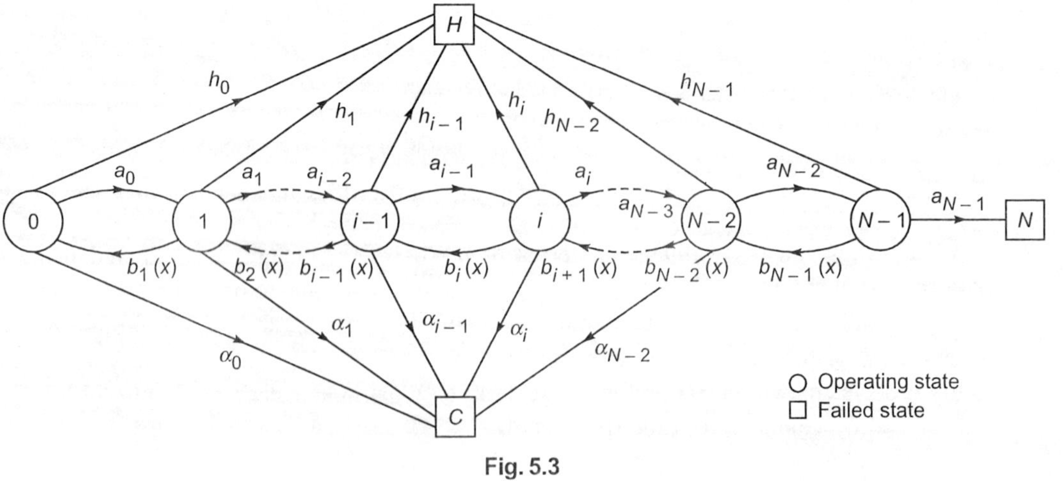

Fig. 5.3

The differential-difference equations associated with the transition diagram (by probability considerations) are

$$\frac{dp_0(t)}{dt} + A_0 p_0(t) = \int b_1(x)\, p_1(x, t)\,dx \tag{5.1.21}$$

$$\left(\frac{\partial}{\partial x} + \frac{\partial}{\partial t} + b_i(x) + A_i\right) p_i(x, t) = b_{i+1}(x)\, p_{i+1}(x, t) + a_{i-1}\, p_{i-1}(x, t),\ i = 1, 2, \ldots, N-2 \tag{5.1.22}$$

$$\left(\frac{\partial}{\partial x} + \frac{\partial}{\partial t} + b_{N-1}(x) + A_{N-1}\right) p_{N-1}(x, t) = a_{N-2}\, p_{N-2}(x, t) \tag{5.1.23}$$

$$\frac{dp_N(t)}{dt} = a_{N-1}\int p_{N-1}(x, t)\,dx \tag{5.1.24}$$

$$\frac{dp_c(t)}{dt} = \int\left\{\sum_{i=0}^{N-2} \alpha_i p_i(x,t)\right\} dx \tag{5.1.25}$$

$$\frac{dp_H(t)}{dt} = \int\left\{\sum_{i=0}^{N-1} h_i p_i(x,t)\right\} dx \tag{5.1.26}$$

with initial and boundary conditions

$$p_0(0) = 1,\ p_N(0) = p_C(0) = p_H(0) = p_i(x, 0) = 0$$
$$p_i(0, t) = a_{i-1}\int p_{i-1}(x, t)\,dx, \text{ for all } i \geq 1.$$

Limits of integration are 0 to ∞ and $p_i(t) = \int p_i(x, t)dx$, $i = 0, 1, 2, \ldots$

Equations (5.1.21), (5.1.24)—(5.1.26) are ordinary first order differential equations while eqns. (5.1.22) and (5.1.23) are partial differential equations of first order (Lagrange's type) and may be solved easily. The solutions are given by

$$p_0(t) = [1 + \int (\int b_i(x)\, p_i(x, t)dx)e^{A_0 t}dt]e^{-A_0 t}$$
$$p_i(x, t) = [\int\{b_{i+1}(x)\, p_{i+1}(x, t) + a_{i-1}\, p_{i-1}(x, t)\}e^{+A_i x + \int b_i(x)dx}dx + a_{i-1}\int p_{i-1}(x, t-x)dx].$$
$$e^{-A_i x} - \int b_i(x)dx,\ i = 1, 2, \ldots, N-2$$
$$p_{N-1}(x, t) = a_{N-2}\,[\int p_{N-2}(x, t)e^{xA_{N-1}} + \int b_{N-1}(x)dx\,dx + \int p_{N-2}(x, t-x)dx]\, e^{-xA_{N-1}\int b_{N-1}(x)dx}$$

$$p_N(t) = a_{N-1}\int\{\int p_{N-1}(x, t)dx\}dt;\ p_c(t) = \int\left\{\sum_{i=0}^{N-2}\alpha_i p_i(x,t)\,dx\right\}dt,$$

$$p_H(t) = \int\left\{\int\sum_{i=0}^{N-1} h_i p_i(x,t)\,dx\right\}dt$$

where, $A_i = a_i + \alpha_i + h_i$, $i = 0, 1, 2, \ldots, N-2$ and $A_{N-1} = a_{N-1} + h_{N-1}$.

Availability functions $A(t)$ and mean time to system failure are

$$A(t) = \int \left\{ \sum_{i=0}^{N-1} p_i(x,t) \right\} dx, \text{ MTSF } = \int R(t)dt$$

taking $p_0(t) = \int p_0(x, t)dx$ and $R(t)$ equal to $A(t)$ when there is no repair.

Steady-state probabilities are obtained by taking $\frac{d}{dt} \to 0$ as $t \to \infty$ and the probabilities independent of t. The system availability function $A(\infty) = \int \left\{ \sum p_i(x) \right\} dx$.

Results for error free, non-repairable, without common cause failures and with constant repair rate may be obtained by putting $h_i = 0$, $b_i(x) = 0$, $\alpha_i = 0$ and $b_i(x) = b$, respectively in the foregoing analysis.

5.2 ON-SURFACE TRANSIT SYSTEM

Reliability is an important consideration in the planning, design and operation of transit systems. Lower reliability means increased unscheduled maintenance and decreased equipment availability. If the total system availability is low, more vehicles will be required to meet the demand but even with more vehicles, the system performance may not be satisfactory. More vehicles can increase system availability but do not decrease the incidence of system failures. Transit system reliability reflects the ability of a transit system to keep operating schedules.

The normally operating vehicle performance can be degraded either due to hardware failures or human errors. If the degradation of the vehicle is serious, it is taken to the repair workshop, otherwise, the vehicle continues its operation even though it is degraded. The operating vehicle may fail either from its degraded state or from its normal operational state. In either case, the vehicle failure can be either due to hardware failures or due to human errors. The failed vehicle is towed to the repair workshop. The fully repaired vehicle goes back into its normal service. This interesting transit system is considered by Dhillon and Rayapati (1985) making the following assumptions and notations:

(i) failure, repair, towing and other transition rates are constant,
(ii) human error rates are constant,
(iii) failures and repairs are statistically independent,
(iv) fully repaired vehicle is as good as new, and
(v) at $t = 0$, system is good.

$p_i(t)$ probability density that the transit system is in state i at time t for $i = 0, 1, 2, 3, 4, 5$.

λ_1 vehicle degradation rate due to hardware failures.
λ_2 hardware failure rate of the vehicle from state 1 to state 3,
λ_3 hardware failure rate of the vehicle from state 2 to state 3.
λ_4 hardware failure rate of the vehicle from state 0 to state 3,
λ_5 vehicle degradation rate due to human errors,
λ_6 failure rate of the vehicle due to human errors from state 2 to state 4,
λ_7 failure rate of the vehicle due to human errors from state 1 to state 4,

λ_8 failure rate of the vehicle due to human errors from state 0 to state 4,

λ_i transition rate from state 'i' to state 5, for i = 9, 10, 11, 12. λ_9 and λ_{10} are towing rates from states 3 and 4 respectively to state 5.

μ repair rate of the vehicle from state 5 to state 0.

Using the notations and assumptions given above, the transition diagram is as shown in Fig. 5.4.

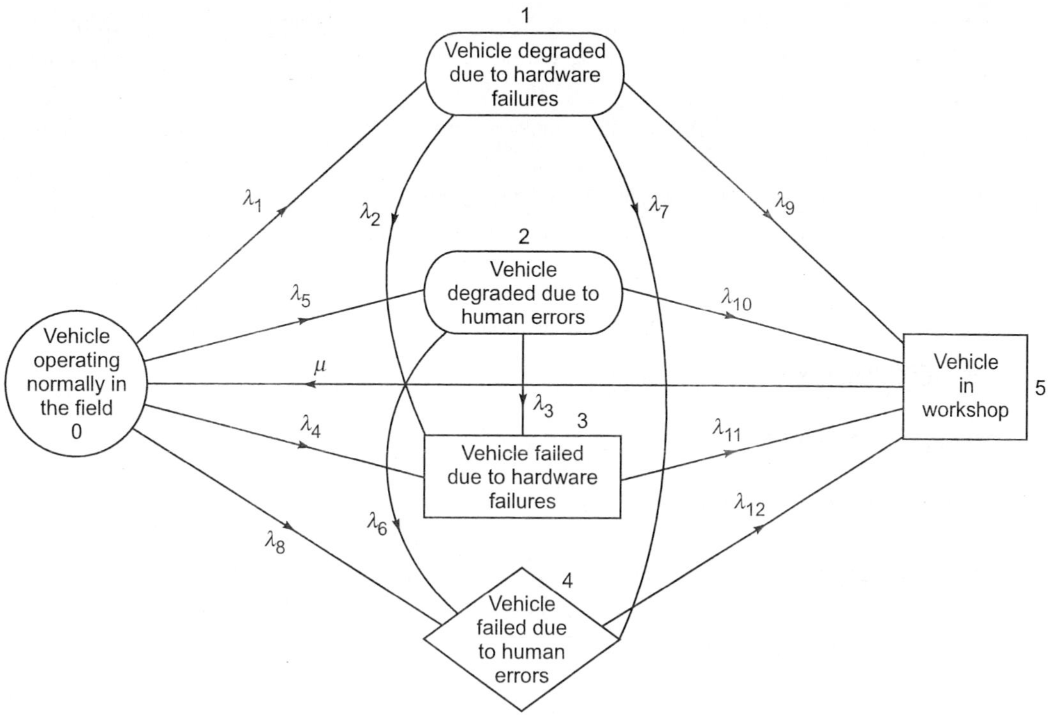

Fig. 5.4

Probability considerations give the following differential-difference equations associated with Fig. 5.4:

$$\frac{dp_0(t)}{dt} + (\lambda_1 + \lambda_4 + \lambda_5 + \lambda_8)\, p_0(t) - \mu\, p_5(t) \tag{5.2.1}$$

$$\frac{dp_1(t)}{dt} + (\lambda_2 + \lambda_7 + \lambda_9)\, p_1(t) = \lambda_1\, p_0(t) \tag{5.2.2}$$

$$\frac{dp_2(t)}{dt} + (\lambda_3 + \lambda_6 + \lambda_{10})\, p_2(t) = \lambda_5\, p_0(t) \tag{5.2.3}$$

$$\frac{dp_3(t)}{dt} + \lambda_{11}\, p_3(t) = \lambda_4\, p_0(t) + \lambda_2\, p_1(t) + \lambda_3\, p_2(t) \tag{5.2.4}$$

$$\frac{dp_4(t)}{dt} + \lambda_{12}\, p_4(t) = \lambda_8\, p_0(t) + \lambda_7\, p_1(t) + \lambda_6 p_2(t) \tag{5.2.5}$$

$$\frac{dp_5(t)}{dt} + \mu p_5(t) = \sum_{i=1}^{4} \lambda_{i+8} p_i(t) \tag{5.2.6}$$

with initial condition (i.e., at $t = 0$)

$$p_0(0) = 1 \text{ and } p_i(0) = 0 \text{ for } i \neq 0.$$

Setting the derivatives equal to zero in eqns. (5.2.1)—(5.2.6) and solving recursively, all the steady-state probabilities are obtained in terms of p_0 which is evaluated using the normalizing condition $\sum_{i=0}^{5} p_i = 1$. The steady-state availability of the transit system is $A_v = p_0 + p_1 + p_2$. Putting $\mu = 0$ in eqns. (5.2.1)—(5.2.6) and taking Laplace transforms of the equations and solving for $p_0(s)$, $p_1(s)$, $p_2(s)$, Laplace transforms of $p_0(t)$, $p_1(t)$ and $p_2(t)$, respectively, we obtain Laplace transform of reliability function as

$$R(s) = L\,R(t) = p_0(s) + p_1(s) + p_2(s).$$

Inversion of $R(s)$ gives $R(t)$.

5.3 PRIORITY SYSTEMS

There are systems whereupon failure in a complex system the repair of an unit/subsystem is given preference over the other unit/system depending upon the importance of the unit/subsystem. Such systems are called priority systems. When a component of lower priority is under repair and a component of higher priority arrives for repair, then the lower priority component is preempted and higher priority component is repaired. After completion of the service of higher priority component, the lower priority component is given service. If the repair of lower priority component is restarted as a freshly failed component neglecting the time spent in its repair before the arrival of higher priority component, this is called 'preemptive repeat repair' discipline. On the other hand, if the repair of lower priority component is started from the point where it was left at the time of arrival of the higher priority component, this is called 'preemptive resume repair' discipline. In this section, we consider such systems making use of different techniques so that the readers may apply these methods suitably to other systems.

(a) ***Two-Unit System:*** A two-unit standby system with imperfect switch and preemptive repeat repair policy is considered. As an example, a boiler in a thermal power plant can, besides the normal and complete failure, operate in partial failure mode with a part of tubing having failed, while the impaired part of the system is isolated and repaired. A similar situation can be seen in a power distribution network. The system is considered by Gupta et al., [1983] using regenerative point technique.

The system consists of two units—one operating and the other in cold standby. A unit has normal, partial failure and total failure modes while the switching device has two modes—good and bad. When failure of an active unit occurs, the switch disconnects it and connects the standby unit. If the switch is not good, the standby unit cannot be connected for continuous operation of the system. The repair of the switch is given preference over the repair of the failed unit. We further assume that a partially failed unit passes into the state of total failure. When the unit is in partial failure state, it is working as well as being repaired. Repaired unit enters the normal state. If during the repair, the unit fails totally, the service of the unit starts afresh and the time already spent in repair goes to waste. Further, if the operating unit fails partially, while the other unit is in total failure state and is in repair, the former gets priority of preemptive repeat type. The unit repaired from the total failure state goes to normal state. Failures from operable state and partial failure state are exponentially distributed while the repair times of the switch and the units have general distributions. Defining the following notations:

E set of states $\{S_i\}$, $i = 0$ to 5.

O, operative; S, standby; P, partial failure; T, total failure; S_r, switch in repair

α_1, α_2 : respective failure rates of an operative and a partially failed unit

β_1, β_2 : respective repair rates of partially failed unit and total failed unit

$g_1(t)$, $G_1(t)$: probability density function (pdf) and cumulative distribution functions (cdf) of the repair time of a partially failed unit.

$g_2(t)$, $G_2(t)$: pdf and cdf of the repair time of a totally failed unit.

$G_s(t)$: cdf of the repair time of the switch

p : probability that the switch operates successfully

p_{ij} : direct transition probability from S_i to S_j

μ_i : mean sojourn time in state S_i

$q_{ij}(t)$: pdf of direct transition time from state S_i to S_j

$A_i(t)$: P[System is up at time $t \mid E_0 = S_i$]

$M_i(t)$: probability that the system up initially in state S_i continues in that state at least until time t

$B_i(t)$: cdf of the time to system failure $\mid E_0 = S_i$

$m_2 = \int t\, dG_2(t)$, $m_s = \int t\, dG_s(t)$

© Stieltjes convolution

Limits of integration are from 0 to ∞ unless stated otherwise.

Following are the system states:

S_0 : one unit operative and other unit is in standby;

S_1 : one unit in partial failure and under repair, the other in standby;

S_2 : switching device is under repair and the standby remains unswitched while the other unit is totally failed;

S_3: one unit operative, the other is totally failed and under repair;

S_4: operative unit is in partial failure state and under repair while the service of the totally failed unit is interrupted;

S_5 : both units are in total failure state.

Transition diagram for the system is shown in Fig. 5.5.

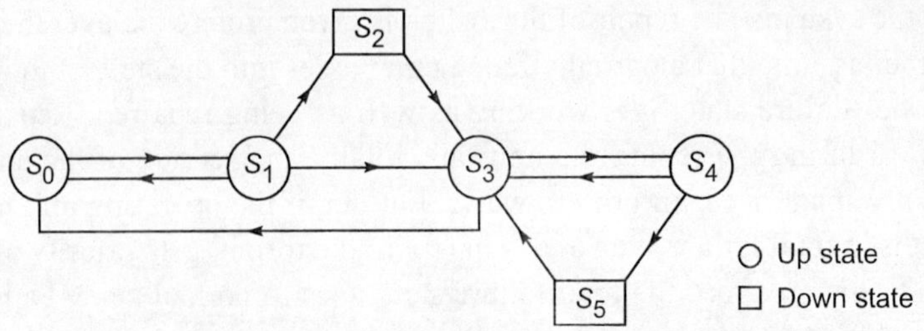

Fig. 5.5

The transition probabilities associated with the transition diagram are:

$$p_{01} = p_{23} = p_{53} = 1 \; ; p_{10} = p_{43} = \int e^{-\alpha_2 t} g_1(t)dt = g_1^*(\alpha_2) = \phi, \text{ say}$$

$$p_{12} = q\int \alpha_2 e^{-\alpha_2 t} \overline{G}(t)dt = q\,\overline{\phi}\,; \; p_{13} = p\int \alpha_2 e^{-\alpha_2 t}\overline{G}_1(t)dt = p\,\overline{\phi}\,;$$

$$p_{30} = \int e^{-\alpha_1 t} g_2(t)dt = g_2^*(\alpha_1) = \psi, \text{ say; } p_{34} = \int \alpha_1 e^{-\alpha_1 t}\overline{G}_2(t)dt = \overline{\psi}\,;$$

$$p_{45} = \int \alpha_2 e^{-\alpha_2 t}\overline{G}_1(t)dt = \phi. \tag{5.3.1}$$

Sojourn times are $\mu_0 = \dfrac{1}{\alpha_1}$; $\mu_3 = \int e^{-\alpha_1 t}\overline{G}_2(t)dt = \overline{\psi}/\alpha_1$; $\mu_1 = \mu_4 = \int e^{-\alpha_2 t}\overline{G}_1(t)dt = \dfrac{\overline{\phi}}{\alpha_2}$.

(5.3.2)

To determine the distribution of first passage time to the states S_2 and S_5 when the system completely fails, we regard these states as absorbing, so that the regenerative states are S_0, S_1, S_3 and S_4. Probabilistic considerations give the equations

$$B_0(t) = Q_{01}(t) © B_1(t); \; B_1(t) = Q_{12}(t) + Q_{10}(t) © B_0(t) + Q_{13}(t) © B_3(t);$$

$$B_3(t) = Q_{30}(t) © B_0(t) + Q_{34}(t) © B_4(t) \; ; \; B_4(t) = Q_{45}(t) + Q_{43}(t) © B_3(t).$$

where, $Q_{ij}(t) = P(X_{n+1} = j,\, T_{n+1} - T_n \le t \mid X_n = i.)$

$$\text{MTSF} = \frac{(1 - p_{34}\,p_{43})(\mu_0 + p_{01}\,\mu_1) + p_{01}\,p_{13}(\mu_3 + p_{34}\,\mu_4)}{(1 - p_{01}\,p_{10})(1 - p_{34}\,p_{43}) - p_{01}\,p_{13}\,p_{30}} = \left.\frac{dB_0(s)}{ds}\right|_{s=0}$$

Using (5.3.1) and (5.3.2), we have

$$\text{MTSF} = \frac{(\alpha_2 + \alpha_1\overline{\phi})(1 - \overline{\psi}\phi + p\,\overline{\phi}\,\overline{\psi})}{\alpha_1\alpha_2\,\overline{\phi}\,(1 - \overline{\psi}\phi - p\psi)}$$

Also, we have

$$M_0(t) = e^{-\alpha_1 t} \; ; \; M_1(t) = M_4(t) = e^{-\alpha_2 t}\overline{G}_1(t); \; M_3(t) = e^{-\alpha_1 t} G_2(t).$$

$$A_0(t) = M_0(t) + q_{01}(t) © A_1(t) \; ; \; A_1(t) = M_1(t) + q_{10}(t) © A_0(t) + q_{12}(t) © A_2(t) + q_{13}(t) © A_3(t)$$

$$A_2(t) = q_{23}(t) © A_3(t) \; ; \; A_3(t) = M_3(t) + q_{30}(t) © A_0(t) + q_{34}(t) © A_4(t)$$

$$A_4(t) = M_4(t) + q_{43}(t) © A_3(t) + q_{45}(t) © A_5(t) \; ; \; A_5(t) = q_{53}(t) © A_3(t).$$

The steady-state availability

$$A_0 = \underset{t\to\infty}{\text{Lim}}\ A_0(t) = \underset{s\to 0}{\text{Lim}}\ s.\ A_0^*(s) = \frac{a}{a+b}$$

where, $a = (\mu_0 + \mu_1)p_{30} + (1 - p_{10})(\mu_3 + \mu_4 p_{34})$; $b = (1 - p_{10})m_2 p_{34} p_{45} + m_s\, p_{12} p_{30}$.

$$A_0 = \frac{(\alpha_2 + \alpha_1\bar{\phi})(\psi + \bar{\phi}\bar{\psi})}{(\alpha_2 + \alpha_1\bar{\phi})(\psi + \bar{\phi}\bar{\psi}) + (m_2\bar{\psi}\bar{\phi} + m_s q\psi)\bar{\phi}^2\alpha_1\alpha_2}$$

(b) ***System with Imperfect Switch and Waiting Time***

In this section, a system having more complex configuration than the system discussed in 5.3(a) and with imperfect switch is considered. Since it may not be possible to attend the failed system/component immediately for its repair on its failure, waiting time is taken into account. The system consists of two classes L_1 and L_2 working in parallel. Class L_2 consists of N_2 components connected in series (not necessarily identical) while class L_1 is composed having N_1 identical components connected in standby redundancy. There is a single repair facility and L_2 components are always given preference for repair over L_1 class components. Further assume that

(i) Class L_1 is repaired as a whole, i.e., no class L_1 components are repaired till all have failed.

(ii) When class L_1 is under repair and a component of class L_2 fails causing system failure, the repair of L_1 class is preempted. After completion of the repair of class L_2 component, the repair of class L_1 begins afresh, i.e., as the repair of freshly failed class (preemptive repeat priority discipline).

(iii) The failure behaviour of class L_2 components is unaffected by the number of failed L_1 class components, if any.

(iv) Switch-over device is imperfect and the system fails upon the failure of the switch-over device.

(v) Non-failed components cannot fail when the system is failed and all components of class L_2 may fail while only one component of class L_1 can fail at a time.

(vi) If the component of class L_2 arrives for repair during the repair of a component of class L_1, then it will have to wait till the repair of the L_1 class component in repair service is finished up. After that class L_1 is preempted by component of class L_2.

b: probability of successful operation of the switch (failure probability is $1 - b$)

$\eta(s)\Delta$: the first order probability that the repair of switch-over device is done in the interval $(s, s + \Delta)$ conditioned that it has not been repaired upto time s

λ_i: constant failure rate of ith component in class L_2 such that $\lambda = \sum_{i=1}^{N_2} \lambda_i$

$\alpha(t)\Delta$: the first order probability that the failure in class L_1 takes place in the interval $(t, t + \Delta)$ and it had not happened upto time t.

$\beta_i(z)\Delta$: the first order probability that ith component of class L_2 is commissioned for repair in the interval $(z, z + \Delta)$ conditioned that it has been waiting for repair upto time z.

$\delta(v)\Delta$: the first order probability that class L_1 is waiting for repair and the waiting time lies between v and $v + \Delta$.

$\phi(y)$: probability density for class L_1 repair distribution.

$\psi_i(u)$: probability density for class L_2 repair distribution.

$p_{0,k}(t)$: probability density that at time t, the system is operating but k components in L_1 class have failed, $0 \leq k < N_1$. Switch is good.

$p_{0,N_1}(y, t)$: probability density that at time t, the system is degraded due to the failure of all N_1 components. Repair is done and the elapsed repair time on L_1 class is y. Switch is good.

$p_{i,k}(u, t)$: probability density that at time t, the system which is still failed due to the failure of ith component in class L_2, is being repaired, $k(0 \leq k < N_1)$ components of class L_1 have failed and are not being repaired. The elapsed repair time on class L_2 component is u. Switch is good.

$p_{i,N_1}(u, t)$: probability density that at time t, the system is still failed due to failure of ith component of class L_2 and it has preempted the L_1 class which was under repair. Repair of class L_2 component is being done and elapsed repair time is u. At the instant of preemption $u = 0$. Switch is good.

$p_w(v, t)$: probability density that at time t, class L_1 is waiting for repair and the waiting time lies in $(v, v + \Delta)$. Switch is good.

$p_{w,k,i}(z, t)$: probability density that at time t, ith failed component of class L_2 is waiting for repair (due to L_1 in service) and the elapsed waiting time is z. k components of class L_1 are in failed state. Switch is good.

$p_{F,k}(s, t)$: probability density that at time t, the system is in failed state due to imperfect working of switch while k components of L_1 class are in failed state. The switch is being repaired with elapsed repair time s.

General distribution with probability density $H(s)$ is given by $\eta(s)\exp\left(-\int_0^s \eta(s)ds\right)$ with similar expressions for $\alpha(t)$, $\beta_i(z)$, $\delta(v)$, $\phi(y)$, $\psi_i(u)$:

The system considered by Singh [1977, 1980] is analysed using direct integration (without using Laplace transforms). The differential-difference equations for the system are:

$$\frac{d}{dt}p_{0,0}(t) + [\lambda + \alpha(t)]\, p_{0,0}(t) = \int p_{0,N_1}(y, t)\, \phi(y)dy + \sum_{i=1}^{N_2} \int p_{i,0}(u, t)\, \psi_i(u)du \quad (5.3.3)$$

$$\frac{d}{dt}p_{0,k}(t) + [\lambda + \alpha(t)]\, p_{0,k}(t) = \int p_{F,k}(s, t)\, \eta(s)ds + b\alpha(t)p_{0,k-1}(t) + \sum_{i=1}^{N_2} \int p_{i,k}(u, t)\, \psi_i(u)du,$$

$$(1 \leq k \leq N_1 - 1) \quad (5.3.4)$$

$$\frac{\partial}{\partial s}p_{F,k}(s, t) + \frac{\partial}{\partial t}p_{F,k}(s, t) + \eta(s)p_{F,k}(s, t) = 0, \; (1 \leq k \leq N_1 - 1) \quad (5.3.5)$$

$$\frac{\partial}{\partial y}p_{0,N_1}(y, t) + \frac{\partial}{\partial t}p_{0,N_1}(y, t) + [\lambda + \phi(y)]p_{0,N_1}(y, t) = 0, \quad (5.3.6)$$

$$\frac{\partial}{\partial u}p_{i,k}(u, t) + \frac{\partial}{\partial t}p_{i,k}(u, t) + \psi_i(u)p_{i,k}(u, t) = 0, \; (1 \leq i \leq N_2,\; 0 \leq k \leq N_1) \quad (5.3.7)$$

$$\frac{\partial}{\partial z}p_{w,k,i}(z,t)+\frac{\partial}{\partial t}p_{w,k,i}(z,t)+\beta_i(z)p_{w,k,i}(z,t)=0,\ (0\le k\le N_1,\ 1\le i\le N_2) \tag{5.3.8}$$

$$\frac{\partial}{\partial v}p_w(v,t)+\frac{\partial}{\partial t}p_w(v,t)+\delta(v)p_w(v,t)=0 \tag{5.3.9}$$

with following initial and boundary conditions

$$p_{0,k}(0) = 1 \text{ for } k = 0 \text{ and zero otherwise (initial conditions)}$$
$$p_{F,k}(0,t) = (1-b)\ \alpha(t)\ p_{0,k-1}(t),\ 1\le k\le N_1-1$$
$$p_{i,k}(0,t) = \int p_{w,k,i}(z,t)\ \beta_i(z)dz,\ 0\le k\le N_1-1;\ 1\le i\le N_2$$
$$p_{i,N_1}(0,t) = \lambda_i\int p_{0,N_1}(y,t)dy,\ 1\le i\le N_2$$
$$p_{0,N_1}(0,t) = b\ \alpha(t)\ p_{0,N_1-1}(t)+\sum_{i=1}^{N_2}\int p_{i,N_1}(u,t)\ \psi_i(u)du$$
$$p_{w,k,i}(0,t) = \lambda_i\, p_{0,k}(t),\ 1\le i\le N_2\ ;\ 0\le k\le N_1$$
$$p_w(0,t) = b\xi(t)\ p_{0,N_1}-1(t).$$

The limits for integration are from 0 to ∞ until unless it is stated.

Equations (5.3.3) and (5.3.4) are first order ordinary differential equations while eqns. (5.3.5)—(5.3.9) are partial differential equations (Lagrange's type) of first order and first degree. These are solved using the initial and boundary equation and the solutions are as follows:

$$p_{F,k}(s,t)=\phi_1(t-s)\exp\left(-\int_0^s\eta(s)ds\right),\ 1\le k\le N_1-1 \tag{5.3.10}$$

$$p_{0,N_1}(y,t)=\phi_2(t-y)\exp\left(-\lambda y-\int_0^y\phi(y)dy\right) \tag{5.3.11}$$

$$p_{i,k}(u,t)=\phi_3(t-u)\exp\left(-\int_0^u\psi_i(u)du\right),\ 1\le i\le N_2\ ;\ 0\le k\le N_1 \tag{5.3.12}$$

$$p_{w,k,i}(z,t)=\phi_4(t-z)\exp\left(-\int_0^z\beta_i(z)dz\right),\ 1\le i\le N_2\ ;\ 0\le k\le N_1 \tag{5.3.13}$$

$$p_w(v,t)=\phi_5(t-v)\exp\left(-\int_0^v\delta(v)dv\right), \tag{5.3.14}$$

where, ϕ_1, ϕ_2, ϕ_3, ϕ_4, ϕ_5 are arbitrary functions given by

$$\phi_1(t) = p_{F,k}(0,t) = (1-b)\ \alpha(t)\ p_{0,k-1}(t),$$

$$\phi_2(t) = p_{0,N_1}(0,\ t) = b\ \alpha(t)\ p_{0,N_1-1}(t) + \sum_{i=1}^{N_2} \int \phi_3(t-u)\ A_i(u)du,$$

$$\phi_3(t) = p_{i,k}(0,t) = \begin{cases} \int \phi_4(t-z)\beta_i(z)dz = \int \lambda_i\ p_{0,k}(t-z)\beta_i(z)dz, 0 \le k \le N_1 - 1 \\ \int \lambda_i\ p_{0,N_1}(y,t)dy = \int \lambda_i\ \phi_2(t-y)\exp\left(-\lambda y - \int_0^y B(y)dy\right), k = N_1 \end{cases}$$

$$\phi_4(t) = p_{w,k,i}(0,\ t) = \lambda_i\ p_{0,k}(t);\ \phi_5(t) = p_w(0,\ t) = b\ \alpha(t)\ p_{0,N_1-1}(t).$$

Substituting various values from eqns. (5.3.10)—(5.3.14) in eqns. (5.3.3) and (5.3.4) and solving (using the initial conditions), $p_{0,0}(t)$ and $p_{0,k}(t)$ are given by

$$p_{0,0}(t) = \left[1 + \int\left\{\int \phi_2(t-y)B(y)\exp(-\lambda y)dy + \sum_{i=1}^{N_2}\int \phi_3(t-u)A_i(u)du\right\}\cdot \exp\left(\lambda t + \int_0^t \alpha(t)dt\right)dt\right]\cdot\exp\left(-\lambda t - \int_0^t \alpha(t)dt\right) \qquad (5.3.15)$$

$$p_{0,k}(t) = \left[\int\left\{\int \phi_5(t-s)H(s)ds + \sum_{i=1}^{N_2}\int \phi_2(t-u)A_i(u)du + b\alpha(t)p_{0,k-1}(t)\right\}\cdot \exp\left(\lambda t + \int_0^t \alpha(t)\right)dt\right]\exp\left(-\lambda t - \int_0^t \alpha(t)\,dt\right),\ 1 \le k \le N_1 - 1. \qquad (5.3.16)$$

where,

$$A_i(u) = \psi_i(u)\ \exp\left(-\int_0^u \psi_i(u)du\right);\ B(y) = \phi(y)\ \exp\left(-\int_0^y \phi(y)dy\right).$$

Thus, all the probabilities are in terms of $p_{0,N_1-1}(t)$which in turn is evaluated by (5.3.16). $p_{0,0}(t)$ can be evaluated by (5.3.15).

Availability function is given by

$$A(t) = 1 - Q(t)$$

where, $$Q(t) = \int \sum_{k=1}^{N_1-1} p_{F,k}(s,\ t)ds + \int \sum_{k=0}^{N_1}\sum_{i=1}^{N_2} p_{i,k}(u,\ t)du + \int \sum_{k=0}^{N_1}\sum_{i=1}^{N_2} p_{w,k,i}(z,\ t)dz.$$

When switch-over device is perfect, $b = 1$ and $p_{F,K}(s,\ t) = 0,\ \eta(s) = 0$. Putting these values in equations, corresponding results are obtained. For the limiting behaviour or steady state of the system

$$\frac{\partial}{\partial t} \to 0,\ \frac{d}{dt} \to 0,\ p_{0,N_1}(y,\ t) \to p_{0,N_1}(y),\ p_{i,k}(u,\ t) \to p_{i,k}(u),\ p_{0,k}(t) \to p_{0,k};\ p_{w,k,i}(z,\ t) \to p_{w,k,i}(z),$$

$p_w(v, t) \rightarrow p_w(v)$, $p_{F,k}(s, t) \rightarrow p_{F,K}(s)$ as $t \rightarrow \infty$. The partial differential eqns. (5.3.5)—(5.3.9) change to ordinary differential equations which can be solved easily by direct integration.

Special Note:- If the repair policy is preemptive resume instead of preemptive repeat, governing equations are given here for this system—Singh [1977]:

Redundancy is within class L_1 and components of class L_2 are given priority for repair over class L_1. If the components of L_1 are under repair and a failure occurs in L_2 class, then the components of L_1 are preempted and the components of L_2 are taken for repair. After completion of the repair of L_2 class components, the class L_1 is taken for repair again and the repair is started from the point at which it was preempted last. Components of L_1 can be preempted any number of times but on each entry the repair time will be reduced by the time spent in the previous repair. Other assumptions are same as for preemptive repeat case with the only difference that m is the number of L_1 class components in working order. Like $p_{0,N_1}(y, t)$ and $p_{i,k}(u, t)$, $p_R(y, t)$ and $p_{r,m,i}(u, t)$ are defined. In addition to these, define $Q_{r,m,i}(u, v, t)\Delta$ as the probability that m components of L_1 class are in working order at time t and the system is failed due to the failure of ith component of L_2 class. The system is under repair because of the repair of class L_2 component and the repair time of L_2 class component lies between u and $u + \Delta$. Class L_1 has the total preempted time lying between v and $v + \Delta$. Switch is good. Governing equations for the system are

$$\frac{d}{dt}p_{0,m}(t) + [\lambda + \alpha(t) - b]p_{0,m}(t) = \int p_{f,m}(s, t)\ \eta(s)ds + \alpha(t)p_{0,m+1}(t) + \sum_{i=1}^{n} \int p_{r,m,i}(u, t)\ \psi_i(u)du$$

$$+ \sum_{i=1}^{n} \int(\int Q_{r,m,i}\ (u, v, t)\,\psi_i(u)du\,dv)\ (1 \le m \le k - 1) \qquad (5.3.17)$$

$$\frac{d}{dt}p_{0,k}(t) + [\lambda + \alpha(t)]p_{0,k}(t) = \sum_{i=1}^{n} \int p_{r,k,i}(u, t)\ \psi_i(u)du + \int p_R(y, t)\ \phi(y)dy$$

$$+ \sum_{i=1}^{n} \int Q_{r,k,i}\ (u, 0, t)\ \psi_i(u)du \qquad (5.3.18)$$

$$\frac{\partial}{\partial s}p_{f,m}(s, t) + \frac{\partial}{\partial t}p_{f,m}(s, t) + \eta(s)p_{f,m}(s, t) = 0,\ 1 \le m \le k - 1 \qquad (5.3.19)$$

$$\frac{\partial}{\partial z}p_{w,m,i}(z, t) + \frac{\partial}{\partial t}p_{w,m,i}(z, t) + \beta_i(z)p_{w,m,i}(z, t) = 0,\ 1 \le m \le k;\ 1 \le i \le n \qquad (5.3.20)$$

$$\frac{\partial}{\partial y}p_R(y, t) + \frac{\partial}{\partial t}p_R(y, t) + \phi(y)\ p_R(y, t) = 0 \qquad (5.3.21)$$

$$\frac{\partial}{\partial x}p_w(x, t) + \frac{\partial}{\partial t}p_w(x, t) + \delta(x)\ p_w(x, t) = 0 \qquad (5.3.22)$$

$$\frac{\partial}{\partial u}p_{r,m,i}(u, t) + \frac{\partial}{\partial t}p_{r,m,i}(u, t) + \psi_i(u)\ p_{r,m,i}(u, t) = 0,\ 1 \le m \le k;\ 1 \le i \le n \qquad (5.2.23)$$

$$\frac{\partial}{\partial u} Q_{r,m,i}(u, v, t) + \frac{\partial}{\partial t} Q_{r,m,i}(u, v, t) + \psi_i(u)\, Q_{r,m,i}(u, v, t) = 0,\ 1 \le m \le k;\ 1 \le i \le n \qquad (5.2.24)$$

with initial and boundary conditions as follows:

$$p_{0,m}(0) = 1 \text{ for } m = k \text{ and zero for } m \ne k.$$

$$p_{f,m}(0, t) = (1 - b)\ \alpha(t)\ p_{0,m+1}(t),\ 1 \le m \le k - 1,$$

$$p_{r,m,i}(0, t) = \int p_{w,m,i}(z, t)\beta_i(z)dz,\ 1 \le m \le k\ ;\ p_w(0, t) = b\ \alpha(t)\ p_{0,1}(t)$$

$$p_R(0, t) = \int p_w(x, t)\ \delta(x)dx + b\ \alpha(t)p_{0,1}(t);\ p_{w,m,i}(0, t) = \lambda_i\ p_{0,m}(t),\ 1 \le m \le k;$$

$$Q_{r,m,i}(0, v, t) = \int_0^v dv \int p_{r,m,i}(u, t)du,\ 1 \le m \le k - 1;\ Q_{r,k,i}(0,0,t) = \lambda_i p_0, k(t).$$

Again equations (5.3.17) to (5.3.24) are ordinary differential and partial differential equations of first order and first degree and can be solved using the initial and boundary conditions.

A very interesting industrial problem for duplex casting system under preemptive resume repair policy is analyzed by Gupta et al., [2004].

5.4 MISSION RELIABILITY

The probability that the system fulfils the mission during the time interval (0, t), is called mission reliability $R(t, \tau)$. The system fails if the unit under repair does not become operable in a time less than τ. In this section, two different systems are considered and analysis is done using different methods.

(a) Consider the system consisting of $M + N$ identical units. Out of these $M + N$ units, M are active units while N units are in warm standby. As soon as an operating unit fails, it is switched out. Switch is perfect. Further, we assume that

λ: constant failure rate of operating unit

λ_1: constant failure rate of standby unit, $\lambda_1 < \lambda$,

$\eta(t)$: general repair rate of a unit,

$p_0(t)$: probability density that at time t the system is in operating state,

$p_n(t, x)dx$: probability that at time t the system is in operative state and n units are in repair station. The elapsed repair time of unit under repair is x, $n = 1, 2,\ldots, N$,

$p_{n+1}(t, x, y)dxdy$: probability that at time t, the system is in major breakdown and the elapsed repair time of a unit under repair is x and its elapsed waiting time is y, where $y \le x$ and $0 \le y \le \tau$ (τ is the allowed down time),

$p_f(t)$: probability density that at time t, the system is in failed state. Limits for integration are from zero to ∞ if not indicated.

(i) Each unit after repair is as good as new.

(ii) While the system is in major breakdown, other good units do not fail.

(iii) The system fails if a unit under repair does not become operable in a time less than τ, the constant time, measured from an instant at which major breakdown occurred, where it occurs as soon as $N + 1$ units fail.

Kodama and Takamatsu [1973] considered the system and obtained various results using Laplace transforms. Following are the difference-differential equations associated with the system:

$$\frac{d}{dt}p_0(t)+ (M\lambda + N\lambda_1)\, p_0(t) = \int p_1(t, x)\,\eta(x)dx, \tag{5.4.1}$$

$$\frac{\partial}{\partial t}p_n(t, x) + \frac{\partial}{\partial x}p_n(t, x) + (M\lambda + (N - n)\lambda_1 + \eta(x))p_n(t, x) = (1 - \delta_{n_1})(M\lambda + (N - \mathrm{n} + 1)\lambda_1)$$

$$p_{n-1}(x - t)$$

$$n = 1, 2,\ldots, N \tag{5.4.2}$$

$$\left(\frac{\partial}{\partial t} + \frac{\partial}{\partial x} + \frac{\partial}{\partial y} + \eta(x)\right)p_{N+1}(t, x, y) = 0 \tag{5.4.3}$$

$$\frac{d}{dt}p_f(t) = \int_{\tau}^{\infty} p_{N+1}(t, x, \tau)dx \tag{5.4.4}$$

where, δ_{ij} is the Kronecker delta and the initial and boundary conditions are $p_0(0) = 1$ otherwise zero,

$$p_{N+1}(t, x, 0) = M\lambda\, p_N(t, x) \tag{5.4.5}$$

$$p_n(t, 0) = \delta n_1(M\lambda + N\lambda_1)p_0(t) + (1 - \delta_{nN}) \int_0^t p_{n+1}(t, x)\,\eta(x)dx +$$

$$+ \delta_{nN} \int_0^{(\tau, t)} dy \int_0^t p_{N+1}(t, x, y)\; \eta(x)dx, \; n = 1, 2, \ldots, N \tag{5.4.6}$$

and $\quad (\tau, t)$ = min of τ and t.

Solutions of eqns. (5.4.1) and (5.4.4) (ordinary differential equations) are

$$p_0(t) = [1 + \int (\int p_1(t, x)\,\eta(x)dx)\; \exp(M\lambda + N\lambda_1)t\; dt]\; \exp(-M\lambda - N\lambda_1)t, \tag{5.4.7}$$

$$p_f(\mathrm{t}) = \int\left(\int_{\tau}^{\infty} p_{N+1}(t, x, \tau)\,dx\right)dt \tag{5.4.8}$$

Equations (5.4.2) and (5.4.3) are partial differential equations of first order. The solutions are given by

$$p_n(t, x) = [\phi_1(t - x) + (1 - \delta_{n1})(M\lambda + (N - n + 1)\lambda_1)\int p_{n-1}(t, x).$$
$$\exp\,\{M\lambda x - (N{-}n)\lambda_1 x + \int\eta(x)dx\}\, dx].\exp\{-M\lambda x - (N - n)\lambda_1 x + - \int\eta(x)dx\};$$
$$n = 1, 2,\ldots, N, \tag{5.4.9}$$

$$p_{N+1}(t, x, y) = \phi_2(t - x, x - y)\; \exp(-\eta(x)y) \tag{5.4.10}$$

where, $\phi_1(t) = p_n(t, 0)$ given by eqn. (5.4.6)

$\phi_2(t - x, x) = p_{N+1}(t, x, 0)$ given by (5.4.5)

and $\int p_n(t, x)dx = p_n(t)$.

From the above analysis, it is observed that all the probabilities $p_n(x, t) n = 1, 2, \ldots, N + 1$ are obtained in terms of $p_0(t)$ which is given by (5.4.7). Equation (5.4.7) is an integral equation solvable by usual methods or numerical methods.

Mission reliability function for the system is

$$R(t, \tau) = 1 - p_f(t)$$

and mean time to system failure (MTSF) = $\int R(t, \tau)dt$.

(b) In this subsection, a more practical system considered by Singh [1980] is presented. Singh considers a two-unit system with imperfect switch and general failure and repair distributions. He also considers simultaneous failure like Fukuta and Kodama [1974] taking the following assumptions:

(i) The system consists of two dissimilar units A_1 and A_2 connected in redundancy,
(ii) there is only one repair station,
(iii) when one unit fails and the other is good, the failure rate is $\eta_i(t)$, but when only one unit is good, failure rate is $\eta'_i(t)$, $i = 1, 2$, possibly $\eta'_i(t) \neq \eta_i(t)$,
(iv) when both units fail simultaneously, the failure rate is $\eta_{12}(x)$,
(v) when both units fail simultaneously unit A_1 is repaired first with probability density $\alpha_i(y)$ $(\Sigma \alpha_i(y) = 1)$,
(vi) repair rate for each unit A_i is $\beta_i(y)$, $i = 1, 2$ and the repaired unit is like a new one,
(vii) the switch-over device is imperfect with probability of successful working equal to 'b' and its repair rate is $\zeta(s)$, and
(viii) the system fails if it stays inoperable for more than the allowed down time τ, after both the units fail or the switch-over device fails.

Define

$p_0(t)$: probability density that at time t system is in operable state.

$Q(t)$: probability density that at time t system is in failed state.

$p_{F,i}(y, t)$: probability density that at time t, A_i the failed unit is under repair and its elapsed repair time is y while the other unit is good. Switching device is working.

$p_F(s, t)$: probability density that at time t, the system is in failed state due to the failure of switching device. Both units are good. Switch is under repair and its elapsed repair time is $s (0 \leq s \leq \tau)$.

$Q_{r,w,i}(y, z, t)$: probability density that at time t, the system is failed due to the failure of both units. Unit A_i is under repair with elapsed repair time y and the other unit waits for repair with elapsed waiting time z, $z < y$ and $0 \leq z \leq \tau$. Switch is in working condition.

$Q_{R,w,i}(y, t)$: probability density that at time $t - y$, both the units are failed simultaneously and at time t, unit A_i is under repair while the other unit waits for repair. The elapsed repair time is y, $0 \leq y \leq \tau$. Switch is good.

(τ, t) : minimum of τ and t.

Following are the governing differential-difference equations for the system:

$$\frac{d}{dt}p_0(t) + [\eta_1(t) + \eta_2(t) + \eta_{12}(t) - b]\, p_0(t) = \int_0^t p_F(s,\, t)\, \zeta(s)ds + \sum_{i=1}^{2} \int_0^t p_{F,i}(y,\, t)\beta_i(y)dy, \quad (5.4.11)$$

$$\frac{\partial}{\partial s}p_F(s,\, t) + \frac{\partial}{\partial t}p_F(s,\, t) + \zeta(s)p_F(s,\, t) = 0 \quad (5.4.12)$$

$$\frac{\partial}{\partial y}p_{F,i}(y,\, t) + \frac{\partial}{\partial t}p_{F,i}(y,\, t) + [\eta_{3-i}(t) + \beta_i(y)]p_{F,i}(y,\, t) = 0,\ i = 1,\, 2 \quad (5.4.13)$$

$$\frac{\partial}{\partial y}Q_{r,w,i}(y,\, z,\, t) + \frac{\partial}{\partial z}Q_{r,w,i}(y,\, z,\, t) + \frac{\partial}{\partial t}Q_{r,w,i}^{(y,z,t)} + \beta_i(y)Q_{r,w,i}(y,\, z,\, t) = 0\ i = 1,\, 2 \quad (5.4.14)$$

$$\frac{\partial}{\partial y}Q_{R,w,i}(y,\, t) + \frac{\partial}{\partial t}Q_{R,w,i}(y,\, t) + \beta_i(y)Q_{R,w,i}(y,\, t) = 0,\ i = 1,\, 2 \quad (5.4.15)$$

$$\frac{d}{dt}Q(t) = \begin{cases} 0, \text{for } t < \tau \\ \sum_{i=1}^{2}\left[\int_\tau^t Q_{r,w,i}(y, \tau, t)\,dy + \int_\tau^t Q_{R,w,i}(y,t)\,dy\right] + \int_\tau^t p_F(s,t)ds, \text{ for } t \geq \tau, \end{cases} \quad (5.4.16)$$

The initial and boundary conditions are

$$p_0(0) = 1;\ Q(0) = 0;\ p_F(0,\, t) = (1 - b)p_0(t)\ \eta_i(t),\ i = 1,\, 2$$

$$p_{F,i}(0,\, t) = b\ \eta_i(t)\ p_0(t) + \int_0^\tau dz \int_y^t Q_{r,w,3-i}(y,\, z,\, t)\beta_{3-i}(y)dy +$$

$$\int_0^{(\tau,t)} Q_{R,w,3-i}\,(y,\, t)\ \beta_{3-i}(y)dy,\ i = 1,\, 2$$

$$Q_{r,w,i}(y,\, 0,\, t) = b\eta'_{3-i}(t)p_{F,i}(y,\, t),\ i = 1,\, 2$$

$$Q_{R,w,i}(0,\, t) = \int_0^\tau \alpha_i(y)\eta_{12}(t)p_0(t)dy,\ i = 1,\, 2$$

Equations (5.4.12) to (5.4.15) are partial differential-difference equations (Lagrange's type) while eqns. (5.4.11) and (5.4.16) are first order ordinary differential equations. On solving these equations, we obtain

$$p_F(s,\, t) = \phi_1(t - s)\ \exp\left(-\int_0^s \zeta(s)ds,\right.$$

$$p_{F,i}(y, t) = \phi_2(t-y)\exp\left(-\eta_{3-i}(t)y - \int_0^y \beta_i(y)dy\right), \; i = 1, 2$$

$$Q_{r,w,i}(y, z, t) = \phi_3(y-z, y-t)\exp(-\beta_i(y)z), \; \mathrm{i} = 1, 2$$

$$Q_{R,w,i}(y, t) = \phi_4(t-y)\exp\left(-\int_0^y \beta_i(y)dy\right), \; i = 1, 2$$

$$p_0(t) = \left[1 + \int\left\{\int_0^t \phi_1(t-s)H(s)ds + \sum_{i=1}^{2}\int_0^t \phi_2(t-y)\exp(-\eta_{3-i}(y)y)K_i(y)dy\right\}\right.$$

$$\left.\exp\left(\int_0^t (\eta_1(t)+\eta_2(t)+\eta_{12}(t)-b)dt\right)dt\right].\exp\left(-\int_0^t (\eta_1(t)+\eta_2(t)+\eta_{12}(t)-b)dt\right)$$

$$Q(t) = \left[\sum_{i=1}^{2}\left\{\int_\tau^t \phi_3(y-\tau, y-t)\exp(-\beta_i(y)\tau)dy + \int_\tau^t \phi_4(t-y)\exp\left(\int_0^y \beta_i(y)dy\right)dy\right\}\right.$$

$$\left. + \int_\tau^t \phi_1(t-s)\exp\left(-\int_0^s \zeta(s)ds\right)ds\right]dt$$

where, $\phi_1(t) = p_F(0, t)$; $\phi_2(t) = p_{F,i}(0, t)$; $\phi_3(y, y-t) = Q_{r,w,i}(y, 0, t)$;

$$\phi_4(t) = Q_{R,w,i}(0, t); \; H(s) = \zeta(s)\exp\left(-\int_0^s \zeta(s)ds\right)$$

$$K_i(y) = \beta_i(y)\exp\left(-\int_0^y \beta_i(y)dy\right)$$

The mission reliability function $R(t, \tau) = 1 - Q(t)$.

For perfect switch-over device, putting $b = 1$ and $p_F(s, t) = 0$ in the foregoing results, the probabilities are obtained. Also taking $\frac{d}{dt} = \frac{\partial}{\partial t} = 0$, and probabilities independent of t as $t \to \infty$, steady-state probabilities may be evaluated and hence the long-run availability of the system.

An interesting problem of flexible polymer powder production system for mission reliability has been discussed in detail by Gupta et al., [2005].

5.5 CONTAINER MANUFACTURING

In the foregoing sections of this chapter, special types of systems have been discussed using different methods. Such systems occur in electrical-electronics, telecommunication and computer, etc. systems. In this section, a mechanical system is to be discussed. In Chapters 6-8, special type mechanical systems concerned with process industries are discussed.

In a container used in (sugar industry) manufacturing plant, there are three major parts–(a) the main shell construction (b) end dish construction and (c) fitting of end dishes. These three subsystems work in series hence successful working of all is necessary. Since (b) and (c) parts generally do not fail, the part (a) main shell construction is the most important. The study was carried out by Singh and Mahajan [1998] for main shell construction. The raw material (i.e., long steel plates) used for the construction of shell is supplied from some steel plant. During the manufacturing process, all the steel sheets are cut into different sizes in cutting yard by gas cutting machine. These sheets are taken to the hydraulic pressing machine to shape them before rolling. The rolling machines then shape the sheets into required hollow cylindrical shape. The rolled cylinder is welded with long seem welding machines in order to form an open cylinder, i.e., the main shell of the container. Taking following notations and assumptions, analysis of the shell manufacturing system is carried out.

The shell construction system is composed of the following subsystems:

(i) gas cutting machine A having only one unit subjected to major failure only,
(ii) hydraulic pressing machine B consisting two units in parallel subjected to minor and major failures,
(iii) rolling machine C consisting three units in parallel subjected to minor and major failures,
(iv) long seem welding machine D having three units in parallel subjected to minor and major failures,
(v) there are no simultaneous failures among the subsystems,
(vi) switch-over device is perfect.

$\overline{B}, \overline{C}, \overline{D}$ represent the reduced working states of the subsystems B, C, D.

λ_i: respective mean constant failure rate of subsystems A, B, C, D, $\overline{B}, \overline{C}, \overline{D}$ $i = 1, 2, \ldots, 7$.

μ_i: respective mean constant repair rate of subsystems A, B, C, D, $\overline{B}, \overline{C}, \overline{D}$ $i = 1, 2, \ldots, 7$.

p_i: steady-state probability that the system is in ith state, $i = 1, 2, \ldots, 27$

The transition diagram of the system is shown in Fig. 5.6.

The steady-state probability equations associated with the transition diagram are (by simple probability consideration taking $d/dt = 0$ or using the mnemonic rule):

$$\alpha_1 p_0 = \mu_1 p_1 + \mu_5 p_{21} + \mu_6 p_{22} + \mu_7 p_{23}, \ (\alpha_1 = \lambda_1 + \sum_{5}^{7} \lambda_i);$$

$$\mu_1 p_1 = \lambda_1 p_0 \ ; \ \mu_i p_{i+1} = \lambda_i p_{21}, \ i = 1, 2;$$
$$\mu_1 p_4 = \lambda_1 p_{22} \ ; \ \mu_3 p_5 = \lambda_3 p_{22} \ ; \ \mu_1 p_6 = \lambda_1 p_{23} \ ; \ \mu_4 p_7 = \lambda_4 p_{23};$$
$$\mu_i p_{i+7} = \lambda_1 p_{24}, \ i = 1, 2, 3 \ ; \ \mu_i p_{i+10} = \lambda_i p_{25}, \ i = 1, 2 \ ; \ \mu_4 p_{13} = \lambda_4 p_{25};$$
$$\mu_1 p_{14} = \lambda_1 p_{26} \ ; \ \mu_i p_{i+12} = \lambda_i p_{26} \ , \ i = 3, 4 \ ; \ \mu_i p_{i+16} = \lambda_i p_{27}, \ i = 1, 2, 3, 4;$$

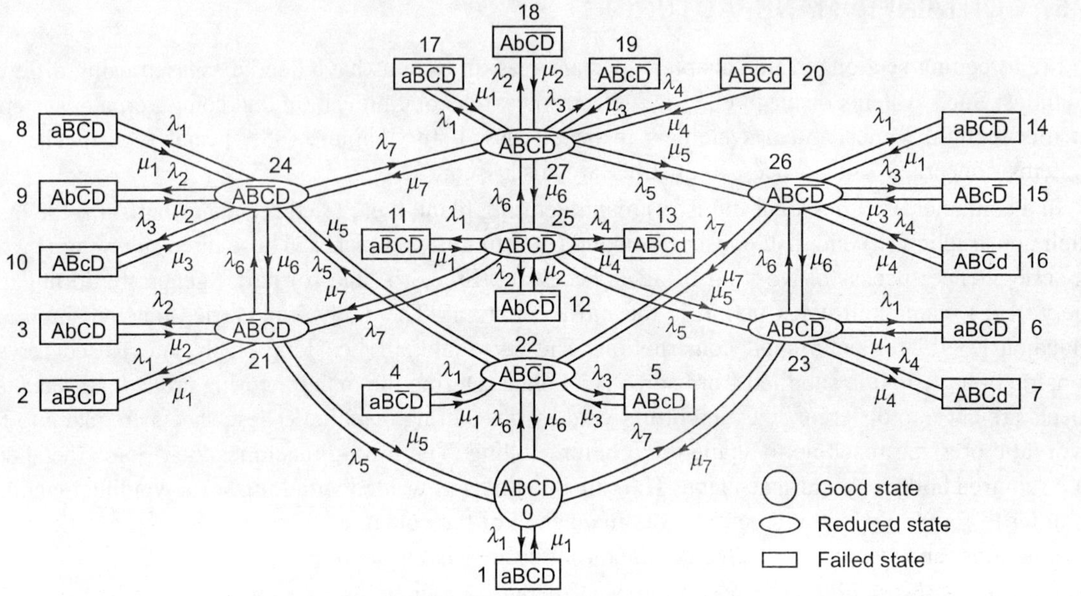

Fig. 5.6

$$\alpha_2 p_{21} = \mu_1 p_2 + \mu_2 p_3 + \mu_6 p_{24} + \mu_7 p_{25} + \lambda_5 p_0, \ (\alpha_2 = \lambda_1 + \lambda_2 + \lambda_6 + \lambda_7 + \mu_5);$$
$$\alpha_3 p_{22} = \mu_1 p_4 + \mu_3 p_5 + \mu_5 p_{24} + \mu_7 p_{26} + \lambda_6 p_0, \ (\alpha_3 = \lambda_1 + \lambda_3 + \lambda_5 + \lambda_7 + \mu_6);$$
$$\alpha_4 p_{23} = \mu_1 p_6 + \mu_4 p_7 + \mu_6 p_{26} + \mu_5 p_{25} + \lambda_7 p_0, \ (\alpha_4 = \lambda_1 + \lambda_4 + \lambda_5 + \lambda_6 + \mu_7);$$

$$\alpha_5 p_{24} = \mu_1 p_8 + \mu_2 p_9 + \mu_3 p_{10} + \mu_7 p_{27} + \lambda_5 p_{22} + \lambda_6 p_{21}, \ (\alpha_5 = \sum_1^3 \lambda_i + \lambda_7 + \mu_5 + \mu_6);$$

$$\alpha_6 p_{25} = \mu_1 p_{11} + \mu_2 p_{12} + \mu_4 p_{13} + \mu_6 p_{27} + \lambda_7 p_{21} + \lambda_5 p_{23}, \ (\alpha_6 = \lambda_1 + \lambda_2 + \lambda_4 + \lambda_6 + \mu_5 + \mu_7);$$

$$\alpha_7 p_{26} = \mu_1 p_{14} + \mu_3 p_{15} + \mu_4 p_{16} + \mu_5 p_{27} + \lambda_6 p_{23} + \lambda_7 p_{22}, \ (\alpha_7 = \lambda_1 + \sum_3^5 \lambda_i + \mu_6 + \mu_7);$$

$$\alpha_8 p_{27} = \sum_1^4 \mu_i p_{16+i} + \lambda_5 p_{26} + \lambda_6 p_{25} + \lambda_7 p_{24}, \ (\alpha_8 = \sum_1^4 \lambda_i + \sum_5^7 \mu_i) \qquad (5.5.1)$$

Solving these equations recursively and using the normalizing condition $\sum_{i=0}^{27} p_i = 1$, we have

$$p_1 = \frac{\lambda_1}{\mu_1} p_0; \ p_{i+1} = \frac{\lambda_i}{\mu_i} K_9 p_0, \ i = 1, 2; \ p_4 = \frac{\lambda_1}{\mu_1} K_{10} p_0; \ p_5 = \frac{\lambda_3}{\mu_3} p_0 K_{10};$$

$$p_6 = \frac{\lambda_1}{\mu_1} p_0 K_{11}; \ p_7 = \frac{\lambda_4}{\mu_4} p_0 K_{11}; \ p_{i+7} = \frac{\lambda_i}{\mu_i} p_0 K_{12}, \ i = 1, 2, 3;$$

$$p_{i+10} = \frac{\lambda_i}{\mu_i} p_0 K_{13},\ i = 1, 2;\ p_{13} = \frac{\lambda_4}{\mu_4} p_0 K_{13};\ p_{14} = \frac{\lambda_1}{\mu_1} p_0 K_{14};\ p_{i+14} = \frac{\lambda_{i+2}}{\mu_{i+2}} p_0 K_{14},\ i = 1, 2;$$

$$p_{i+16} = \frac{\lambda_i}{\mu_i} p_0 K_{15},\ i = 1, 2, 3, 4\ ;\ p_{i+20} = p_0 K_{i+8},\ i = 1, 2, \ldots, 7; \tag{5.5.2}$$

$$p_0 = \left[1 + \frac{\lambda_1}{\mu_1} + \left(\frac{\lambda_1}{\mu_1} + \frac{\lambda_2}{\mu_2}\right) K_9 + \left(\frac{\lambda_1}{\mu_1} + \frac{\lambda_3}{\mu_3}\right) K_{10} + \left(\frac{\lambda_1}{\mu_1} + \frac{\lambda_4}{\mu_4}\right) K_{11} + \right.$$

$$\left(\frac{\lambda_1}{\mu_1} + \frac{\lambda_2}{\mu_2} + \frac{\lambda_3}{\mu_3}\right) K_{12} + \left(\frac{\lambda_1}{\mu_1} + \frac{\lambda_2}{\mu_2} + \frac{\lambda_4}{\mu_4}\right) K_{13} +$$

$$\left.\left(\frac{\lambda_1}{\mu_1} + \frac{\lambda_3}{\mu_3} + \frac{\lambda_4}{\mu_4}\right) K_{14} + \sum_{1}^{4} \frac{\lambda_i}{\mu_i} K_{15} + \sum_{i=9}^{15} K_i \right]^{-1} \tag{5.5.3}$$

where, $A_1 = \lambda_6 + \lambda_7 + \mu_5$; $A_2 = \lambda_5 + \lambda_7 + \mu_6$; $A_3 = \lambda_5 + \lambda_6 + \mu_7$; $A_4 = \lambda_7 + \mu_5 + \mu_6$;

$$A_5 = \lambda_6 + \mu_5 + \mu_7\ ;\ A_6 = \lambda_5 + \mu_6 + \mu_7\ ;\ A_7 = \sum_{5}^{7} \mu_i;\ T_1 = 1 - K_3K_7 - K_1K_8 - K_1K_4K_7 -$$

$$\frac{K_2(K_4K_7 + K_8)(K_3 + K_1K_4)}{1 - K_2K_4}$$

$$T_2 = A_1 - \frac{\mu_6\lambda_6}{A_4} - \frac{\mu_7\lambda_7}{A_5} - \frac{\mu_6\mu_7\lambda_6\lambda_7}{T_5}\left(\frac{1}{A_4} + \frac{1}{A_5}\right)^2;\ T_3 = A_2 - \left(\frac{\mu_5\lambda_5}{A_4}\right) - \left(\frac{\mu_7\lambda_7}{A_6}\right) -$$

$$\frac{\mu_5\mu_7\lambda_5\lambda_7}{T_5}\left(\frac{1}{A_4} + \frac{1}{A_6}\right)^2;$$

$$T_4 = A_3 - \frac{\mu_5\lambda_5}{A_5} - \frac{\mu_6\lambda_6}{A_6} - \frac{\mu_5\mu_6\lambda_5\lambda_6}{T_5}\left(\frac{1}{A_5} + \frac{1}{A_6}\right)^2;\ T_5 = A_7 - \frac{\mu_7\lambda_7}{A_4} - \frac{\lambda_6\mu_6}{A_5} - \frac{\lambda_5\mu_5}{A_6};$$

$$K_1 = \frac{1}{T_4}\left[\frac{\mu_5\lambda_7}{A_5} + \frac{\mu_5\mu_6\lambda_6\lambda_7}{T_5}\left(\frac{1}{A_4} + \frac{1}{A_5}\right)\left(\frac{1}{A_5} + \frac{1}{A_6}\right)\right];$$

$$K_2 = \frac{1}{T_4}\left[\frac{\mu_6\lambda_7}{A_6} + \frac{\mu_5\mu_6\lambda_5\lambda_7}{T_5}\left(\frac{1}{A_4} + \frac{1}{A_6}\right)\left(\frac{1}{A_5} + \frac{1}{A_6}\right)\right];$$

$$K_3 = \frac{1}{T_3}\left[\frac{\mu_5\lambda_6}{A_4} + \frac{\mu_5\mu_7\lambda_6\lambda_7}{T_5}\left(\frac{1}{A_4} + \frac{1}{A_5}\right)\left(\frac{1}{A_4} + \frac{1}{A_6}\right)\right];$$

$$K_4 = \frac{1}{T_3}\left[\frac{\mu_7\lambda_6}{A_6} + \frac{\mu_5\mu_7\lambda_6\lambda_7}{T_5}\left(\frac{1}{A_4} + \frac{1}{A_6}\right)\left(\frac{1}{A_5} + \frac{1}{A_6}\right)\right]$$

$$K_5 = \frac{K_3 + K_1 K_4}{1 - K_2 K_4};\ K_6 = \frac{1}{1 - K_2 K_4}\left(\frac{\lambda_7 K_4}{T_4} + \frac{\lambda_6}{T_3}\right);$$

$$K_7 = \frac{1}{T_2}\left[\frac{\mu_6 \lambda_5}{A_4} + \frac{\mu_6 \mu_7 \lambda_5 \lambda_7}{T_5}\left(\frac{1}{A_4} + \frac{1}{A_6}\right)\left(\frac{1}{A_4} + \frac{1}{A_5}\right)\right]$$

$$K_8 = \frac{1}{T_2}\left[\frac{\mu_7 \lambda_5}{A_5} + \frac{\mu_6 \mu_7 \lambda_5 \lambda_6}{T_5}\left(\frac{1}{A_4} + \frac{1}{A_5}\right)\left(\frac{1}{A_5} + \frac{1}{A_6}\right)\right];\ K_{10} = K_5 K_9 + K_6;$$

$$K_{11} = K_1 K_9 + \frac{\lambda_7}{T_4} + K_2 K_{10};$$

$$K_9 = \frac{1}{T_1}\left[\frac{K_4 K_7 \lambda_7}{T_4} + \frac{\lambda_6 K_7}{T_3} + \frac{\lambda_7 K_8}{T_4} + \frac{\lambda_5}{T_2} + \frac{K_2 K_4 K_7 + K_2 K_8}{1 - K_2 K_4}\left(\frac{\lambda_6}{T_3} + \frac{\lambda_7}{T_4}\right)\right];$$

$$K_{12} = \frac{1}{A_4}\left[K_9\left\{\lambda_6 + \frac{\mu_7 \lambda_6 \lambda_7}{T_5}\left(\frac{1}{A_4} + \frac{1}{A_5}\right)\right\} + K_{10}\left\{\lambda_5 + \frac{\mu_7 \lambda_5 \lambda_7}{T_5}\left(\frac{1}{A_4} + \frac{1}{A_6}\right)\right\}\right.$$
$$\left. + K_{11}\frac{\mu_7 \lambda_5 \lambda_6}{T_5}\left(\frac{1}{A_5} + \frac{1}{A_6}\right)\right];$$

$$K_{13} = \frac{1}{A_5}\left[K_9\left\{\lambda_7 + \frac{\mu_6 \lambda_6 \lambda_7}{T_5}\left(\frac{1}{A_4} + \frac{1}{A_5}\right)\right\} + K_{10}\frac{\mu_6 \lambda_5 \lambda_7}{T_5}\left(\frac{1}{A_4} + \frac{1}{A_6}\right)\right.$$
$$\left. + K_{11}\left\{\lambda_5 + \frac{\mu_6 \lambda_5 \lambda_6}{T_5}\left(\frac{1}{A_5} + \frac{1}{A_6}\right)\right\}\right]$$

$$K_{14} = \frac{1}{A_6}\left[K_9 + \frac{\mu_5 \lambda_6 \lambda_7}{T_5}\left(\frac{1}{A_4} + \frac{1}{A_5}\right) + K_{10}\left\{\lambda_7 + \frac{\mu_5 \lambda_5 \lambda_7}{T_5}\left(\frac{1}{A_4} + \frac{1}{A_6}\right)\right\}\right.$$
$$\left. + K_{11}\left\{\lambda_6 + \frac{\mu_5 \lambda_5 \lambda_6}{T_5}\left(\frac{1}{A_5} + \frac{1}{A_6}\right)\right\}\right]$$

$$K_{15} = \frac{1}{T_5}\left[K_9 \lambda_6 \lambda_7\left(\frac{1}{A_4} + \frac{1}{A_5}\right) + K_{10} \lambda_5 \lambda_7\left(\frac{1}{A_4} + \frac{1}{A_6}\right) + K_{11} \lambda_6 \lambda_5\left(\frac{1}{A_5} + \frac{1}{A_6}\right)\right].$$

Availability of the shell construction system is given by

$$A_v = p_0 + \sum_{i=21}^{27} p_i = \left[1 + \sum_{i=9}^{15} K_i\right] p_0 \tag{5.5.4}$$

where, p_0 is given by (5.5.3).

Singh and Mahajan studied the behaviour of the system for the behaviour change of subsystems/units. As example effect of failure rates of gas cutting machine (λ_1) and pressing machine (λ_2) on availability is shown in the table taking $\lambda_3 = 0.002$, $\lambda_4 = 0.002$, $\lambda_5 = 0.025$, $\lambda_6 = 0.005$, $\lambda_7 = 0.004$, $\mu_1 = 0.2$, $\mu_2 = 0.1$, $\mu_3 = 0.1$, $\mu_4 = 0.2$, $\mu_5 = 0.2$, $\mu_6 = 0.25$, $\mu_7 = 0.25$ in eqn. (5.5.4).

λ_1/λ_2	*0.000*	*0.001*	*0.002*	*0.003*	*0.004*	*0.005*
0.000	0.99945	0.99834	0.99724	0.99613	0.99503	0.99393
0.001	0.99448	0.99338	0.99229	0.99119	0.99010	0.98902
0.002	0.98956	0.98847	0.98739	0.98631	0.99523	0.98415
0.003	0.98469	0.98361	0.98254	0.98147	0.98040	0.97933
0.004	0.97986	0.97880	0.97773	0.97667	0.97562	0.97456
0.005	0.97509	0.97403	0.97298	0.97193	0.97088	0.96983

The table shows that failure rate of the gas cutting machine affects the availability more rapidly than the failure rate of the pressing machine. Similarly, tables for repair rates and comparative tables for repair and failure rates can be formed to study the behaviour of the system under different modes. The above table cautions that the management should be careful about the gas cutting machine.

If there is a standby in gas cutting system, i.e., the system A never fails and there is simultaneous failure in the pressing machine B, then there are sixteen failed states instead of twenty states. Taking $\lambda_1 = 0$ and constant failure rate of B equal to λ^* and constant repair rate μ^*, the transition diagram becomes as shown in Fig. 5.7. The equations associated with the transition diagram may be written and solved recursively. The availability of the system, in this case, is given by

$$A_v = p_0 + \sum_{i=17}^{23} p_i = \left[1 + \sum_{i=9}^{15} K_i\right] p_0$$

where,

$$p_0 = \Bigg[1 + \frac{\lambda^*}{\mu^*} + \frac{\lambda_2}{\mu_2} K_9 + \left(\frac{\lambda^*}{\mu^*} + \frac{\lambda_3}{\mu_3}\right) K_{10} + \left(\frac{\lambda^*}{\mu^*} + \frac{\lambda_4}{\mu_4}\right) K_{11}$$

$$+ \left(\frac{\lambda_2}{\mu_2} + \frac{\lambda_3}{\mu_3}\right) K_{12} + \left(\frac{\lambda_2}{\mu_2} + \frac{\lambda_4}{\mu_4}\right) K_{13} + \left(\frac{\lambda^*}{\mu^*} + \frac{\lambda_3}{\mu_3} + \frac{\lambda_4}{\mu_4}\right) K_{14}$$

$$+ \left(\frac{\lambda_2}{\mu_2} + \frac{\lambda_3}{\mu_3} + \frac{\lambda_4}{\mu_4}\right) K_{15} + \sum_{i=9}^{15} K_i \Bigg]^{-1} \tag{5.5.5}$$

Effect of failure and repair rates on availability of the system may be studied forming tables. Singh and Mahajan provide these tables. If the availability of main shell construction is denoted by Av_1 and the availability of end dish construction system and end dish fitting system be Av_2 and Av_3, respectively, then the overall availability of the container manufacturing system is

$$A(v) = Av_1 \times Av_2 \times Av_3$$

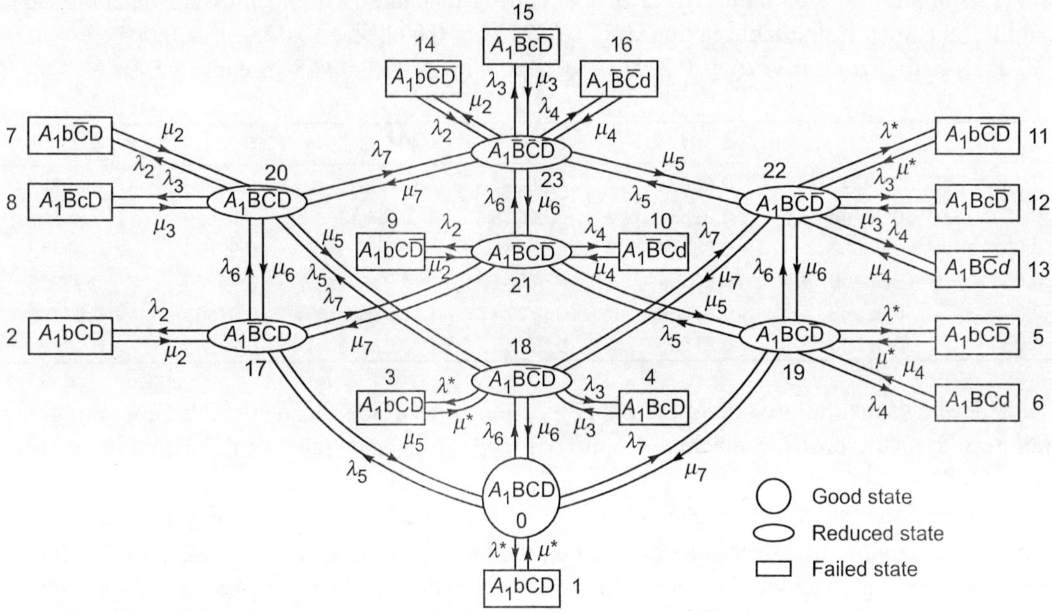

Fig. 5.7

Appendix: When $\mu = 0$,

$$p_0(s) = \frac{1}{s + B_1}; \; p_1(s) = \frac{\lambda_1}{(s + B_1)(s + B_2)}; \; p_2(s) = \frac{\lambda_5}{(s + B_1)(s + B_3)}$$

where, $B_1 = \lambda_1 + \lambda_4 + \lambda_5 + \lambda_8$; $B_2 = \lambda_2 + \lambda_7 + \lambda_9$;
$B_3 = \lambda_6 + \lambda_3 + \lambda_{10}$ and reliability of the transit system is

$$R(t) = \left(1 + \frac{\lambda_1}{B_2 - B_1} + \frac{\lambda_5}{B_3 - B_1}\right) e^{-B_1 t} - \frac{\lambda_1}{B_2 - B_1} e^{-B_2 t} - \frac{\lambda_5}{B_3 - B_1} e^{-B_3 t}.$$

In steady state,

$$p_0 = \frac{1}{1 + B_7}; \; p_1 = \frac{\lambda_2}{B_2} p_0; \; p_2 = \frac{\lambda_5}{B_3} p_0; \; p_3 = B_4 p_0; \; p_4 = B_5 p_0;$$

$$p_5 = B_6 p_0$$

where, B_1, B_2, B_3 are given above and

$$B_4 = \frac{1}{\lambda_{11}}\left(\lambda_4 + \frac{\lambda_1\lambda_2}{B_2} + \frac{\lambda_5\lambda_3}{B_3}\right); \; B_5 = \frac{1}{\lambda_{12}}\left(\lambda_8 + \frac{\lambda_5\lambda_6}{B_3} + \frac{\lambda_1\lambda_7}{B_2}\right)$$

$$B_6 = \frac{1}{\mu}\left(\frac{\lambda_9\lambda_1}{B_2} + \frac{\lambda_{10}\lambda_5}{B_3} + \lambda_{11}B_4 + \lambda_{12}B_5\right); \; B_7 = \frac{\lambda_1}{B_2} + \frac{\lambda_5}{B_3} + B_4 + B_5 + B_6.$$

EXERCISE

1. State k-out-of-n system.
2. What is priority systems and explain its?
3. Write short notes on:
   (a) Mission reliability
   (b) Container manufacturing

6

Chapter

Reliability Technology in Process Industries

There are industries having complex systems which are interconnected in series. The systems, as parts of the industry, produce different materials or parts for a particular product. There is continuous production process in the systems and continuous supply of produced material among the systems. The industrial system, on failure of any one subsystem, becomes standstill. Such industrial systems are known as process industries. In other words, manufacturing of a product is a batch process. Generally, items produced by one unit are further used by another unit to make the item complete for its use or to produce another item out of it. This process is called batch process. An industry adopting system of batch processing is process industry. In general, process industry is defined as an industry where there is a sequence of processes. In a process industry, we cannot bypass intermediate sub-processes. In this chapter, three process industries, namely, paper, cement and utensils are discussed. On similar lines, other process industries for various parameters such as mean time to system failure (MTSF), mean time to repair (MTTR) of the system, reliability and availability of the system may be analyzed.

6.1 THE PAPER INDUSTRY

The raw material for the production of paper is soft and hardwood and bamboo (in general). The raw material is chopped into small pieces of approximately uniform size (called chips) and transported to the store by the use of compressed air. A chain conveyor feeding system carries these chips from the store to digesters whenever required. These chips are cooked using $NaOH + Na_2S$ and steam at 8 kg/cm^2 pressure and 175°C temperature. The chips after cooking are converted into pulp. The pulp is pneumatically transported to storage tanks from where it is transported for further processing through fiberizer and refiner. The pulp is filtered through filters and washed with water in three/four stages to remove the cooking chemicals. The washed pulp discharged from the washer is stored in a surge tank. The pulp is next processed by bleaching and screening for the production of white paper while for the production of brown coarse grade of paper the pulp is screened directly without going through bleaching system. Chlorine gas is passed through pulp tank for bleaching. The screening is done to separate out the oversize and odd shape particles or debris. The pulp is then passed through a cleaner which separates heavy material from the pulp and then the pulp is sent to paper rolling machine. In rolling machine, the pulp is spread evenly over an endless belt made of meshed wire running between breast and couch rolls. The paper in the form of sheets produced by the rolling machine is sent to the dryers to smoothen and iron out any irregularities. The dried paper sheet is finally rolled in the form of rolls and sent for final packing.

The paper industry/production system consists of six subsystems—(a) feeding (b) pulping (c) washing (d) bleaching (e) screening (f) paper formation/production:

(a) Feeding System—It consists of a chain conveyor for carrying chips from store to digesters and blower with blowing units for pneumatic conveying of chips to the digesters.

(b) Pulping System—It consists of digesters for cooking the pulp using NaOH, Na_2S and steam, knotter (fiberizer) to remove the knots from the cooked pulp, decker for removing the black liquor from the cooked pulp and refiner (opener) to open the knots.

(c) Washing System—It consists of screening unit for separating the unwanted foreign material from the pulp, cleaner for removing heavy material from the pulp and washer for removing chemicals through washing.

(d) Bleaching System—It consists of filter for filtering the unbleached pulp and opener to open the fibres.

(e) Screening System—It is composed of four subsystems; filter to remove black liquor, screen for removing the knots and other undesirable material, cleaner and mixer for cleaning the fibres and mixing of freshwater with the pulp and washer to wash the pulp for brightness.

(f) Paper Formation System—It consists of fibre decomposition and water suction unit, pressing unit for ironing and smoothening the paper sheets and dryers for removing the moisture content from the paper sheets.

All the systems in the paper mill work in series. The six systems introduced above are described and analyzed to obtain overall capability of the paper industry as follows:

(a) Feeding System

In this system (i) the blower A supplies compressed air which is used for pneumatic conveying the chips to the digesters, (ii) standby unit B carries wooden chips from store to the digesters using compressed air (iii) the chips carrying system comprised of three units in series, namely, chain D_1, conveyor D_2 and bucket elevator D_3. The failure of the blower A system causes failure of the feeding system. The capacity of B is small and it is used whenever an extra demand of chips occurs or in case of sudden failure of the blower A. The schematic diagram and transition diagram of the feeding system are given in Figs. 6.1 and 6.2, respectively.

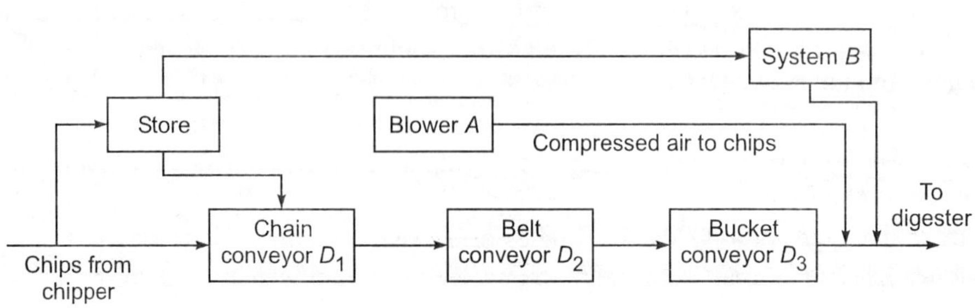

Fig. 6.1

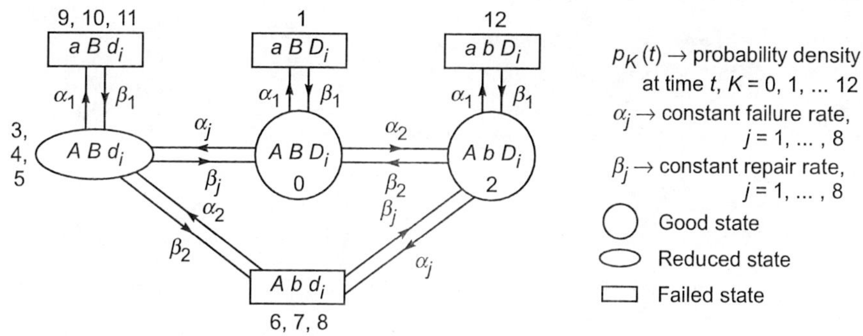

Fig. 6.2

The differential-difference equations associated with the transition diagram (using simple probabilistic arguments) are:

$$\left(\frac{d}{dt}+\sum_{1}^{5}\alpha_i\right) p_0(t) = \sum_{1}^{2} \beta_i p_i(t) + \sum_{3}^{5} \beta_j p_j(t) \qquad (6.1.1)$$

$$\left(\frac{d}{dt}+\alpha_1+\beta_2+\sum_{6}^{8}\alpha_i\right) p_2(t) = \alpha_2 p_0(t) + \beta_1 p_{12}(t) + \sum_{3}^{5} \beta_j p_{j+3}^{(t)} \qquad (6.1.2)$$

$$\left(\frac{d}{dt}+\alpha_1+\alpha_2+\beta_j\right) p_j(t) = \beta_2 p_{j+3}^{(t)} + \beta_1 p_{j+6}^{(t)} + \alpha_j p_0(t),\ j = 3,\ 4,\ 5 \qquad (6.1.3)$$

$$\left(\frac{d}{dt}+\beta_1\right) p_i(t) = \alpha_1 p_k(t), \text{ for } k = 0,\ i = 1\ ;\ k = 2,\ i = 12\ ;\ k = j,\ i = j + 6,$$

$$j = 3,\ 4,\ 5 \qquad (6.1.4)$$

$$\left(\frac{d}{dt}+\beta_2+\beta_j\right) p_{j+3}^{(t)} = \alpha_j p_2(t) + \alpha_2 p_j(t),\ j = 3,\ 4,\ 5 \qquad (6.1.5)$$

with initial conditions $p_0(0) = 1$ otherwise zero.

Taking Laplace transforms of eqns. (6.1.1)–(6.1.5) (using initial conditions) and solving recursively, the probabilities are obtained in terms of $p_0(s)$. The availability function $Av_1(s)$ is given by

$$Av_1(s) = L\left[p_0(t) + p_2(t) + \sum_{3}^{5} p_j(t)\right] = p_0(s) + p_2(s) + \sum_{3}^{5} p_j(s)$$

where, $p_0(s) = [s + x_3 - y_3]^{-1}$;

$$p_2(s) = \frac{\alpha_2}{s + x_2 - y_2}\left[1 + \sum_{3}^{5} \frac{\beta_j^2}{(s+\beta_2+\beta_j)(s+x_1-y_1)}\right] p_0(s);$$

$$p_j(s) = \left[\frac{\alpha_2\beta_2\alpha_j}{(s+\beta_2+\beta_j)(s+x_1-y_1)(s+x_2-y_2)}\right.$$

$$\cdot\left\{1+\sum_{3}^{5}\frac{\beta_j^2}{(s+\beta_2+\beta_j)(s+x_1-y_1)}\right\}+\frac{\beta_j}{(s+x_1-y_1)}\Bigg]\,p_0(s),\ j = 3,\ 4,\ 5.$$

$$x_1 = \alpha_1 + \alpha_2 + \beta_j;\ x_2 = \alpha_1 + \beta_2 + \sum_{3}^{5}\alpha_j\ ;\ x_3 = \sum_{1}^{2}\alpha_i + \sum_{3}^{5}\alpha_j\ ;$$

$$y_1 = \frac{\alpha_1\beta_1}{s+\beta_1} + \frac{\alpha_2\beta_2}{s+\beta_2+\beta_j}\ ;$$

$$y_2 = \sum_{3}^{5}\frac{\alpha_j\beta_j}{s+\beta_2+\beta_j}\left[1+\frac{\alpha_2\beta_2}{(s+\beta_2+\beta_j)(s+x_1-y_1)}\right]+\frac{\alpha_1\beta_1}{s+\beta_1}\ ;$$

$$y_3 = \frac{\alpha_1\beta_1}{s+\beta_1} + \beta_2 M_1 + \sum_{3}^{5}\beta_j M_2\ ;\ p_2(s) = M_1\,p_0(s)\ ;\ p_j(s) = M_2\,p_0(s);$$

s is Laplace parameter.

Mean time to system failure (MTTF) $= \lim_{s\to 0} s\,Av_1(s)$.

The steady-state probabilities are obtained taking $\frac{d}{dt} = 0$, $p_k(t) = p_k$ as $t \to \infty$ in eqns. (6.1.1)–(6.1.5) and solving recursively. All the probabilities are obtained in terms of p_0 which, in turn, is obtained using normalizing condition $\sum_{k=0}^{12} p_k = 1$. Steady-state availability is given by

$$Av_1 = p_0 + p_2 + \sum_{3}^{5} p_j = \frac{1+L_1}{L_3} + \sum_{3}^{5}\frac{\alpha_2\alpha_j}{L_3}$$

where,

$$L_1 = \frac{\alpha_2 + \sum[\alpha_2\alpha_j/(\beta_2+\beta_j+\alpha_2)]}{\beta_2 + \sum\alpha_j - \sum\frac{\alpha_j\beta_j}{\beta_2+\beta_j} - \sum\frac{\alpha_2\beta_2\beta_j}{(\beta_2+\beta_j)(\alpha_2+\beta_2+\beta_j)}}$$

$$L_2 = [\beta_2(1 + L_1) + \beta_j]/\beta_j(\alpha_2 + \beta_2 + \beta_j)\ ;$$

$$L_3 = \left[\left(1+\frac{\alpha_1}{\beta_1}\right)\left(1+L_1+\sum L_2\alpha_j\right)+\sum\left(\alpha_2 L_2\alpha_j+\frac{L_i\alpha_j}{\beta_2+\beta_j}\right)\right];\ j = 3,\ 4,\ 5$$

Analysis of the system for general repair time is given by Kumar et. al. [1988].

(b) Pulping System

The pulping system, an important part of the mill, consists of four subsystems:

(i) The digester D_1 : Here a mixture of wooden chips and NaOH + Na_2S (1 : 3.5 ratio) is heated by steam at 175°C. Failure of digester stops the cooking process and hence leads to system failure.

(ii) The decker D_2 : It is used to remove the black liquor from the pulp. Failure of this causes complete failure of the process of black liquor removal. With only one or two working units, it is possible to produce low quality paper which is uneconomical.

(iii) The knotter D_3 : It consists of one main unit and one standby. Complete failure occurs on failure of both the units. It is used to tear, cut and plough the fibres.

(iv) The opener D_4 : It consists of one main unit and one standby. Complete failure of this subsystem occurs when both the units fail. It is used to break the walls of the fibres into ribbons ensuring the availability of large surface area for bonding.

The block diagram and transition diagram for the system are shown in Fig. 6.3 and Fig. 6.4, respectively.

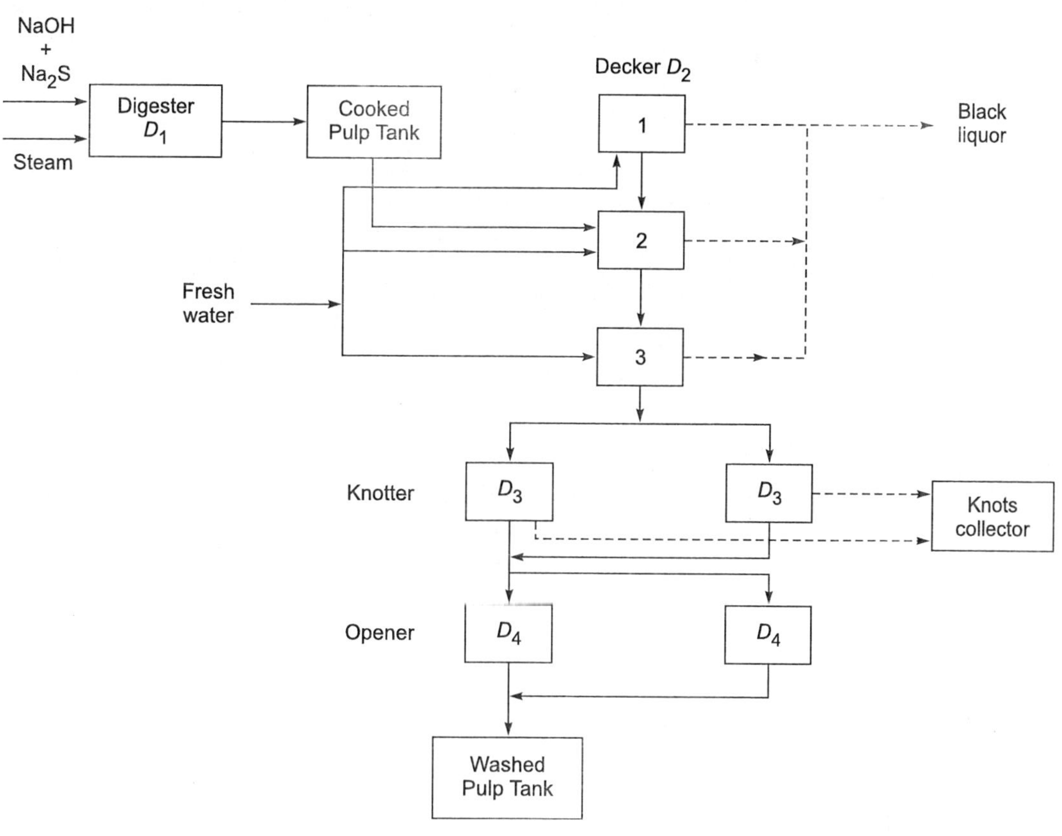

Fig. 6.3

Taking $p_j(t)$, probability density at time t in jth state of the system and α_j, β_j the constant failure and repair rates for jth state, the transition diagram for the system is

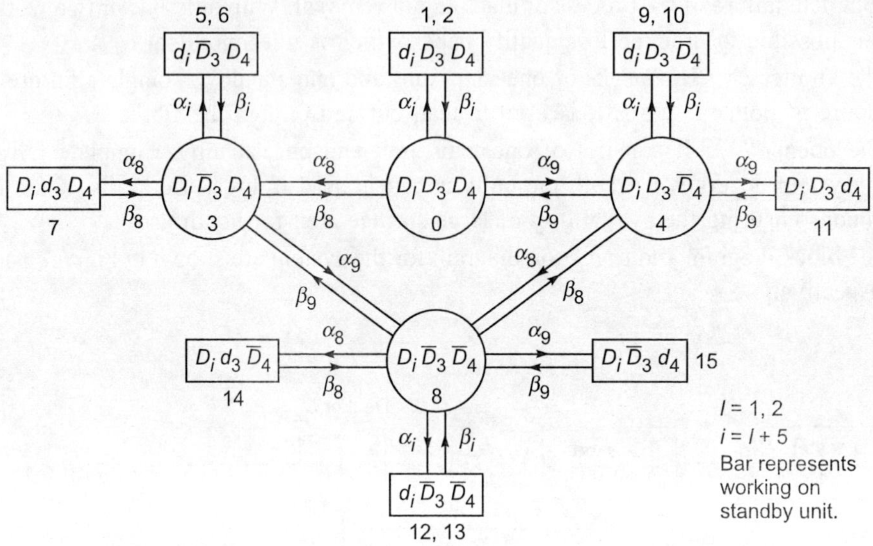

Fig. 6.4

The governing differential-difference equations associated with the transition diagram are

$$\left(\frac{d}{dt} + \sum_{6}^{9} \alpha_j\right) p_0(t) = \sum_{6}^{9} \beta_j p_{j-5}(t) \tag{6.1.6}$$

$$\left(\frac{d}{dt} + \beta_8 + \sum_{6}^{9} \alpha_j\right) p_3(t) = \sum_{6}^{9} \beta_j\, p_{j-1}(t) + \alpha_8\, p_0(t) \tag{6.1.7}$$

$$\left(\frac{d}{dt} + \beta_9 + \sum_{6}^{9} \alpha_j\right) p_4(t) = \alpha_9\, p_0(t) + \beta_6\, p_9(t) + \beta_7\, p_{10}(t) + \beta_8\, p_8(t) + \beta_9\, p_{11}(t), \tag{6.1.8}$$

$$\left(\frac{d}{dt} + \beta_8 + \beta_9 + \sum_{6}^{9} \alpha_j\right) p_8(t) = \sum_{6}^{9} \beta_j\, p_{j+6}(t) + \alpha_8\, p_4(t) + \alpha_9\, p_3(t), \tag{6.1.9}$$

$$\left(\frac{d}{dt} + \beta_j\right) p_i(t) = \alpha_j\, p_k(t) \tag{6.1.10}$$

with initial conditions $p_0(0) = 1$ otherwise zero.

In eqn. (6.1.10) take for $j = 6, 7$, $k = 0$, $i = j - 5$; $k = 3$, $i = j - 1$; $k = 4$, $i = j + 3$; $k = 8$, $i = j + 6$, for $j = 8$, $k = 3$, $i = j - 1$; $k = 8$, $i = j + 6$, and for $j = 9$, $k = 4$, $i = j + 2$; $k = 8$, $i = j + 6$.

The above (6.1.6)–(6.1.10) are first order differential equations and can be solved easily using Laplace transforms. Availability function $Av_2(t) = p_0(t) + p_3(t) + p_4(t) + p_8(t)$

$= L^{-1}[p_0(s) + p_3(s) + p_4(s) + p_8(s)]$. Taking $\frac{d}{dt} = 0$, $p_j(t) = p_j$ for $t \to \infty$, steady state probabilities may be evaluated using the normalizing condition $\sum_{j=0}^{15} p_j = 1$. Here only the equations are given avoiding the detailed solution which is simple in this case. In the analysis of other subsystems, only the schematic and/or transition diagrams along with the governing differential-difference equations are given. Since the equations are simple and may be solved using any method, the details are not given. The results similar to the feeding system may be obtained for availability and mean time to system failure, etc.

(c) Washing System

The washing system comprise the following :

(i) The screen F_1, having two units in series. The subsystem fails completely when any one of these units fails. It is used to remove the oversize, uncooked and odd shape fibres from pulp through straining.

(ii) The cleaner F_2, having '*l*' units in parallel such that failure of any one or more units reduces the cleaning efficiency of the subsystem which affects the quality of the paper and hence the profit. This can be repaired by unskilled workers in a short time period.

(iii) The decker F_3, having one main unit and one standby unit. Failure of this subsystem occurs only when both the units fail.

Denoting by $p_j(t)$ as the probability density at time t in jth state, λ_j, α_j, the constant failure and β_j constant repair rates, the transition diagram is given in Fig. 6.5.

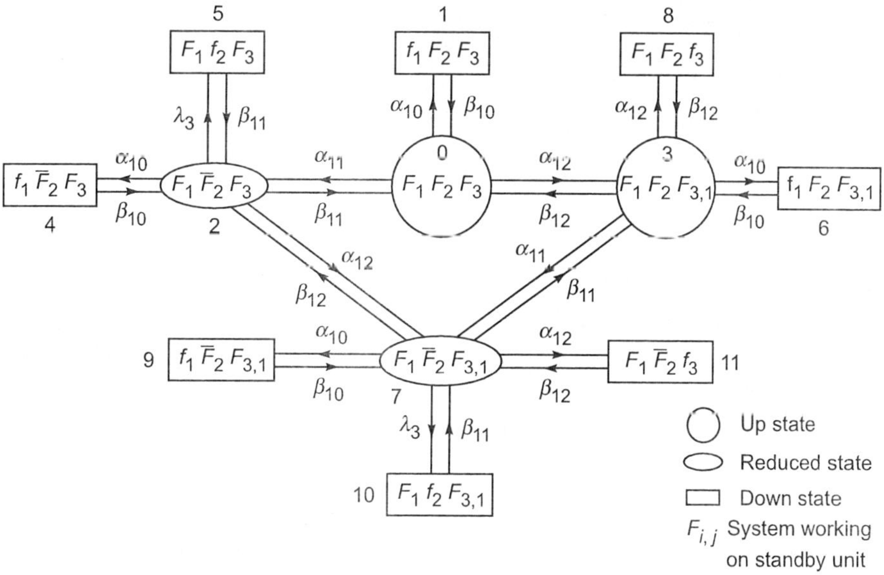

Fig. 6.5

Following the notations given in transition diagram, the differential-difference equations associated are:

$$\left(\frac{d}{dt}+\sum_{10}^{12}\alpha_j\right)p_0(t)=\sum_{10}^{12}\beta_j p_{j-9}(t), \tag{6.1.11}$$

$$\left(\frac{d}{dt}+\alpha_{10}+\alpha_{12}+\lambda_3+\beta_{11}\right)p_2(t)=\alpha_{11}p_0(t)+\beta_{11}p_5(t)+\beta_{12}p_7(t)+\beta_{10}p_4(t), \tag{6.1.12}$$

$$\left(\frac{d}{dt}+\sum_{10}^{12}\alpha_j+\beta_{12}\right)p_3(t)=\sum_{10}^{12}\beta_j p_{j-4}(t)+\alpha_{12}p_0(t), \tag{6.1.13}$$

$$\left(\frac{d}{dt}+\alpha_{10}+\lambda_3+\lambda_{12}+\beta_{11}+\beta_{12}\right)p_7(t)=\alpha_{12}p_2(t)+\alpha_{11}p_3(t)+\sum_{10}^{12}\beta_j p_{j-1}(t), \tag{6.1.14}$$

$$\left(\frac{d}{dt}+\beta_j\right)p_i(t)=\alpha_j p_k(t) \tag{6.1.15}$$

$$\left(\frac{d}{dt}+\beta_{11}\right)p_i(t)=\lambda_3 p_k(t) \tag{6.1.16}$$

with initial conditions $p_0(0) = 1$ otherwise $= 0$.

In eqn. (6.1.15), for $j = 10$, $k = 0$, $i = j - 9$; $k = 2$, $i = j - 6$; $k = 3$, $i = j - 4$; $k = 7$, $i = j - 1$, $j = 12$, $k = 3$, $i = j - 4$; $k = 7$, $i = j - 1$,

and in eqn. (6.1.16) for $k = 2$, $i = 5$; $k = 7$, $i = 10$.

Availability function $Av_3(t) = p_0(t) + p_2(t) + p_3(t) + p_7(t) = L^{-1}[p_0(s) + p_2(s) + p_3(s) + p_7(s)]$

and MTTF $= \lim_{s\to 0} s\,Av_3(s)$, where $L\{Av_3(t)\} = Av_3(s)$.

(d) Bleaching System

This system consists of two subsystems only:

(i) The filter G_1 consisting of two units and is said to be failed when one of the two units is failed (due to fall in quality). Its main function is to remove chlorine and unbleached mass from the pulp received from the washing system.

(ii) The opener G_2, comprised of two units and is said to be failed when one of the two units is failed (due to fall in quality). Its main function is to open up the fibres through combing action.

G_1 is similar as $\overline{G_1}$ and G_2 is similar to $\overline{G_2}$. Also G_1 is in series with G_2, $\overline{G_1}$ is in series with $\overline{G_2}$. The transition diagram for the system is given in Fig. 6.6.

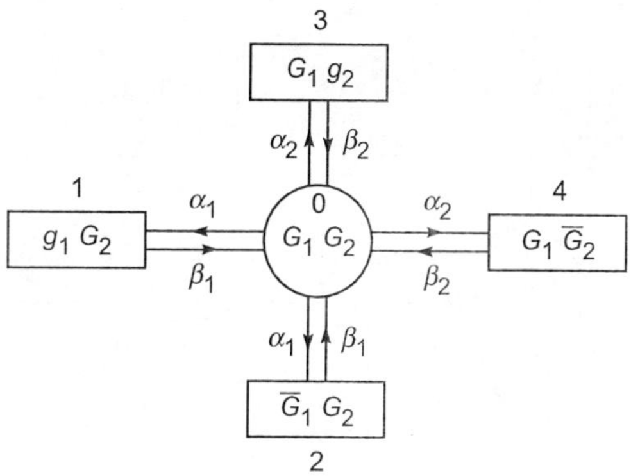

Fig. 6.6

The differential-difference equations associated with the transition diagram are

$$\left(\frac{d}{dt} + 2\sum_{1}^{2} \alpha_j\right) p_0(t) = 2\sum_{1}^{2} \beta_j p_j(t) \tag{6.1.17}$$

$$\left(\frac{d}{dt} + \beta_j\right) p_j(t) = \alpha_1\, p_0(t), \quad j = 1, 2 \tag{6.1.18}$$

$$\left(\frac{d}{dt} + \beta_j\right) p_j(t) = \alpha_2\, p_0(t), j = 3, 4 \tag{6.1.19}$$

with initial conditions $p_0(0) = 1$ otherwise $= 0$.

Availability function $Av_4(t) = p_0(t) = L^{-1}p_0(s)$

$$\text{MTTF} = \lim_{s\to 0} s\, Av_4(s) = [\beta_1\, \beta_2 + 2\alpha_1\beta_2 + 2\alpha_2\beta_1]^{-1}$$

(e) Screening System

The screening system comprises the following:

(i) The filter S_1, works for removal of black liquor from the pulp. Its failure causes failure of the system.

(ii) The screen S_2, removes the knots and other undesirable materials from the pulp. Failure of screen causes complete failure of the system.

(iii) The cleaner S_3, consists of three units in parallel. The failure of any one unit reduces the efficiency of the plant. Complete failure of cleaner reduces the efficiency of the plant but the system remains operative. Manual operation is possible during the repair. Water is mixed here with the pulp by centrifugal action.

(iv) The decker S_4, reduces the blackness of the pulp. The failure of decker causes the complete failure of the system.

The transition diagram based on the above description and assumptions is shown in Fig. 6.7.

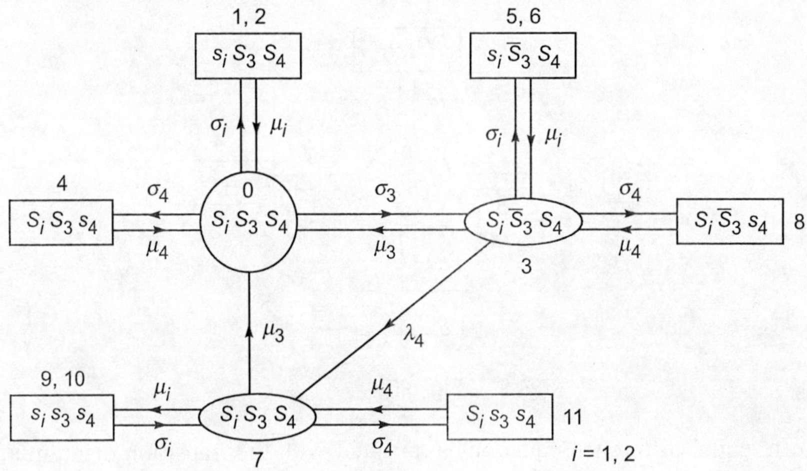

Fig. 6.7

Taking σ_i as constant failure rate and μ_i as constant repair rate, probability considerations give the following differential-difference equations associated with the transition diagram:

$$\left(\frac{d}{dt} + \sum_1^4 \sigma_j\right) p_0(t) = \sum_1^4 \mu_j p_j(t), \tag{6.1.20}$$

$$\left(\frac{d}{dt} + \mu_3 + \sigma_4 + \lambda_4 + \sum_1^2 \sigma_i\right) p_3(t) = \sigma_3 p_0(t) + \mu_4 p_8(t) + \sum_1^2 \mu_j p_{j+4}(t), \tag{6.1.21}$$

$$\left(\frac{d}{dt} + \mu_3 + \sigma_4 + \sum_1^2 \sigma_j\right) p_7(t) = \lambda_4 p_3(t) + \mu_4 p_{11}(t) + \sum_1^2 \mu_j p_{j+8}(t) \tag{6.1.22}$$

$$\left(\frac{d}{dt} + \mu_j\right) p_i(t) = \sigma_j p_k(t) \tag{6.1.23}$$

with initial conditions $p_0(0) = 1$ otherwise zero.

In eqn. (6.1.23), for $j = 1, 2$, $k = 0$, $i = j$; $k = 3$, $i = j + 4$; $k = 7$, $i = j + 8$; $j = 4$, $k = 0$, $i = j$; $k = 3$, $i = j + 4$; $k = 7$, $i = j + 7$.

The availability function $Av_5(t) = p_0(t) + p_3(t) + p_7(t)$ and

$$\text{MTTF} = \lim_{s \to 0} s\, Av_5(s) = 2(\lambda_4 + \mu_3)^2[\mu_3(\lambda_4 + \mu_3)^2 + \lambda_4 \mu_3^2 + \lambda_4^2(\lambda_4 + 2\mu_3)]^{-1}$$

(f) Paper Formation System

It is the important system of a paper mill consisting of the following subsystems:

(i) The wire mat F_1, used for depositing the suspended fibre on the top of the wire mesh and sucking the water from the pulp by suction. It controls the width of paper sheet. Failure of F_1 causes failure of the system.

(ii) Synthetic belt F_2, which provides the support to run the fibre mat through press and drying sections. Failure of F_2 causes failure of the system.

(iii) Rollers F_M, consisting of M rollers in series. Failure of any one fails the system. These rollers help the wire mat and the synthetic belt to roll on them smoothly.

(iv) Vaccum pump F_K, comprises four units in parallel. It is used for sucking water from the pulp through wire mat. It has two standby units. The system fails if three or more units fail at a time.

All the above four subsystems (F_1, F_2, F_M, F_K) require common steam supply. The failure of steam supply causes failure of the system. The failure of steam supply causes the failure of other system (s) which is run by steam and is in repair. If after time t the steam supply is failed the system is failed. The schematic diagram and transition diagram of the system are shown in Figs. 6.8 and 6.9, respectively.

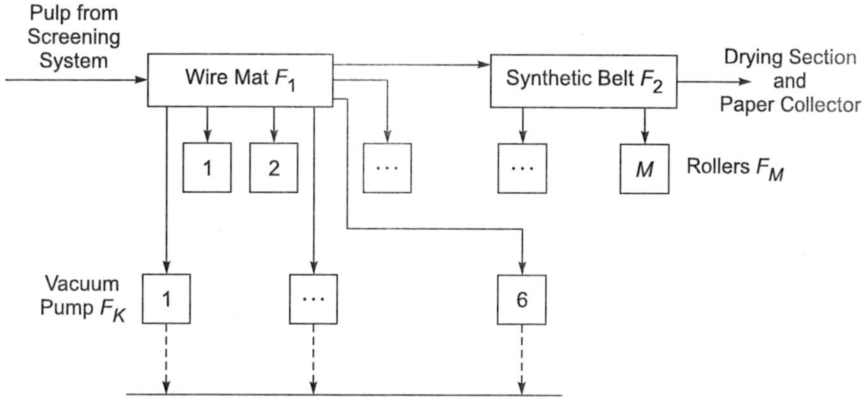

Fig. 6.8

Probability considerations give the following differential-difference equations associated with the transition diagram (6.9) :

$$\left(\frac{d}{dt}+\sum_{7}^{9}\sigma_i+\lambda_5+\lambda_6+\sum_{1}^{2}\sigma_i\right)p_0(t)=\sum_{5}^{8}\mu_i\,p_{i-4}(t)+\mu_9\{p_{(SF)_1}(t)+p_{(SF)_2}(t)\} \tag{6.1.24}$$

$$\left(\frac{d}{dt}+\mu_8+\lambda_5+\sum_{5}^{9}\sigma_i\right)p_4(t)=\sigma_8\,p_0(t)+\sum_{5}^{8}\mu_i\,p_i(t)+\mu_9\,p_{(SF)_3}(t) \tag{6.1.25}$$

$$\left(\frac{d}{dt}+\mu_8+\sigma_9+\sum_{7}^{10}\upsilon_i\right)p_8(t)=\lambda_5\,p_0(t)+\sigma_8\,p_4(t)+\sum_{5}^{8}\mu_i\,p_{i+4}(t)+\mu_9\,p_{(SF)_4}(t) \tag{6.1.26}$$

$$\left(\frac{d}{dt}+\mu_i+\lambda_5\right)p_n(t)=\sigma_i\,p_k(t),\ \text{for } i=5,6,7,\ k=0,\ n=i-4; \tag{6.1.27}$$

$$k=4,\ n=i;\ k=8,\ n=i+4$$

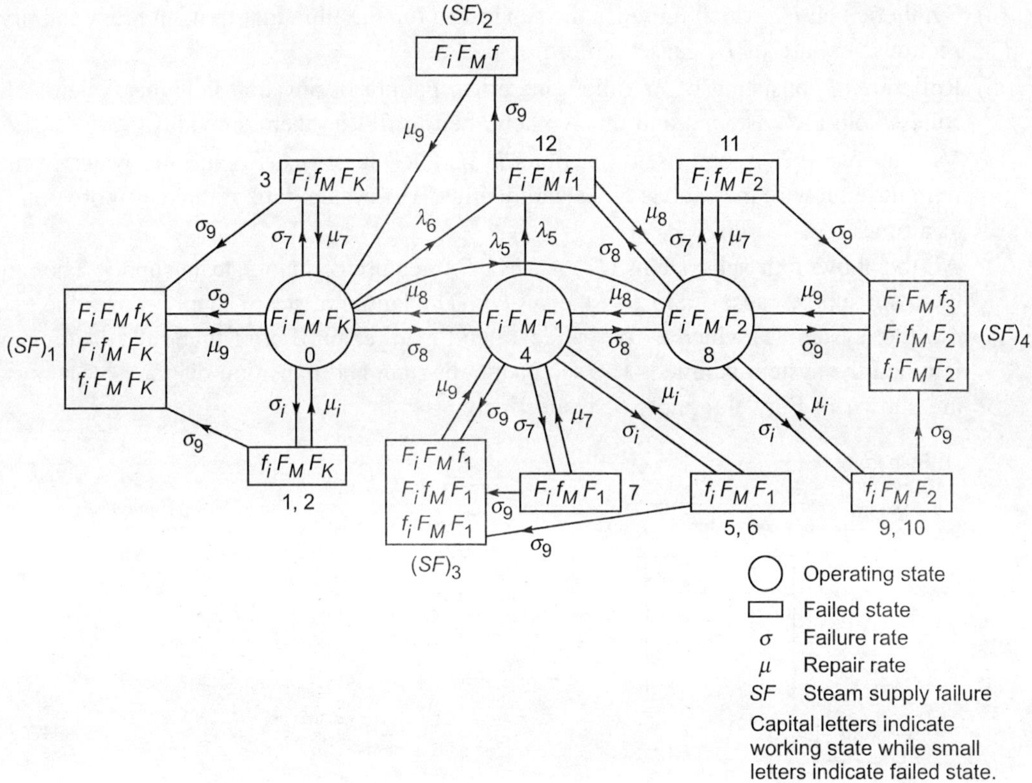

Fig. 6.9

$$\left(\frac{d}{dt} + \sigma_9 + \mu_8\right) p_{12}(t) = \lambda_6\, p_0(t) + \lambda_5\, p_4(t) + \sigma_8\, p_8(t) \tag{6.1.28}$$

$$\left(\frac{d}{dt} + \mu_9\right) p_{(SF)_i}(t) = \sigma_9\, p_j(t),\ i = 1, 2, 3, 4\ j = 0, 1, 2, \ldots, 12 \tag{6.1.29}$$

with initial conditions $p_0(0) = 1$ otherwise zero.

In eqn. (6.1.29) for $i = 1, j = 0, 1, 2, 3$; $i = 2, j = 12$; $i = 3, j = 4, 5, 6, 7$; $i = 4, j = 8, 9, 10, 11$.

The availability function $Av_6(t)$ and MTTF are given by

$$Av_6(t) = p_0(t) + p_4(t) + p_8(t) \text{ and MTTF} = \lim_{s \to 0} s\, Av_6(s).$$

where, $Av_6(s) = L[p_0(t) + p_4(t) + p_8(t)] = p_0(s) + p_4(s) + p_8(s)$, s is Laplace transform parameter.

Since all the systems discussed above from (*a*) to (*f*) are working in series in a paper industry, the overall availability of the paper mill can be evaluated from the following formula:

$$Av = Av_1(t) \cdot Av_2(t) \cdot Av_3(t) \cdot Av_4(t) \cdot Av_5(t) \cdot A_6(t).$$

Steady-state behaviour of the systems may be studied as discussed for feeding system in 6.1(a). Since, in case of process industries the management always is interested in long-run behaviour of the industrial system, most of the authors studied the systems for long-run availability of the system. Changing the failure and repair rates in the above study, tables and graphs may be prepared for availability and mean time to system failure both for transient and steady-state behaviour of the industry, which are helpful for ready reference. Various measures such as preventive maintenance, etc. can be adopted to increase the availability of the system. For more details, about the study of paper industry readers are referred to see works of Kumar et al., [1988–1993]. Considering different practical facets of the subsystems valuable parametric knowledge about the industry can be obtained.

6.2 THE CEMENT INDUSTRY

Cement, an important material for our present life, is obtained by burning and crushing the clay and lime carbonate stones. Dry and wet, the two processes are used to manufacture the cement. In dry process, clay and limestone are dried, crushed and grinded finely. These are mixed together in calculated proportions and the process is called blending. The mixed powder is supplied to a huge rotating kiln set up in an inclined position (angle 15°). The kiln has two ends, the charge or feed end and the fuel end. The temperature at ends remains between 400°C and 1800°C. The charge end undergoes dehydration, dissociation and decomposition as it slides downwards due to rotation of kiln about its own axis. At the lower end, the burnt product comes out in the form of green lump-shaped, very hot material called clinker. The clinkers are first cooled in special chambers called clinker coolers and 2-3% of gypsum is added to the clinkers. This mixture is very finely pulverized. The fine is the pulverization, the better is the quality of the cement. The powdered cement is stored in silos from where it is drawn for packing and transportation. In comparison to dry process, the wet process is more convenient but costly than dry process. In wet process, the grinding of the raw materials is done separately and in the presence of water so that instead of powder a fine dispersion of limestone (limestone slurry) and clay (clay slurry) are obtained. These are stored in separate tanks. These are blended in proper proportions and the blended slurry is fed into rotary kiln. In wet process, the kilns are longer and bigger in dimensions. In this section manufacturing of the cement by dry process is analyzed for availability. The process is described as follows [Ref. Gupta et al. (2005)]:

(i) The boulders are transported in huge dumpers and dumped into the hopper of the crusher.

(ii) The jaw crusher (A), a single unit system, crushes the big stones into small pieces. It is subjected to major failure only. After this the hammer mill (B) crushes the small pieces of stones into further small sizes. The hammer mill is also single unit system subject to major failure only.

(iii) The clay materials found in the quarry are also dumped into the crusher and stacked along with the limestone. The crushed materials are checked for calcium carbonate, lime, alumina, ferrous oxide and silica contents. Any component found short in the materials is added separately. The additive material and crushed limestone are conveyed to the storage hoppers. The raw materials are fed to the raw mill by means of conveyor and proportioned with the help of weigh feeders. The materials are grinded to the desired fineness in raw mill (or raw mixer) C. The fine powder so obtained from raw mill is blown upwards, collected in cyclones and fed to the giant sized continuous blending and storage silo by use of aeropole. System C consists of small steel balls subject to minor and major failures.

(iv) The material is dropped merely by gravity from the blending to the storage silo thereby conserving power. The material is pumped into the preheater by an aeropole. This material, from the bottom of the preheater, is fed to the rotary kiln. The burning is carried out in rotary kiln formed of steel tubes. The slurry is injected at the upper end of kiln. Hot gases or flames are forced through the lower end of kiln. The upper end of kiln is called dry zone and the water from slurry is evaporated in this zone. In the next section of kiln, carbon dioxide from slurry is evaporated and small lumps (called nodules) are formed at this stage. The nodules gradually roll down passing through zones (known as burning zone) where the temperature is about 1400°C to 1500°C. In the burning zone, the calcined product is formed and nodules are converted into small, hard, dark, greenish, blue balls known as the clinkers or raw cement. The clinkers are of size 3 mm to 20 mm. This section is called as clinker section (*D*) and consists of two units in series subject to major failure.

(v) The clinkers obtained from clinker section (*D*) are grinded finely in ball mills and tube mills. During the grinding, a small quantity of gypsum (3 to 4%) is added to control the setting. This section is called the cement mill (*E*) consisting of a number of small steel balls and is subjected to minor and major failures.

(vi) Cement needs very careful packing and storage arrangements. It is stored in specially designed concrete storage tanks called silos. The packing system (*F*) consists of two units working in series subject to only major failure.

Schematic diagram of the cement industry is shown in Fig. 6.10.

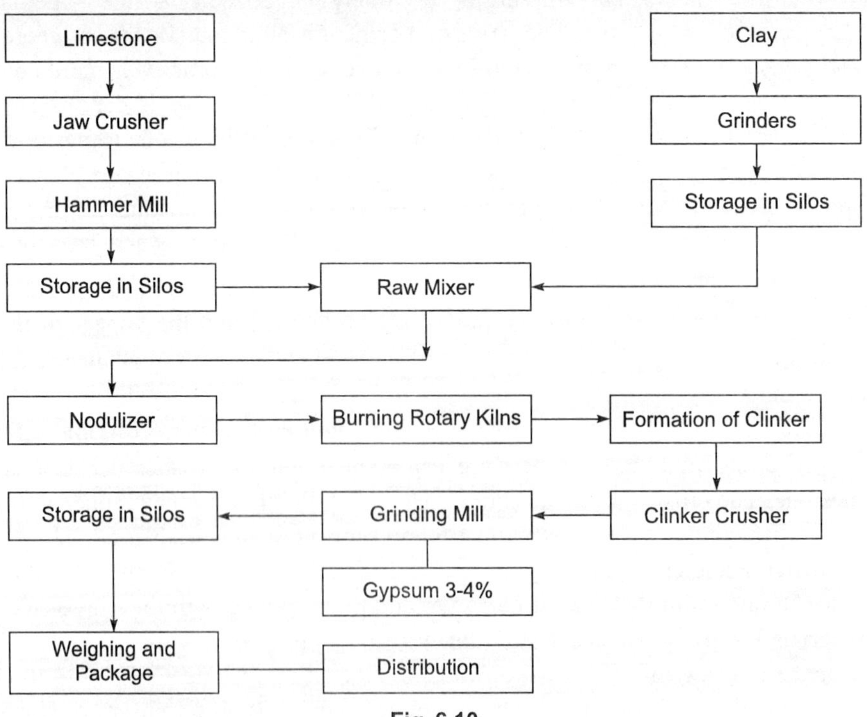

Fig. 6.10

Due to minor failure the system works in reduced state. Let us denote the reduced states of C and E by $\bar{C}$ and $\bar{E}$ and the failed states of A, B, C, D, E, F by small letters a, b, c, d, e, f, respectively. Assuming further that the repairs and failures are independent of each other, no simultaneous failure among subsystems (A, B, C, D, E, F), subsystems C and E can fail completely only through reduced states, repaired components are like new and repair of C and E in reduced state is allowed only upto a certain limit. Further assume that

α_i: respective constant failure rate of subsystems $A, B, D, F, \bar{C}, \bar{E}, C, E$, $i = 1, \ldots, 8$.

β_i: respective constant repair rate of subsystems $A, B, D, F, \bar{C}, \bar{E}, C, E$, $i = 1, \ldots, 8$.

$p_i(t)$: probability density that at time t the system is in ith state, $i = 1, 2, \ldots, 24$.

Following the above notations and assumptions, the transition diagram is shown in Fig. 6.11.

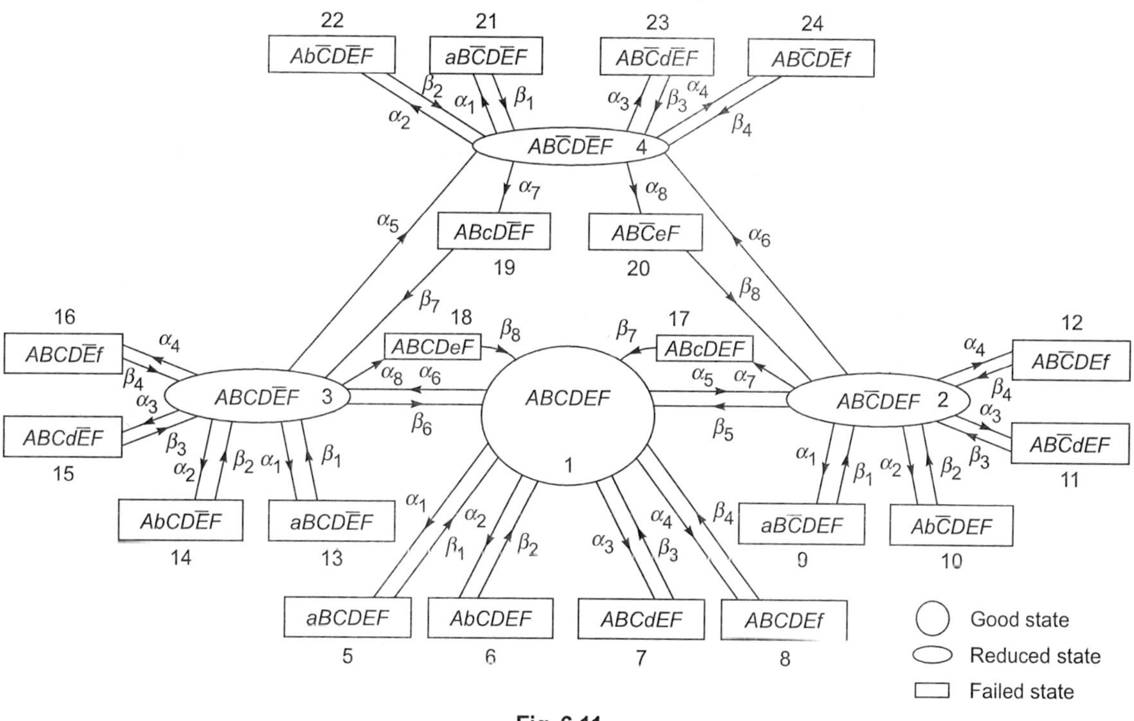

Fig. 6.11

Probability considerations give the following differential equations associated with the system:

$$\left(\frac{d}{dt}+\sum_{1}^{6}\alpha_i\right)p_1(t)=\sum_{1}^{4}\beta_i p_{i+4}(t)+\beta_5 p_2(t)+\beta_6 p_3(t)+\beta_7 p_{17}(t)+\beta_8 p_{18}(t), \tag{6.2.1}$$

$$\left(\frac{d}{dt}+\sum_{1}^{4}\alpha_i+\alpha_6+\alpha_7+\beta_5\right)p_2(t)=\sum_{1}^{4}\beta_i p_{i+8}(t)+\beta_8 p_{20}(t)+\alpha_5 p_1(t), \tag{6.2.2}$$

$$\left(\frac{d}{dt}+\sum_{1}^{5}\alpha_i+\alpha_8+\beta_6\right)p_3(t)=\sum_{1}^{4}\beta_i p_{i+12}(t)+\beta_7 p_{19}(t)+\alpha_6 p_1(t) \tag{6.2.3}$$

$$\left(\frac{d}{dt}+\sum_{1}^{4}\alpha_i+\alpha_7+\alpha_8\right)p_4(t)=\sum_{1}^{4}\beta_i p_{i+20}(t)+\alpha_5 p_3(t)+\alpha_6 p_2(t) \tag{6.2.4}$$

$$\left(\frac{d}{dt}+\beta_i\right)p_{4+i}(t)=\alpha_i p_1(t),\ i=1,2,3,4 \tag{6.2.5}$$

$$\left(\frac{d}{dt}+\beta_i\right)p_{8+i}(t)=\alpha_i p_2(t),\ i=1,2,3,4 \tag{6.2.6}$$

$$\left(\frac{d}{dt}+\beta_i\right)p_{12+i}(t)=\alpha_i p_3(t),\ i=1,2,3,4 \tag{6.2.7}$$

$$\left(\frac{d}{dt}+\beta_7\right)p_{17}(t)=\alpha_7 p_2(t)\ ;\ \left(\frac{d}{dt}+\beta_8\right)p_{18}(t)=\alpha_8 p_3(t) \tag{6.2.8}$$

$$\left(\frac{d}{dt}+\beta_7\right)p_{19}(t)=\alpha_3 p_4(t)\ ;\ \left(\frac{d}{dt}+\beta_8\right)p_{20}(t)=\alpha_8 p_4(t) \tag{6.2.9}$$

$$\left(\frac{d}{dt}+\beta_i\right)p_{20+i}(t)=\alpha_i p_4(t),\ i=1,2,3,4 \tag{6.2.10}$$

with initial conditions $p_1(0) = 1$ and zero otherwise.

Taking Laplace transforms of eqns. (6.2.1)–(6.2.10) and using the initial condition, the probability transforms are obtained as follows:

$p_{1+i}(s) = b_i p_1(s),\ i = 1, 2, 3,\ ;\ p_{4+i}(s) = C_i p_1(s),\ i = 1, 2, 3, 4;$
$p_{8+i}(s) = C_i b_1 p_1(s),\ i = 1, 2, 3, 4;\ p_{12+i}(s) = C_i b_2 p_1(s),\ i = 1, 2, 3, 4;$
$p_{j+10}(s) = C_j b_i p_1(s),\ j = 7, 8$ and $i = 1, 2\ ;\ p_{j+12}(s) = C_j b_3 p_1(s),\ j = 7, 8\ ;$
$p_{20+i}(s) = C_i b_3 p_1(s),\ i = 1, 2, 3, 4$ and s is Laplace parameter, where

$$C_i=\frac{\alpha_i}{\beta_i+s},\ i=1,2,3,4;\ C_j=\frac{\alpha_j}{s+\beta_j},\ j=7,8\ ;$$

$$A_1=\frac{\alpha_7\beta_7}{s+\beta_7}\ ;\ A_2=\frac{\alpha_8\beta_8}{s+\beta_8}\ ;\ K=s-\frac{\alpha_1\beta_1}{s+\beta_1}-\frac{\alpha_2\beta_2}{s+\beta_2}-\frac{\alpha_3\beta_3}{s+\beta_3}-\frac{\alpha_4\beta_4}{s+\beta_4}j$$

$$K_1=\frac{(K+\lambda_4)\alpha_5}{(K+\lambda_2)(K+\lambda_4)-A_2\alpha_6}\ ;$$

$$K_2=\frac{A_2\alpha_5}{(K+\lambda_2)(K+\lambda_4)-A_2\alpha_6}\ ;\ K_3=\frac{(K+\lambda_4)\alpha_6}{(K+\lambda_3)(K+\lambda_4)-A_1\alpha_5}\ ;$$

$$K_4=\frac{A_1\alpha_5}{(K+\lambda_3)(K+\lambda_4)-A_1\alpha_5}\ ;$$

$$b_1 = \frac{K_1 + K_2 + K_3}{1 - K_2K_4}\ ;\ b_2 = \frac{K_1 + K_3 + K_4}{1 - K_2K_4}\ ;$$

$$b_3 = \frac{\alpha_5(K_1 + K_3 + K_4) + \alpha_6(K_1 + K_2 + K_3)}{(K + \lambda_4)(1 - K_2K_4)}\ ;$$

$$\lambda_1 = \sum_1^6 \alpha_i,\ \lambda_2 = \sum_1^4 \alpha_i + \alpha_6 + \alpha_7 + \beta_5\ ;\ \lambda_3 = \sum_1^5 \alpha_i + \alpha_8 + \beta_6\ ;\ \lambda_4 = \sum_1^4 \alpha_i + \alpha_7 + \alpha_8.$$

Laplace transform of availability function

$$Av(s) = L\,[p_1(t) + p_2(t) + p_3(t) + p_4(t)] = \sum_1^4 p_i(s) = (1 + b_1 + b_2 + b_3)p_1(s)$$

where, $p_1(s) = [(K + \lambda_1) - b_1(\beta_5 + A_1) - b_2(\beta_6 + A_2)]^{-1}$.

For steady-state analysis, take $\frac{d}{dt} \to 0$, $p_i(t) \to p_i$ as $t \to \infty$ and solve the resulting equations recursively. All the probabilities are obtained in terms of p_1 which is obtained by using the normalizing condition, i.e., $\sum_i^{24} \beta_i = 1$. The long-run availability

$$Av = p_1 + p_2 + p_3 + p_4 = [1 + L_1 + L_3 + L_4 + L_2L_3]\ p_1$$

where,

$$p_1 = \left[\left(1 + \frac{\alpha_1}{\beta_1} + \frac{\alpha_2}{\beta_2} + \frac{\alpha_3}{\beta_3} + \frac{\alpha_4}{\beta_4}\right)(1 + L_1 + L_3 + L_4 + L_2L_3) + \frac{\alpha_7}{\beta_7}(L_3 + L_4) + \frac{\alpha_8}{\beta_8}(L_1 + L_2L_3 + L_4)\right]^{-1}$$

and

$$L_1 = \frac{\alpha_6(\alpha_7 + \alpha_8)}{[(\alpha_7 + \alpha_8)(\alpha_5 + \alpha_8 + \beta_6) - \alpha_5\alpha_7]}\ ;$$

$$L_2 = \frac{\alpha_6\alpha_7}{[(\alpha_7 + \alpha_8)(\alpha_5 + \alpha_8 + \beta_6) - \alpha_5\alpha_7]}\ ;$$

$$L_3 = \frac{\alpha_5(\alpha_7 + \alpha_8 + L_1\alpha_8)}{[(\alpha_7 + \alpha_8)(\alpha_6 + \alpha_7 + \beta_5) - \alpha_6\alpha_8] - \alpha_5\alpha_8L_2}\ ;$$

$$L_4 = \frac{\alpha_5(L_2L_3 + L_1) + \alpha_6L_3}{\alpha_7 + \alpha_8}.$$

Gupta et al., [2005] solved the first order differential equations (6.2.1)–(6.2.10), using initial condition, by Runge–Kutta method. They calculated the availability and MTBF changing repair/failure rate and studied the effect of failure and repair rates of different subsystems on the availability and MTBF. For long-run availability, two tables are given to demonstrate the effect of failure and repair rates on availability.

(a) Effect of failure rates of raw mixer (α_7) and clinker machine (α_3) on long-run availability:

Taking $\alpha_1 = 0.03$, $\alpha_2 = 0.03$, $\alpha_4 = 0.07$, $\alpha_5 = 0.03$, $\alpha_6 = 0.03$, $\alpha_8 = 0.003$, $\beta_1 = 6.0$, $\beta_2 = 6.0$, $\beta_3 = 4.8$, $\beta_4 = 12$, $\beta_5 = 24$, $\beta_6 = 24$, $\beta_7 = 0.3$, $\beta_8 = 0.3$

$\alpha_3 \rightarrow$ $\alpha_7 \downarrow$	0.01	0.02	0.03	0.04	0.05
0.001	0.982147	0.980141	0.978144	0.976155	0.974174
0.002	0.982142	0.980136	0.978139	0.976150	0.974169
0.003	0.982137	0.980132	0.978135	0.976145	0.974164
0.004	0.982132	0.980127	0.978130	0.976141	0.974160
0.005	0.982128	0.980123	0.978126	0.976137	0.974156

The table reveals that increase in failure rate of raw mixer (α_7) affects long-run availability by approximately 0.8% and failure rate of clinker section (α_3) affects by approximately 0.002%. It is obvious from the table that raw mixer plays an important role in the availability of the system in comparison to clinker section.

(b) Effect of failure rate (α_1) and repair rate (β_1) of jaw crusher on long-run availability.

Taking $\alpha_2 = 0.03$, $\alpha_3 = 0.03$, $\alpha_4 = 0.07$, $\alpha_5 = 0.03$, $\alpha_6 = 0.03$, $\alpha_7 = 0.003$, $\alpha_8 = 0.003$, $\beta_2 = 6$, $\beta_3 = 4.8$, $\beta_4 = 12$, $\beta_5 = 24$, $\beta_6 = 24$, $\beta_7 = 0.3$, $\beta_8 = 0.3$

$\alpha_1 \rightarrow$ $\beta_1 \downarrow$	0.01	0.02	0.03	0.04	0.05
2	0.978134	0.973374	0.968659	0.963991	0.959367
4	0.980532	0.978039	0.975748	0.973374	0.971011
6	0.981334	0.979732	0.978134	0.976554	0.974956
8	0.981735	0.980436	0.979332	0.978134	0.976940
10	0.981976	0.981013	0.980052	0.979092	0.978134

It is clear from the table that by increasing the failure rate (α_1) of jaw crusher, availability of the system is affected from 0.4% to 2% with the increase in repair rate from 2 to 10 units. If the increase in repair rate (β_1) of jaw crusher is increased, the availability is affected from 0.4% to 2% with the increase in failure rate (α_1) from 0.01 to 0.05.

6.3 THE UTENSILS INDUSTRY

The utensils industry consists of two major parts, namely, (*a*) sheet formation and (*b*) utensils formation. The raw material for an utensils industry is taken from steel manufacturing industries. The raw material, i.e., flat steel sheets are taken to the cutter to cut them into pieces as per requirements. These sheet pieces are put into the furnace for heating at temperature 800°C. The sheets are then turned up into the hot rolling machine where they are pressed to make them as thick as per requirement. The red hot pressed sheets are softened by roller furnisher machine. After this, the sheets are put into cold rolling machines in order to get the smooth and fine sheets for manufacturing the utensils.

(a) The Sheet Formation System

The system consists of five subsystems—two cutters (A_i, $i = 1, 2$), one working and the other in standby; furnace (B), one unit subjected to major failure; hot rolling machine (C) consisting of two units in parallel subjected to minor and major failures; roller furnisher (D), one unit system subjected to only major failure; and cold rolling machine (E) having two units in parallel, subjected to minor and major failures. Minor failure reduces the working capacity of the system but the system remains in operative condition. The schematic diagram for the sheet formation system is shown in Fig. 6.12.

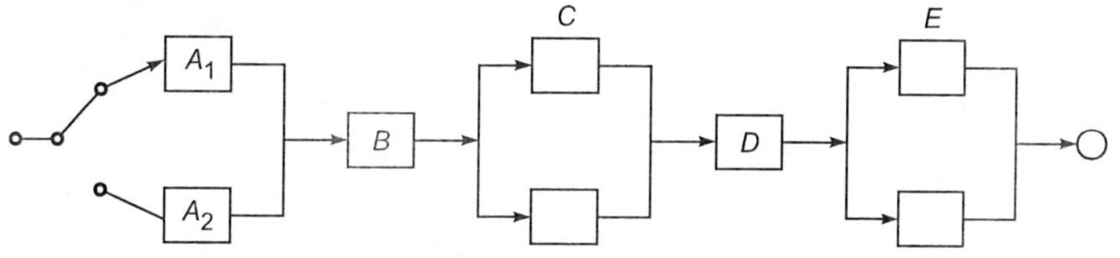

Fig. 6.12

Assuming that the switch-over device is perfect, subsystem A_i never fails and there is no simultaneous failure among the subsystems, following transition diagram for the system is obtained:

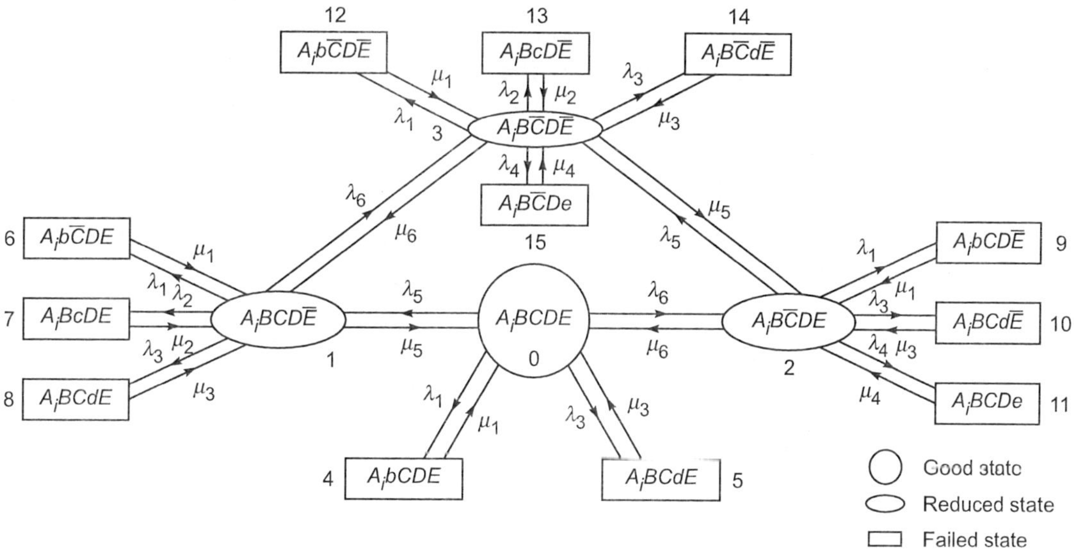

Fig. 6.13

In Fig. 6.13, $\overline{C}$, $\overline{E}$ are used for the reduced states of the subsystems C, E respectively while C, E are used for good states. λ_i (i = 1, 2, ..., 6) denotes the respective constant failure rate of the subsystems B, C, D, E, C and $\overline{E}$ while μ_i (i = 1, 2, ..., 6) is the corresponding repair rate. Denoting $p_i(t)$ as the probability that the system is in ith state at time t, the following differential-difference equations associated with Fig. 6.13 are obtained:

$$\left(\frac{d}{dt} + \alpha_1\right) p_0(t) = \mu_1\, p_4(t) + \mu_3\, p_5(t) + \mu_5\, p_1(ts) + \mu_6\, p_2(t), \tag{6.3.1}$$

$$\left(\frac{d}{dt}+\alpha_2\right) p_1(t) = \sum_1^3 \mu_i\, p_{i+5}(t) + \lambda_5\, p_0(t) + \mu_6\, p_3(t), \tag{6.3.2}$$

$$\left(\frac{d}{dt}+\alpha_3\right) p_2(t) = \mu_1\, p_9(t) + \mu_3\, p_{10}(t) + \mu_4\, p_{11}(t) + \mu_5\, p_3(t) + \lambda_6\, p_0(t) \tag{6.3.3}$$

$$\left(\frac{d}{dt}+\alpha_4\right) p_3(t) = \sum_1^4 \mu_1\, p_{i+11}(t) + \lambda_6\, p_1(t) + \lambda_5\, p_2(t), \tag{6.3.4}$$

$$\left(\frac{d}{dt}+\alpha_1\right) p_4(t) = \lambda_1\, p_0(t)\ ;\ \left(\frac{d}{dt}+\mu_3\right) p_5(t) = \lambda_3\, p_0(t), \tag{6.3.5}$$

$$\left(\frac{d}{dt}+\mu_i\right) p_{5+i} = \lambda_i\, p_1(t), \quad i = 1, 2, 3 \tag{6.3.6}$$

$$\left(\frac{d}{dt}+\mu_1\right) p_9(t) = \lambda_1\, p_2(t)\ ;\ \left(\frac{d}{dt}+\mu_{2+j}\right) p_{9+j}(t) = \lambda_{2+j}\, p_2(t),\ j = 1, 2 \tag{6.3.7}$$

$$\left(\frac{d}{dt}+\mu_K\right) p_{k+11}(t) = \lambda_k\, p_3(t),\ K = 1, 2, 3, 4 \tag{6.3.8}$$

where,

$$\alpha_1 = \lambda_1 + \lambda_3 + \lambda_5 + \lambda_6$$

$$\alpha_2 = \sum_1^3 \lambda_i + \mu_5 + \lambda_6$$

$$\alpha_3 = \lambda_1 + \sum_3^5 \lambda_i + \mu_6$$

$$\alpha_4 = \sum_1^4 \lambda_i + \mu_5 + \mu_6$$

Assuming that initially (i.e., $t = 0$) the system is good, the initial condition is $p_0(0) = 1$ otherwise zero. Equations (6.3.1)–(6.3.8) may be solved using Laplace transforms or by other methods given in Chapter 4. Availability function for the system is

$$Av(t) = \sum_{i=0}^{3} p_i(t).$$

For steady state, taking $\frac{d}{dt} = 0$, $p_i(t) = p_i$ when $t \to \infty$ and solving the resulting equations from (6.3.1) to (6.3.8) recursively and using the normalizing condition $\sum_{i=0}^{15} p_i = 1$, the availability is given by

$$Av = \sum_{i=0}^{3} p_i = \left[1 + \sum_0^3 K_i\right] p_0$$

where, $K_1 = \lambda_5(\mu_5 + \mu_6)(\lambda_5 + \lambda_6 + \mu_5 + \mu_6)/[\mu_5(\mu_5 + \mu_6 + \lambda_6)(\lambda_5 + \mu_5 + \mu_6) - \mu_5\,\lambda_5\,\lambda_6]$,

$K_2 = [\lambda_6(1 + K_1)\mu_5 + \lambda_6\,\mu_6]/\mu_6(\lambda_5 + \mu_5 + \mu_6)$;

$K_3 = (\lambda_6\,K_1 + \lambda_5\,K_2)/(\mu_5 + \mu_6)$, and

$$p_0 = \left[1+\sum_1^3 K_i + \frac{\lambda_1}{\mu_1} + \frac{\lambda_3}{\mu_3} + \sum_1^3 \frac{\lambda_i}{\mu_i}K_1 + \frac{\lambda_1}{\mu_1}K_2 + \sum_1^2 \frac{\lambda_{2+i}}{\mu_{2+i}}K_2 + \sum_1^4 \frac{\lambda_i}{\mu_i}K_3\right]^{-1}.$$

The effect of failure rates and repair rates of subsystems on availability and MTBF may be studied forming availability tables as given in Section 6.2.

(b) The Utensils Formation System

This system consists of four subsystems (i) two cutters (A_i, $i = 1, 2$) one working and other in standby, (ii) the pressing machine (B) one unit only, (iii) the spinning machines (C) four units working in parallel and (iv) the buffing machines (D) two units working in parallel. Subsystem A_i never fails (due to standby) while the other subsystems are subjected to minor and major failures. Taking λ_i, respective mean constant failure rate of B, C, D, $\overline{C}$, $\overline{D}$ ($i = 1, 2, \ldots, 5$) and μ_i ($i = 1, 2, \ldots, 5$), the corresponding mean constant repair rates of B, C, D, $\overline{C}$, $\overline{D}$ the block diagram and transition diagram are given in Figs. 6.14 and 6.15, respectively. Assumptions and notations are similar as described in 6.3(a).

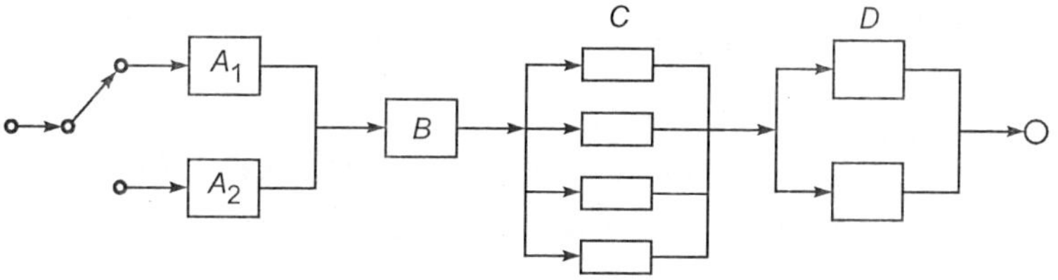

Fig. 6.14

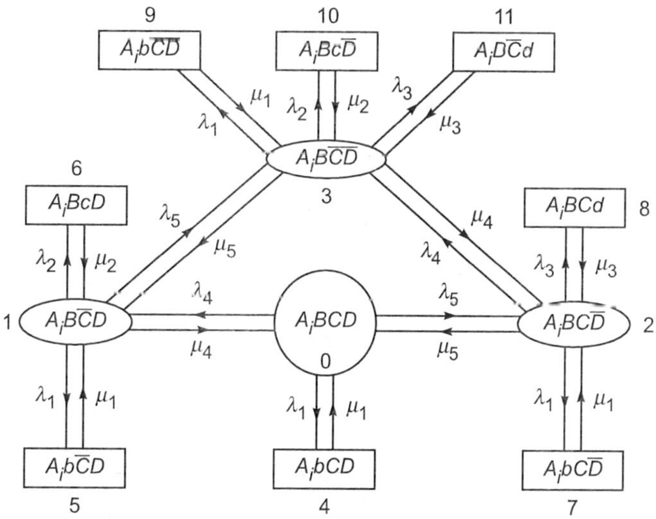

Fig. 6.15

Following the assumptions and notations, the differential-difference equations associated with the transition diagram (using probability consideration) are given as follows:

$$\left(\frac{d}{dt}+\alpha_1\right) p_0(t) = \mu_1 p_4(t) + \mu_4 p_1(t) + \mu_5 p_2(t), \tag{6.3.9}$$

$$\left(\frac{d}{dt}+\alpha_2\right) p_1(t) = \mu_1 p_5(t) + \mu_2 p_6(t) + \mu_5 p_3(t) + \lambda_4 p_0(t), \tag{6.3.10}$$

$$\left(\frac{d}{dt}+\alpha_3\right) p_2(t) = \mu_1 p_7(t) + \mu_3 p_8(t) + \mu_4 p_3(t) + \lambda_5 p_0(t), \tag{6.3.11}$$

$$\left(\frac{d}{dt}+\alpha_4\right) p_3(t) = \sum_{1}^{3} \mu_1 p_{i+8}(t) + \lambda_4 p_2(t) + \lambda_5 p_1(t), \tag{6.3.12}$$

$$\left(\frac{d}{dt}+\mu_1\right) p_4(t) = \lambda_1 p_0(t)\ ;\ \left(\frac{d}{dt}+\mu_i\right) p_{i+4}(t) = \lambda_i p_1(t),\ i = 1, 2 \tag{6.3.13}$$

$$\left(\frac{d}{dt}+\mu_1\right) p_7(t) = \lambda_1 p_2(t)\ ;\ \left(\frac{d}{dt}+\mu_3\right) p_8(t) = \lambda_3 p_2(t) \tag{6.3.14}$$

$$\left(\frac{d}{dt}+\mu_i\right) p_{i+8}(t) = \lambda_i p_3(t),\ i = 1, 2, 3 \tag{6.3.15}$$

with initial condition $p_0(0) = 1$ otherwise equal to zero.

Equations (6.3.9)–(6.3.15) are first order differential equations and may be solved using Laplace transforms or by numerical methods or by any other method discussed in Chapter 4. The system availability

$$Av(t) = \sum_{i=0}^{3} p_i(t)$$

Taking $\frac{d}{dt} \to 0$, $p_i(t) \to p_i$ as $t \to \infty$, the steady-state probabilities are obtained solving recursively the equations resulting from (6.3.9)–(6.3.15) and using the normalizing condition $\sum_{i=0}^{11} p_i = 1$. The availability of the system

$$Av = \sum_{i=0}^{3} p_i = \left[1+\sum_{i=1}^{3} K_i\right] p_0$$

where,

$K_1 = [\lambda_4(\lambda_4 + \lambda_5 + \mu_4 + \mu_5)(\mu_4 + \mu_5)]/\mu_4[(\lambda_5 + \mu_4 + \mu_5)(\lambda_4 + \mu_4 + \mu_5) - \lambda_4\lambda_5]$,
$K_2 = \lambda_2[(1 + K_1)(\mu_4 + \mu_5)]/\mu_5(\lambda_4 + \mu_4 + \mu_5)$,
$K_3 = (\lambda_4 K_2 + \lambda_5 K_1)/(\mu_4 + \mu_5)$ and

$$p_0 = \left[1+\sum_{1}^{3} K_i + \frac{\lambda_1}{\mu_1} + \sum_{1}^{2}\frac{\lambda_i}{\mu_i}K_1 + \left(\frac{\lambda_1}{\mu_1}+\frac{\lambda_3}{\mu_3}\right)K_2 + \sum_{1}^{3}\frac{\lambda_i}{\mu_i}K_3\right]^{-1}$$

Behavioural study of the system may be carried out preparing availability tables.

6.4 AVAILABILITY ANALYSIS OF THE CATTLE FEED PLANT

6.4.1 Introduction

Cattle feeding schedules are very customary and its assembling procedure is similarly straightforward. The primary point of the procedure is to guarantee that the last blend contains consistently same extent of the fixings even in littlest conceivable example of state, one gram as initially planned. This is fundamental in light of the fact that the animal gets uniform every day (or occasional) doses of supplements as recommended by the logical standard feedings. Obviously, to achieve consistently in fixation, a blend of fixings is to be ground to fine work and blended completely. This blended powder can be utilized legitimately for sustaining. On the other hand, the blended ground material can be pelletized because animals, particularly cud-biting sort, as to granulate the pellets while eating rather than simply gulping powdered feed.

In the present chapter, we manage the cattle feed plant. The plant is investigated for time needy & steady-state availability. The overseeing differential conditions have been created utilizing Markov birth-death procedure and settled utilizing starting conditions and as a unique case similar investigation of both Matrix method & Matlab-tool for assessment of time subordinate accessibility in terms of arrangement quality and simplicity of execution is additionally talked about.

6.4.2 The System

The cattle feed plant primarily comprises seven subsystems, in particular elevator, grinder, hopper, mixer, winch, palletiser, screw transport. First the elevator lifts the material & places it into grinder. Processor pounds the crude material and afterward the material is put into the hopper. Hopper is utilized for the capacity and cooling of material by fans present in hopper. The material is put into the mixer for appropriate blending of specific added substances in determined proportion. This mixture is lifted by winch. Winch puts this blend into palletiser. Palletiser enables the blend to push ahead and goes through the gaps and delivers the obligatory shape to blend. At long last screw conveyor cools the item with the assistance of fans present in it and conveys the readied item to the store where it is packed for final delivery.

The cattle feed plant comprises the accompanying seven principle subsystems:

1. **Elevator (A)** comprises one unit. The framework comes up short when this unit comes up short.
2. **Grinder (B)** comprises one unit. It is exposed to significant disappointment as it were.
3. **Hopper (C)** comprises one unit. It is exposed to significant disappointment as it were.
4. **Mixer (D)** comprises two units, one working and the other is in cold standby. The cold standby unit is of lower limit. The framework takes a shot at backup unit in diminished limit. Complete disappointment happens when the two units come up short.
5. **Winch (E)** comprises one unit. The framework fizzles when this unit comes up short.
6. **Palletiser (F)** comprises two units, one working and the other is in cold standby. The cold standby unit is of lower limit. The framework takes a shot at backup unit in diminished limit. Complete disappointment happens when the two units come up short.
7. **Screw conveyor (G)** comprises one unit. The framework comes up short when this subsystem falls flat.

6.4.3 Mathematical Modeling

Assumptions

1. A repair unit is execution savvy, all around great.
2. The subsystems D & F flop through diminished states.
3. Switch-over gadgets are flawless.

Notations

A, B, C, D, E, F, G : Indicate the respective working states.

$\bar{D}$, $\bar{F}$: Indicate the respective reduced states.

a, b, c, d, e, f, g : Indicate the respective failed states.

In view of the above notations & assumptions, the progress chart (Fig. 6.16) is given as pursues.

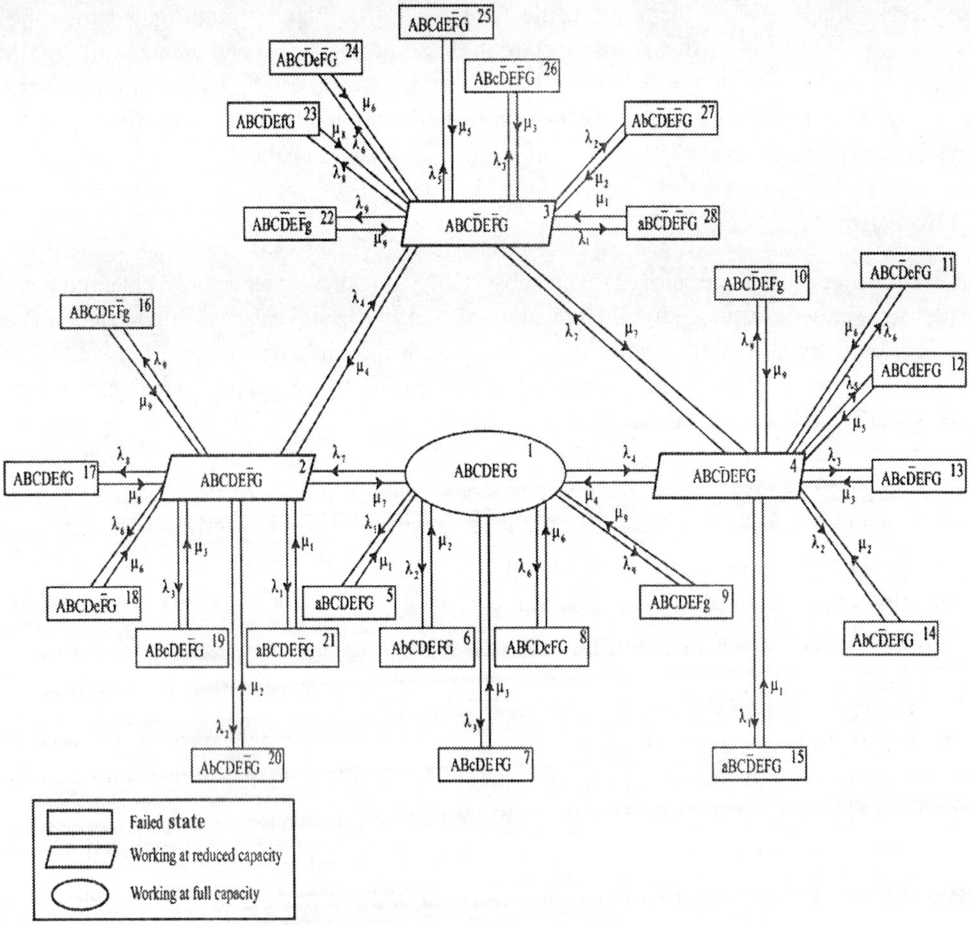

Fig. 6.16 Transition diagram of cattle feed plant.

By utilizing Markov modeling and the governing differential conditions related with the graph are determined as pursues

$$p_1'(t) + a_1p_1(t) = \mu_1p_5(t) + \mu_4p_4(t) + \mu_2p_6(t) + \mu_6p_8(t) + \mu_3p_7(t) + \mu_9p_9(t) + \mu_7p_2(t) \quad (6.4.1)$$

$$p_2'(t) + a_2p_2(t) = \mu_1p_{21}(t) + \mu_9p_{16}(t) + \mu_6p_{18}(t) + \mu_2p_{20}(t) + \mu_4p_3(t) + \mu_8p_{17}(t) + \lambda_7p_1(t) + \mu_3p_{19}(t) \quad (6.4.2)$$

$$p_3'(t) + a_3p_3(t) = \mu_1p_{28}(t) + \mu_8p_{23}(t) + \mu_3p_{26}(t) + \mu_2p_{27}(t) + \mu_5p_{25}(t) + \lambda_4p_2(t) + \mu_6p_{24}(t) + \mu_9p_{22}(t) + \lambda_7p_4(t) \quad (6.4.3)$$

$$p_4'(t) + a_4p_4(t) = \mu_6p_{11}(t) + \mu_1p_{15}(t) + \mu_7p_3(t) + \mu_3p_{13}(t) + \mu_2p_{14}(t) + \mu_5p_{12}(t) + \lambda_4p_1(t) + \mu_9p_{10}(t) \quad (6.4.4)$$

$$p_i'(t) + \mu_jp_i(t) = \lambda_jp_1(t)$$
$$i = 5, 7, 9, 6, 8; j = 1, 3, 9, 2, 6; \quad (6.4.5)$$

$$p_i'(t) + \mu_j p_i(t) = \lambda_j p_2(t)$$
$$i = 16, 18, 20, 17, 19, 21; j = 9, 6, 2, 8, 3, 1; \quad (6.4.6)$$

$$p_i'(t) + \mu_jp_i(t) = \lambda_jp_3(t)$$
$$i = 22, 24, 26, 28, 23, 25, 27; j = 9, 6, 3, 1, 8, 5, 2; \quad (6.4.7)$$

$$p_i'(t) + \mu_jp_i(t) = \lambda_jp_4(t)$$
$$i = 10, 12, 14, 11, 13, 15; j = 9, 5, 2, 6, 3, 1; \quad (6.4.8)$$

where

$$a_1 = \lambda_1 + \lambda_3 + \lambda_9 + \lambda_7 + \lambda_2 + \lambda_6 + \lambda_4,\ a_2 = \lambda_1 + \lambda_3 + \lambda_6 + \lambda_9 + \lambda_2 + \lambda_4 + \lambda_8 + \mu_7$$
$$a_3 = \lambda_1 + \lambda_3 + \lambda_6 + \lambda_9 + \lambda_2 + \lambda_5 + \lambda_8 + \mu_{7+} + \mu_4,\ a_4 = \lambda_1 + \lambda_3 + \lambda_6 + \lambda_9 + \lambda_2 + \lambda_5 + \lambda_7 + \mu_4$$

The conditions (6.4.1) to (6.4.8) are to be explained utilizing the accompanying initial condition.

$$p(t) = [1\ 0\ 0\ 0\ 0\ 0\ 0\ 0\ 0\ 0\ 0 - - - - 0]_{1 \times 28}$$

where

$$p(t) = [p_1(t)\ p_2(t)\ p_3(t) - - - - - - - - - - - p_{28}(t)]_{1 \times 28}$$

We use grid strategy and calculation dependent on ode45 for illuminating these conditions.

6.4.4 Time Dependent Availability Analysis by Matlab

Step 1

(a) Input values of time at which accessibility is to be determined
T span = [0 20 40 60 80 100 120 140 160].

(b) Identify starting condition as
$P(t = 0) = [1\ 0\ 0\ 0\ 0\ 0\ 0\ 0\ 0\ 0 - - - - 0]_{1 \times 28}$

(c) Read no of working stage ($m = 4$)

Step 2

Set scalar relative blunder resistance RelTol ($1e - 4$) and a vector of supreme mistake resilience AbsTol ($1e - 5$)

Step 3

(a) Get the estimation of disappointment and repair rate λ_i, μ_i for $i = 1.....9$
$\lambda_1 = 0.001$, $\lambda_9 = 0.003$, $\lambda_4 = 0.0025$, $\lambda_7 = 0.005$, $\lambda_2 = 0.001$, $\lambda_3 = 0.004$, $\lambda_5 = 0.0025$, $\lambda_8 = 0.005$, $\lambda_6 = 0.006$,

$\mu_1 = 0.01$, $\mu_9 = 0.03$, $\mu_2 = 0.04$, $\mu_4 = 0.02$, $\mu_6 = 0.06$, $\mu_8 = 0.05$, $\mu_5 = 0.02$, $\mu_7 = 0.05$, $\mu_3 = 0.02$,

As these qualities are coefficients of arrangement of differential condition.

(b) Input the arrangement of condition (odefun) (6.4.1 to 6.4.8)

Step 4: Call ode45 (with odefun, tspan, $P(0)$ as information contention)

Step 5: Appoint the aggregate of (working stage) $p_3(t), p_2(t), p_1(t), p_4(t)$ $av(t) = p_3(t) + p_2(t) + p_1(t) + p_4(t)$ for various estimations of time as referenced in T span.

Step 6: Print the estimation of accessible relating to various time esteems in T sapn. Plot the esteem accessibility regarding time (T span)

(T span = [0 20 40 60 80 100 120 140 160]).

Results: Conditions are unraveled utilizing Matlab tool and results are demonstrated as follows.

Table 6.1: Variation of availability with respect to time

Time	Availability
0	1
20	0.81001
40	0.73369
60	0.6983
80	0.67965
100	0.66882
120	0.66208
140	0.65767
160	0.65467

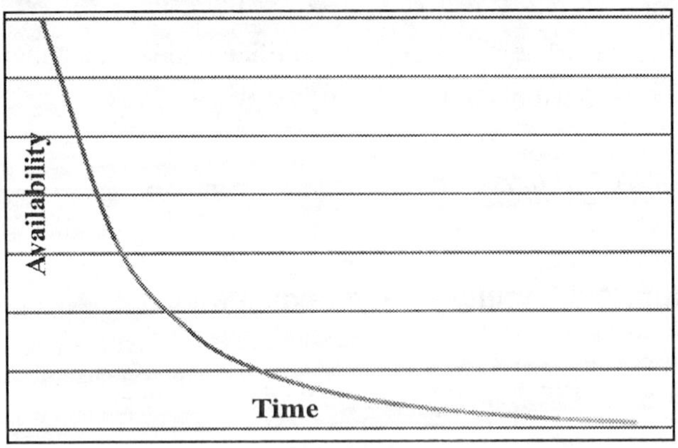

Fig. 6.17

6.4.5 Time Dependent Availability Analysis: Matrix Method

By utilizing matrix method, we have:

- Let assume $n = 28$, $m = 4$
- Get the estimation of disappointment and repair rate λ_i, μ_i for $i = 1.....9$

$\lambda_1 = 0.001$, $\lambda_9 = 0.003$, $\lambda_4 = 0.0025$, $\lambda_7 = 0.005$, $\lambda_2 = 0.001$, $\lambda_3 = 0.004$, $\lambda_5 = 0.0025$, $\lambda_8 = 0.005$, $\lambda_6 = 0.006$, $\mu_1 = 0.01$, $\mu_9 = 0.03$, $\mu_2 = 0.04$, $\mu_4 = 0.02$, $\mu_6 = 0.06$, $\mu_8 = 0.05$, $\mu_5 = 0.02$, $\mu_7 = 0.05$, $\mu_3 = 0.02$,

Matrix A

$A =$

$$
\begin{bmatrix}
-a_1 & \mu_7 & 0 & \mu_4 & \mu_1 & \mu_2 & \mu_3 & \mu_6 & \mu_9 & 0 & 0 & 0 & 0 & 0 & 0 & 0 & 0 & 0 & 0 & 0 & 0 & 0 & 0 & 0 & 0 & 0 & 0 & 0 \\
\lambda_7 & -a_2 & \mu_4 & 0 & 0 & 0 & 0 & 0 & 0 & 0 & 0 & 0 & 0 & 0 & 0 & \mu_9 & \mu_8 & \mu_6 & \mu_3 & \mu_2 & \mu_1 & 0 & 0 & 0 & 0 & 0 & 0 & 0 \\
0 & \lambda_4 & -a_3 & \lambda_7 & 0 & 0 & 0 & 0 & 0 & 0 & 0 & 0 & 0 & 0 & 0 & 0 & 0 & 0 & 0 & 0 & 0 & \mu_9 & \mu_8 & \mu_6 & \mu_5 & \mu_3 & \mu_2 & \mu_1 \\
\lambda_4 & 0 & \mu_7 & -a_4 & 0 & 0 & 0 & 0 & 0 & \mu_9 & \mu_6 & \mu_5 & \mu_3 & \mu_2 & \mu_1 & 0 & 0 & 0 & 0 & 0 & 0 & 0 & 0 & 0 & 0 & 0 & 0 & 0 \\
\lambda_1 & 0 & 0 & 0 & -\mu_1 & 0 & 0 & 0 & 0 & 0 & 0 & 0 & 0 & 0 & 0 & 0 & 0 & 0 & 0 & 0 & 0 & 0 & 0 & 0 & 0 & 0 & 0 & 0 \\
\lambda_2 & 0 & 0 & 0 & 0 & -\mu_2 & 0 & 0 & 0 & 0 & 0 & 0 & 0 & 0 & 0 & 0 & 0 & 0 & 0 & 0 & 0 & 0 & 0 & 0 & 0 & 0 & 0 & 0 \\
\lambda_3 & 0 & 0 & 0 & 0 & 0 & -\mu_3 & 0 & 0 & 0 & 0 & 0 & 0 & 0 & 0 & 0 & 0 & 0 & 0 & 0 & 0 & 0 & 0 & 0 & 0 & 0 & 0 & 0 \\
\lambda_6 & 0 & 0 & 0 & 0 & 0 & 0 & -\mu_6 & 0 & 0 & 0 & 0 & 0 & 0 & 0 & 0 & 0 & 0 & 0 & 0 & 0 & 0 & 0 & 0 & 0 & 0 & 0 & 0 \\
\lambda_9 & 0 & 0 & 0 & 0 & 0 & 0 & 0 & -\mu_9 & 0 & 0 & 0 & 0 & 0 & 0 & 0 & 0 & 0 & 0 & 0 & 0 & 0 & 0 & 0 & 0 & 0 & 0 & 0 \\
0 & 0 & 0 & \lambda_9 & 0 & 0 & 0 & 0 & 0 & -\mu_9 & 0 & 0 & 0 & 0 & 0 & 0 & 0 & 0 & 0 & 0 & 0 & 0 & 0 & 0 & 0 & 0 & 0 & 0 \\
0 & 0 & 0 & \lambda_6 & 0 & 0 & 0 & 0 & 0 & 0 & -\mu_6 & 0 & 0 & 0 & 0 & 0 & 0 & 0 & 0 & 0 & 0 & 0 & 0 & 0 & 0 & 0 & 0 & 0 \\
0 & 0 & 0 & \lambda_5 & 0 & 0 & 0 & 0 & 0 & 0 & 0 & -\mu_5 & 0 & 0 & 0 & 0 & 0 & 0 & 0 & 0 & 0 & 0 & 0 & 0 & 0 & 0 & 0 & 0 \\
0 & 0 & 0 & \lambda_3 & 0 & 0 & 0 & 0 & 0 & 0 & 0 & 0 & -\mu_3 & 0 & 0 & 0 & 0 & 0 & 0 & 0 & 0 & 0 & 0 & 0 & 0 & 0 & 0 & 0 \\
0 & 0 & 0 & \lambda_2 & 0 & 0 & 0 & 0 & 0 & 0 & 0 & 0 & 0 & -\mu_2 & 0 & 0 & 0 & 0 & 0 & 0 & 0 & 0 & 0 & 0 & 0 & 0 & 0 & 0 \\
0 & 0 & 0 & \lambda_1 & 0 & 0 & 0 & 0 & 0 & 0 & 0 & 0 & 0 & 0 & -\mu_1 & 0 & 0 & 0 & 0 & 0 & 0 & 0 & 0 & 0 & 0 & 0 & 0 & 0 \\
0 & \lambda_9 & 0 & 0 & 0 & 0 & 0 & 0 & 0 & 0 & 0 & 0 & 0 & 0 & 0 & -\mu_9 & 0 & 0 & 0 & 0 & 0 & 0 & 0 & 0 & 0 & 0 & 0 & 0 \\
0 & \lambda_8 & 0 & 0 & 0 & 0 & 0 & 0 & 0 & 0 & 0 & 0 & 0 & 0 & 0 & 0 & -\mu_8 & 0 & 0 & 0 & 0 & 0 & 0 & 0 & 0 & 0 & 0 & 0 \\
0 & \lambda_6 & 0 & 0 & 0 & 0 & 0 & 0 & 0 & 0 & 0 & 0 & 0 & 0 & 0 & 0 & 0 & -\mu_6 & 0 & 0 & 0 & 0 & 0 & 0 & 0 & 0 & 0 & 0 \\
0 & \lambda_3 & 0 & 0 & 0 & 0 & 0 & 0 & 0 & 0 & 0 & 0 & 0 & 0 & 0 & 0 & 0 & 0 & -\mu_3 & 0 & 0 & 0 & 0 & 0 & 0 & 0 & 0 & 0 \\
0 & \lambda_2 & 0 & 0 & 0 & 0 & 0 & 0 & 0 & 0 & 0 & 0 & 0 & 0 & 0 & 0 & 0 & 0 & 0 & -\mu_2 & 0 & 0 & 0 & 0 & 0 & 0 & 0 & 0 \\
0 & \lambda_1 & 0 & 0 & 0 & 0 & 0 & 0 & 0 & 0 & 0 & 0 & 0 & 0 & 0 & 0 & 0 & 0 & 0 & 0 & -\mu_1 & 0 & 0 & 0 & 0 & 0 & 0 & 0 \\
0 & 0 & \lambda_9 & 0 & 0 & 0 & 0 & 0 & 0 & 0 & 0 & 0 & 0 & 0 & 0 & 0 & 0 & 0 & 0 & 0 & 0 & -\mu_9 & 0 & 0 & 0 & 0 & 0 & 0 \\
0 & 0 & \lambda_8 & 0 & 0 & 0 & 0 & 0 & 0 & 0 & 0 & 0 & 0 & 0 & 0 & 0 & 0 & 0 & 0 & 0 & 0 & 0 & -\mu_8 & 0 & 0 & 0 & 0 & 0 \\
0 & 0 & \lambda_6 & 0 & 0 & 0 & 0 & 0 & 0 & 0 & 0 & 0 & 0 & 0 & 0 & 0 & 0 & 0 & 0 & 0 & 0 & 0 & 0 & -\mu_6 & 0 & 0 & 0 & 0 \\
0 & 0 & \lambda_5 & 0 & 0 & 0 & 0 & 0 & 0 & 0 & 0 & 0 & 0 & 0 & 0 & 0 & 0 & 0 & 0 & 0 & 0 & 0 & 0 & 0 & -\mu_5 & 0 & 0 & 0 \\
0 & 0 & \lambda_3 & 0 & 0 & 0 & 0 & 0 & 0 & 0 & 0 & 0 & 0 & 0 & 0 & 0 & 0 & 0 & 0 & 0 & 0 & 0 & 0 & 0 & 0 & -\mu_3 & 0 & 0 \\
0 & 0 & \lambda_2 & 0 & 0 & 0 & 0 & 0 & 0 & 0 & 0 & 0 & 0 & 0 & 0 & 0 & 0 & 0 & 0 & 0 & 0 & 0 & 0 & 0 & 0 & 0 & -\mu_2 & 0 \\
0 & 0 & \lambda_1 & 0 & 0 & 0 & 0 & 0 & 0 & 0 & 0 & 0 & 0 & 0 & 0 & 0 & 0 & 0 & 0 & 0 & 0 & 0 & 0 & 0 & 0 & 0 & 0 & -\mu_1
\end{bmatrix}
$$

Results

To ascertain the estimation availability a *c*-program is built up. The tables & figure of time dependent availability are demonstrated as follows.

Table 6.2: Variation of availability with time

Time	Availability
0	1
10	0.8887
20	0.8189
30	0.7739
40	0.7441
50	0.7237
60	0.7093
70	0.6986
80	0.6895
90	0.6773
100	0.6695

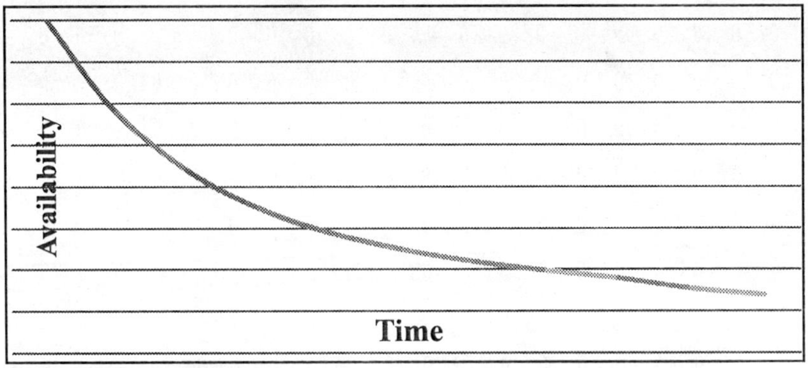

Fig. 6.18 Variation of availability with time.

6.4.6 Steady-State Availability Analysis

Cattle feed plant is required to be accessible for long length of time. Along these lines, long run or steady state probability of the framework is gotten by putting

$d/dt \to 0$ & $P_i(t) \to P_i$ as $t \to \infty$ as in above conditions. The estimations of steady state probabilities are acquired as far as p_1 as pursues:

$$p_2 = k_1 p_1 \quad p_3 = k_2 p_1 \quad p_4 = k_3 p_1 \quad p_5 = c_1 p_1$$

$$p_6 = c_2 p_1 \quad p_7 = c_3 p_1 \quad p_8 = c_6 p_1 \quad p_9 = c_9 p_1$$

$$p_{10} = c_9 k_3 p_1 \quad p_{11} = c_6 k_3 p_1 \quad p_{12} = c_5 k_3 p_1 \quad p_{13} = c_3 k_3 p_1$$

$p_{14} = c_2 k_3 p_1$ $p_{15} = c_1 k_3 p_1$ $p_{16} = c_9 k_1 p_1$ $p_{17} = c_8 k_1 p_1$

$p_{18} = c_6 k_1 p_1$ $p_{18} = c_6 k_1 p_1$ $p_{19} = c_3 k_1 p_1$ $p_{20} = c_2 k_1 p_1$

$p_{21} = c_1 k_1 p_1$ $p_{22} = c_9 k_2 p_1$ $p_{23} = c_8 k_2 p_1$ $p_{24} = c_6 k_2 p_1$

$p_{25} = c_5 k_2 p_1$ $p_{26} = c_3 k_2 p_1$ $p_{27} = c_2 k_2 p_1$ $p_{28} = c_1 k_2 p_1$

where

$$k_1 = \lambda_4 + \lambda_7 - \mu_4 k_3 \big/ \mu_7 \qquad k_3 = \lambda_4^2 + \lambda_4\mu_7 + \lambda_7\lambda_4 + \lambda_4\mu_4 \big/ \mu_4\lambda_4 + \mu_4\mu_7 + \lambda_7\mu_4 + \mu_4^2$$

$$k_3 = \lambda_4^2 + \lambda_4\mu_7 + \lambda_7\lambda_4 + \lambda_4\mu_4 \big/ \mu_4\lambda_4 + \mu_4\mu_7 + \lambda_7\mu_4 + \mu_4^2 \qquad k_2 = ((\lambda_4 + \mu_7)\,k_1 - \lambda_7) \big/ \mu_4$$

$$k_2 = ((\lambda_4 + \mu_7)\,k_1 - \lambda_7) \big/ \mu_4 \quad c_i = \lambda_i \big/ \mu_i \quad \text{where } (1 <= i <= 9)$$

Probability of full limit working state (introductory state), p_1 is controlled by utilizing normalizing condition, for example,

$$\sum_{i=1}^{28} p_i = 1$$

$$p_1\,(1 + k_1 + k_2 + k_3 + c_1 + c_2 + c_3 + c_6 + c_9 + c_9k_3 + c_6k_3 + c_5k_3 + c_3k_3 + c_2k_3 + c_1k_3 + c_9k_1 + c_8k_1 + c_6k_1 + c_3k_1 + c_2k_1 + c_1k_1 + c_9k_2 + c_8k_2 + c_6k_2 + c_5k_2 + c_3k_2 + c_2k_2 + c_1k_2\,) = 1$$

$$p_1 * N = 1$$

$$P_1 = 1/N$$

Presently, steady-state accessibility of cattle feed plant is given by

$A_v = p_1 + p_2 + p_3 + p_4$

Previously mentioned advances can be abridged as in stream outline as given in Fig. 6.19.

Performance Analysis

Tables (6.3-6.9) speak to the estimations of availabilities for different mixes of repair and failure rates of subsystems of a cattle feed plant. These tables uncover the different estimations of accessibility for various mixes of repairs and failures. Based on potential blends (λ_i, μ_i) for example, ideal support technique, we may choose the greatest estimation of accessibility for every subsystem. Along these lines, the ideal estimations of failures and repair rates might be chosen as needs be for every subsystem of the cattle feed plant.

Table 6.3 Effect of failure & repair rates of elevator on availability

λ_1 \ μ_1	**0.01**	**0.02**	**0.03**	**0.04**	**0.05**	**Constant values**
0.001	0.6469	0.6685	0.6761	0.6799	0.6822	$\lambda_2 = 0.005,\ \mu_2 = 0.05;$ $\lambda_3 = 0.003,\ \mu_3 = 0.03;$
0.002	0.6076	0.6469	0.6612	0.6685	0.6730	$\lambda_4 = 0.0025,\ \mu_4 = 0.02;$
0.003	0.5728	0.6266	0.6469	0.6575	0.6641	$\lambda_5 = 0.0025,\ \mu_5 = 0.02;$ $\lambda_6 = 0.004,\ \mu_6 = 0.04;$
0.004	0.5417	0.6076	0.6332	0.6469	0.6554	$\lambda_7 = 0.007,\ \mu_7 = 0.07;$ $\lambda_8 = 0.007,\ \mu_8 = 0.07,$
0.005	0.5139	0.5897	0.6202	0.6366	0.6469	$\lambda_9 = 0.006,\ \mu_9 = 0.06;$

Start

Read estimation of ($\lambda_i\ \mu_i$ ($1 <= i <= 9$))

$$k_3 = \frac{\lambda_4^2 + \lambda_4\mu_7 + \lambda_7\lambda_4 + \lambda_4\mu_4}{\mu_4\lambda_4 + \mu_4\mu_7 + \lambda_7\mu_4 + \mu_4^2},$$

$$k_1 = \frac{\lambda_4 + \lambda_7 - \mu_4 k_3}{\mu_7},$$

$$k_2 = \frac{((\lambda_4 + \mu_7)k_1 - \lambda_7)}{\mu_4}$$

$$c_i = \frac{\lambda_i}{\mu_i} \quad \text{for } i = 1 \ldots 9$$

$$N = (1 + k_1 + k_2 + k_3 + c_1 + c_2 + c_3 + c_6 + c_9 + c_9k_3 + c_6k_3 + c_5k_3 + c_3k_3 + c_2k_3 + c_1k_3 + c_9k_1 + c_8k_1 + c_6k_1 + c_3k_1 + c_2k_1 + c_1k_1 + c_9k_2 + c_8k_2 + c_6k_2 + c_5k_2 + c_3k_2 + c_2k_2 + c_1k_2)$$

$$P_1 = \frac{1}{N},\ p_2 = k_1p_1,\ p_3 = k_2p_1,\ p_4 = k_3p_1$$

$$A_v = p_1 + p_2 + p_3 + p_4$$

Print the value of A_v

Stop

Fig. 6.19 Computational procedure to calculate steady-state availability.

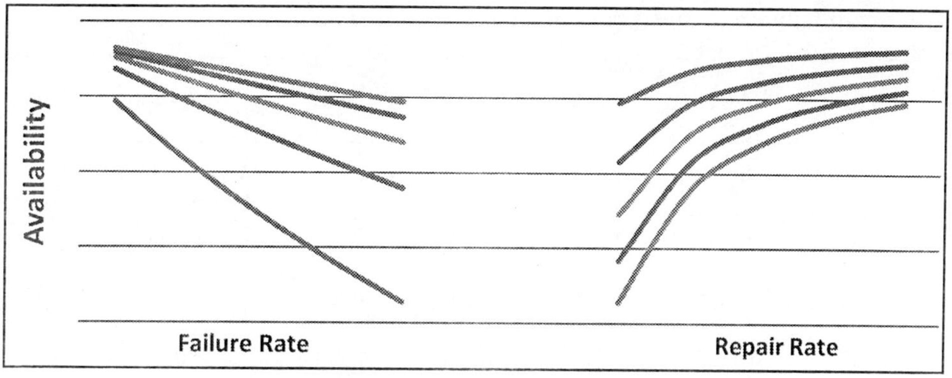

Fig. 6.20 Effect of failure and repair rates of elevator on availability.

Table 6.4 Effect of failure and repair rates of grinder on availability

λ_2 \ μ_2	0.05	0.06	0.07	0.08	0.09	Constant values
0.005	0.6469	0.6540	0.6591	0.6630	0.6751	$\lambda_1 = 0.001$, $\mu_1 = 0.01$; $\lambda_3 = 0.003$, $\mu_3 = 0.03$;
0.010	0.5990	0.6202	0.6294	0.6366	0.6567	$\lambda_4 = 0.0025$, $\mu_4 = 0.02$;
0.015	0.5538	0.5907	0.5991	0.6122	0.6391	$\lambda_5 = 0.0025$, $\mu_5 = 0.02$; $\lambda_6 = 0.004$, $\mu_6 = 0.04$;
0.020	0.5220	0.5620	0.5780	0.5897	0.6245	$\lambda_7 = 0.007$, $\mu_7 = 0.07$;
0.025	0.4939	0.5369	0.5546	0.5687	0.5890	$\lambda_8 = 0.007$, $\mu_8 = 0.07$, $\lambda_9 = 0.006$, $\mu_9 = 0.06$;

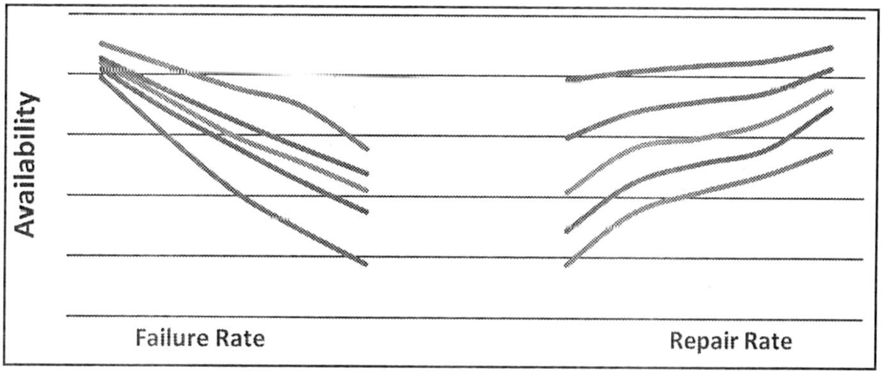

Fig. 6.21 Effect of failure and repair rates of grinder on availability.

Table 6.5 Effect of failure and repair rates of hopper on availability

λ_3 \ μ_3	**0.03**	**0.04**	**0.05**	**0.06**	**0.07**	**Constant values**
0.003	0.6469	0.6575	0.6641	0.6685	0.6717	$\lambda_1 = 0.001,\ \mu_1 = 0.01$; $\lambda_2 = 0.005,\ \mu_2 = 0.05$;
0.006	0.6077	0.6268	0.6388	0.6471	0.6531	$\lambda_4 = 0.0025, \mu_4 = 0.02$; $\lambda_5 = 0.0025, \mu_5 = 0.02$;
0.009	0.5731	0.5988	0.6154	0.6290	0.6355	$\lambda_6 = 0.004,\ \mu_6 = 0.04$;
0.012	0.5421	0.5732	0.5936	0.6081	0.6188	$\lambda_7 = 0.007,\ \mu_7 = 0.07$; $\lambda_8 = 0.007,\ \mu_8 = 0.07$,
0.015	0.5143	0.5497	0.5733	0.5903	0.6030	$\lambda_9 = 0.006,\ \mu_9 = 0.06$;

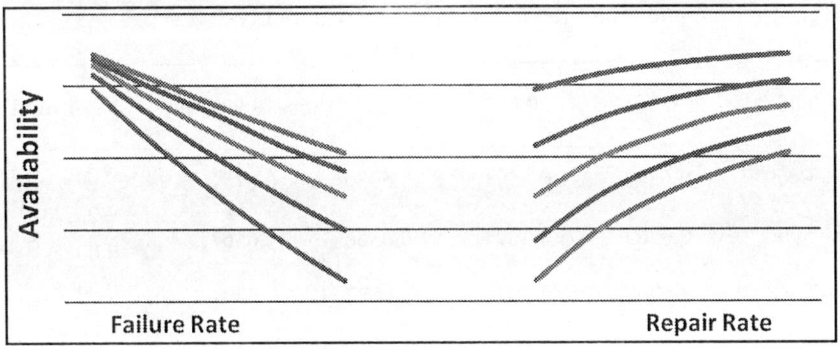

Fig. 6.22 Effect of failure and repair rates of hopper on availability.

Table 6.6 Effect of failure and repair rates of mixer on availability

λ_4 \ μ_4	**0.02**	**0.04**	**0.06**	**0.08**	**0.10**	**Constant values**
0.0025	0.6469	0.6528	0.6553	0.6566	0.6575	$\lambda_1 = 0.001,\ \mu_1 = 0.01$; $\lambda_2 = 0.005,\ \mu_2 = 0.05$;
0.0050	0.6381	0.6468	0.6506	0.6527	0.6541	$\lambda_3 = 0.003,\ \mu_3 = 0.03$;
0.0075	0.6317	0.6418	0.6465	0.6492	0.6510	$\lambda_5 = 0.0025, \mu_5 = 0.02$; $\lambda_6 = 0.004,\ \mu_6 = 0.04$;
0.0100	0.6208	0.6375	0.6429	0.6461	0.6482	$\lambda_7 = 0.007,\ \mu_7 = 0.07$; $\lambda_8 = 0.007,\ \mu_8 = 0.07$,
0.0125	0.6229	0.6339	0.6396	0.6432	0.6456	$\lambda_9 = 0.006,\ \mu_9 = 0.06$;

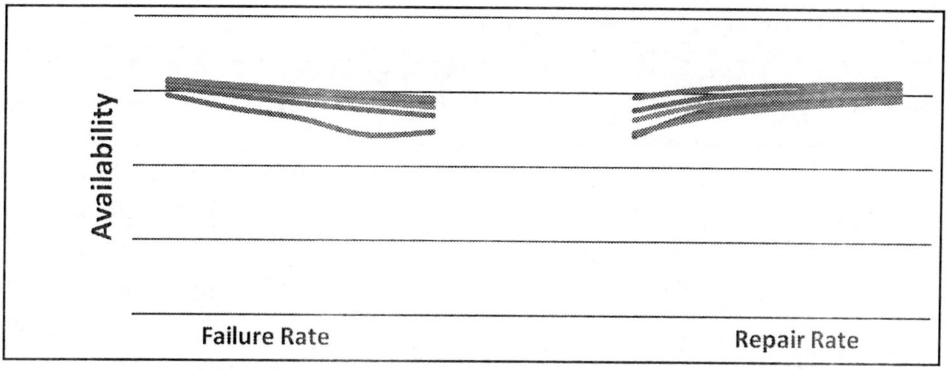

Fig. 6.23 Effect of failure and repair rates of mixer on availability.

Table 6.7 Effect of failure and repair rates of winch on availability

λ_6 \ μ_6	**0.04**	**0.07**	**0.10**	**0.13**	**0.16**	**Constant values**
0.004	0.6469	0.6654	0.6730	0.6772	0.6799	$\lambda_1 = 0.001,\ \mu_1 = 0.01$;
0.008	0.6175	0.6410	0.6554	0.6634	0.6685	$\lambda_2 = 0.005,\ \mu_2 = 0.05$; $\lambda_3 = 0.003,\ \mu_3 = 0.03$;
0.012	0.5629	0.6183	0.6386	0.6501	0.6575	$\lambda_4 = 0.0025,\ \mu_4 = 0.02$; $\lambda_5 = 0.0025,\ \mu_5 = 0.02$;
0.016	0.5318	0.5972	0.6227	0.6374	0.6469	$\lambda_7 = 0.007,\ \mu_7 = 0.07$;
0.020	0.5230	0.5775	0.6076	0.6251	0.6366	$\lambda_8 = 0.007,\ \mu_8 = 0.07$, $\lambda_9 = 0.006,\ \mu_9 = 0.06$;

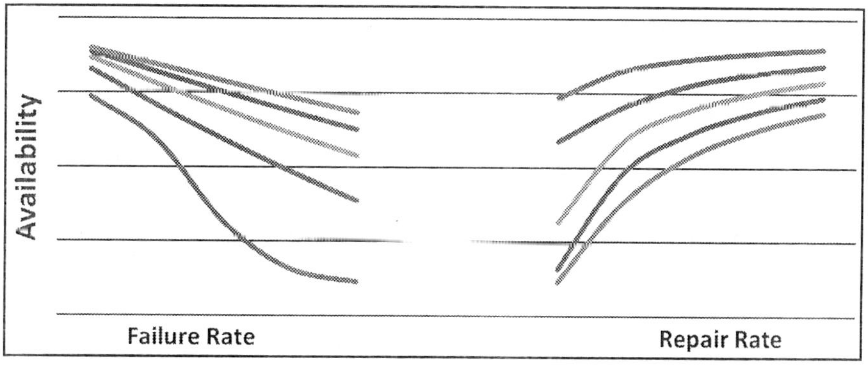

Fig. 6.24 Effect of failure and repair rates of winch on availability.

Table 6.8 Effect of failure and repair rates of palletiser on availability

λ_7 \ μ_7	**0.07**	**0.08**	**0.09**	**0.10**	**0.11**	**Constant values**
0.007	0.645	0.649	0.656	0.659	0.663	$\lambda_1 = 0.001$, $\mu_1 = 0.01$;
0.014	0.642	0.647	0.651	0.655	0.66	$\lambda_2 = 0.005$, $\mu_2 = 0.05$; $\lambda_3 = 0.003$, $\mu_3 = 0.03$;
0.021	0.635	0.642	0.648	0.65	0.657	$\lambda_4 = 0.0025$, $\mu_4 = 0.02$;
0.028	0.633	0.641	0.645	0.645	0.65	$\lambda_5 = 0.0025$, $\mu_5 = 0.02$; $\lambda_6 = 0.004$, $\mu_6 = 0.04$;
0.035	0.63	0.639	0.642	0.643	0.645	$\lambda_8 = 0.007$, $\mu_8 = 0.07$, $\lambda_9 = 0.006$, $\mu_9 = 0.06$;

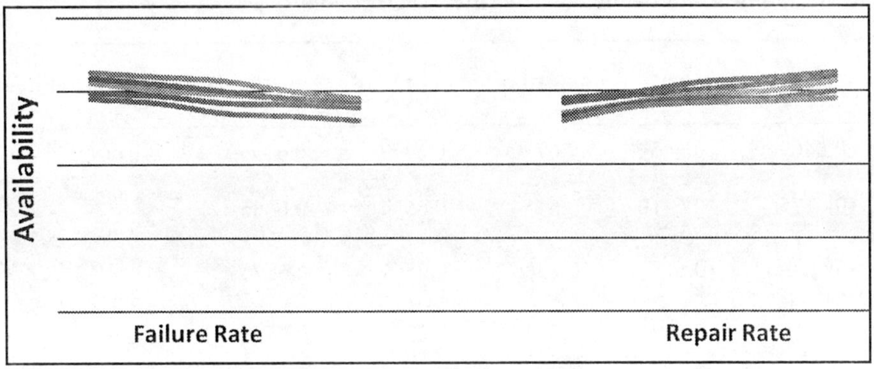

Fig. 6.25 Effect of failure and repair rates of palletiser on availability.

Table 6.9 Effect of failure and repair rates of screw conveyor on availability

λ_9 \ μ_9	**0.06**	**0.07**	**0.08**	**0.09**	**0.1**	**Constant values**
0.006	0.6469	0.6529	0.6575	0.6612	0.664	$\lambda_1 = 0.001$, $\mu_1 = 0.01$;
0.012	0.6176	0.6183	0.6266	0.6332	0.638	$\lambda_2 = 0.005$, $\mu_2 = 0.05$; $\lambda_3 = 0.003$, $\mu_3 = 0.03$;
0.018	0.5828	0.5872	0.5985	0.6072	0.615	$\lambda_4 = 0.0025$, $\mu_4 = 0.02$;
0.024	0.5517	0.5591	0.5728	0.5839	0.593	$\lambda_5 = 0.0025$, $\mu_5 = 0.02$; $\lambda_6 = 0.004$, $\mu_6 = 0.04$;
0.030	0.5239	0.5335	0.5429	0.5620	0.572	$\lambda_7 = 0.007$, $\mu_7 = 0.07$; $\lambda = 0.007$, $\mu_8 = 0.07$,

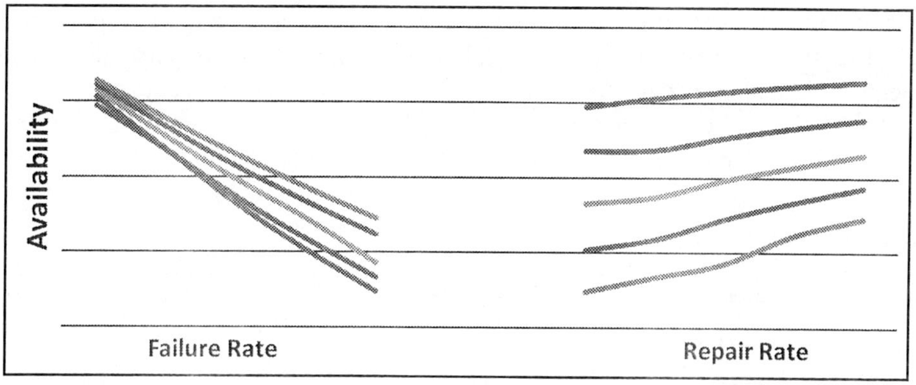

Fig. 6.26 Effect of failure and repair rates of screw conveyor on availability.

Table 6.3 demonstrates the impact of disappointment and repair rates of elevator on accessibility of cattle feed plant. It is seen that for some known estimations of disappointment and repair rate of grinder, hopper, mixer, winch, palletiser, screw transport as disappointment rate (λ_1) of elevator increments from 0.001 (once in 1000 hr) to 0.005 (once in 200 hr), the framework accessibility diminishes impressively by 13.30%. Additionally, as repair rate (μ_1) of elevator increments from 0.01 (once in 100 hr) to 0.05 (once in 20 hr), the framework accessibility increments by 3.53%.

Table 6.4, it is additionally seen that as disappointment rate (λ_2) of grinder increments from 0.005 (once in 200 hr) to 0.025 (once in 40 hr), the framework accessibility diminishes by 15.30%. Essentially, when repair rate (μ_2) of grinder from 0.05 (once in 20 hr) to 0.09 (once in 11.1 hr), at that point the framework accessibility increments by 2.82%.

Table 6.5 likewise uncovers the variety of framework accessibility with change in disappointment rates (λ_3) and fix rates (μ_3) of the hopper. As disappointment rate (λ_3) increments from 0.003 (once in 333.33 hr) to 0.015 (once in 66.66 hr), the framework accessibility diminishes by 13.26%. So also, when fixed rate (μ_3) increments from 0.03 (once in 33.3 hr) to 0.07 (once in 14.28 hr), at that point the framework accessibility increments by 2.48%.

Table 6.6 demonstrates the variety in accessibility of cattle feed plant with the adjustment in repair and failure rates of subsystem mixer (λ_4, μ_4). Further, Table 6.6 uncovers that disappointment rate (λ_4) increments from 0.0025 (once in 400 hr) to 0.0125 (once in 80 hr), the framework accessibility decreases by 2.40%. Essentially, when repair rate (μ_4) increments from 0.02 (once in 50 hr) to 0.10 (once in 10 hr), at that point the framework accessibility increments by about 1.06%.

Table 6.7 additionally uncovers the variety of framework accessibility with change in failure (λ_6) and repair (μ_6) rates of winch. As disappointment rate (λ_6) increments from 0.004 (once in 250 hr) to 0.02 (once in 50 hr), the framework accessibility decreases by 12.39%. Thus, when repair rate (μ_6) increments from 0.04 (once in 25 hr) to 0.16 (once in 6.25 hr), at that point the framework accessibility increments by about 3.30%.

Table 6.8 additionally uncovers the variety of framework accessibility with change in failure (λ_7) and repair (μ_7) rates of palletiser. As disappointment rate (λ_7) increments from 0.007 (once in 142.8 hr) to 0.035 (once in 28.57 hr), the framework accessibility decreases by 1.5%. So also, when repair rate (μ_7) increments from 0.07 (once in 14.28 hr) to 0.11 (once in 9.09 hr), at that point the framework accessibility increments by about 1.80%.

Table 6.9 demonstrates the variety in accessibility of cattle feed plant with the adjustment in repair and failure rates of subsystem screw transport (λ_9, μ_9). Further, Table 6.9 uncovers that disappointment rate (λ_9) increments from 0.006 (once in 166.66 hr) to 0.030 (once in 33.33 hr), the framework accessibility diminishes by 12.30%. Likewise, when repair rate (λ_9) increments from 0.06 (once in 16.66 hr) to 0.1 (once in 10 hr), at that point the framework accessibility increments by about 1.72%.

Table 6.10 Order of repair work

Subsystem	Increase in Availability	Rank
Elevator	3.53%	1st
Winch	3.30%	2nd
Grinder	2.85%	3rd
Hopper	2.58%	4th
Palletiser	1.78%	5th
Screw conveyor	1.72%	6th
Mixer	1.06%	7th

Table 6.10 demonstrates that elevator is the most basic subsystem to the extent support is concerned. In this way, elevator ought to be given top need as the impact of its repair rate on the accessibility is a lot higher than that of winch, grinder, hopper, palletiser, screw transport, mixer. In this manner, based on repair rates, the support need ought to be given according to Table 6.10.

CONCLUDING REMARKS

In section 6.1, the paper industry, a large complex system in process industry, is discussed. In sections 6.2 and 6.3, two other industries, which also belong to process industrial systems, are discussed. In section 6.4, the reliability of cattle feed plant, a large complex system in process industry is calculated using matrix method. In the foregoing analysis of these industries, we used Markov process and matrix methods. If the failure rate or repair rate or both are time dependent, then the systems lose their Markovian character. In such cases, we get partial differential equation which can be solved by using supplementary variable technique (chapter 4).

EXERCISE

1. Explain the paper industry plant in process industry.
2. What is the difference between feeding system, pulping system, and washing system?
3. Write short notes on:
   (a) Digester
   (b) Decker
   (c) Bleaching system
   (d) Paper formation system

7

Chapter

Reliability Technology Applied to Agro-based Industries

There are industries which are totally based on agriculture sector, e.g., sugar industry, butter-oil (ghee) manufacturing industry, etc. The sugar industry is totally dependent upon agriculture because the raw material for it (sugar cane) is supplied from agriculture sector and it is directly affected by any change in the production of sugar cane. The bagasse obtained from crushing of cane sugar is used to produce paper in paper industry. Butter-oil is prepared from milk obtained from cow, buffalo, etc. who are dependent upon agriculture sector. Thus, the butter-oil industry is directly dependent upon agriculture. The fertilizer industry produces different types of fertilizers used to increase the crop production. The production of fertilizer affects the demand of agriculture sector for fertilizers and vice versa. Thus, fertilizer industry is also dependent upon agriculture sector. Also there are other industries such as plywood industry, bread manufacturing industry, cotton textile industry, agriculture implements manufacturing industry, etc. where directly or indirectly the agriculture sector is mainly effective. In this chapter, three industries – sugar, fertilizer and butter-oil manufacturing are analyzed using reliability theory.

7.1 THE SUGAR INDUSTRY

In a sugar plant, the raw material (sugar cane) is transported by tractor-trolleys, trucks, rail, etc. The sugar cane so obtained is fed to the cutters by a chain conveyor system for chopping. The chopped pieces are sent to crushers by another chain conveyor system where they are crushed by a series of crushers so as to squeeze the sugar cane for maximum possible extraction of juice. The squeezed sugar cane (bagasse) is used up in the plant boiler as fuel and also used in paper plant as raw material for paper manufacturing. The juice so obtained contains impurities like small pieces of bagasse, mud, etc. and hence it is passed through a number of filters in series for refining. The juice is heated up by steam to attain a definite temperature and pH value. The heated juice is stored in a tank where sulphur dioxide gas is passed through it for separation of mud. The juice is next sent through a clarifier where further separation of mud by gravity process takes place. The clear juice so obtained is stored in a evaporator where the juice water evaporates and then passes through sulphitor for final refining. The concentrated and refined juice is taken to cooking pans where the remaining juice water gets evaporated, converting the juice to viscous fluid. The semi-solid form of the juice is transferred to crystallizers to evaporate the remaining water. The heating in crystallizers is done by long duration slow heating. Thus, the juice is converted into dark brown magma (molasses) containing yellow sugar crystals in suspended form. The molasses is passed through centrifuges to separate out sugar crystals from it. The yellowish colour of the sugar crystals is washed out by a chemical process. The rest of the crystal free magma is recycled through sulphitor, evaporator, cooking pans, crystallizers, driers and centrifuges to achieve higher output. Finally, the crystals are cooled and collected in graded bags. If there is any part of sugar in powder form, it is again processed through crystallization process.

The process of sugar production is a complex continuous process. The schematic diagram of the process is shown in Fig. 7.1. The whole industrial system is divided into three main subsystems for analysis, namely, feeding, refining and crystallization. In this section, the three subsystems are discussed under some assumptions using suitable notations.

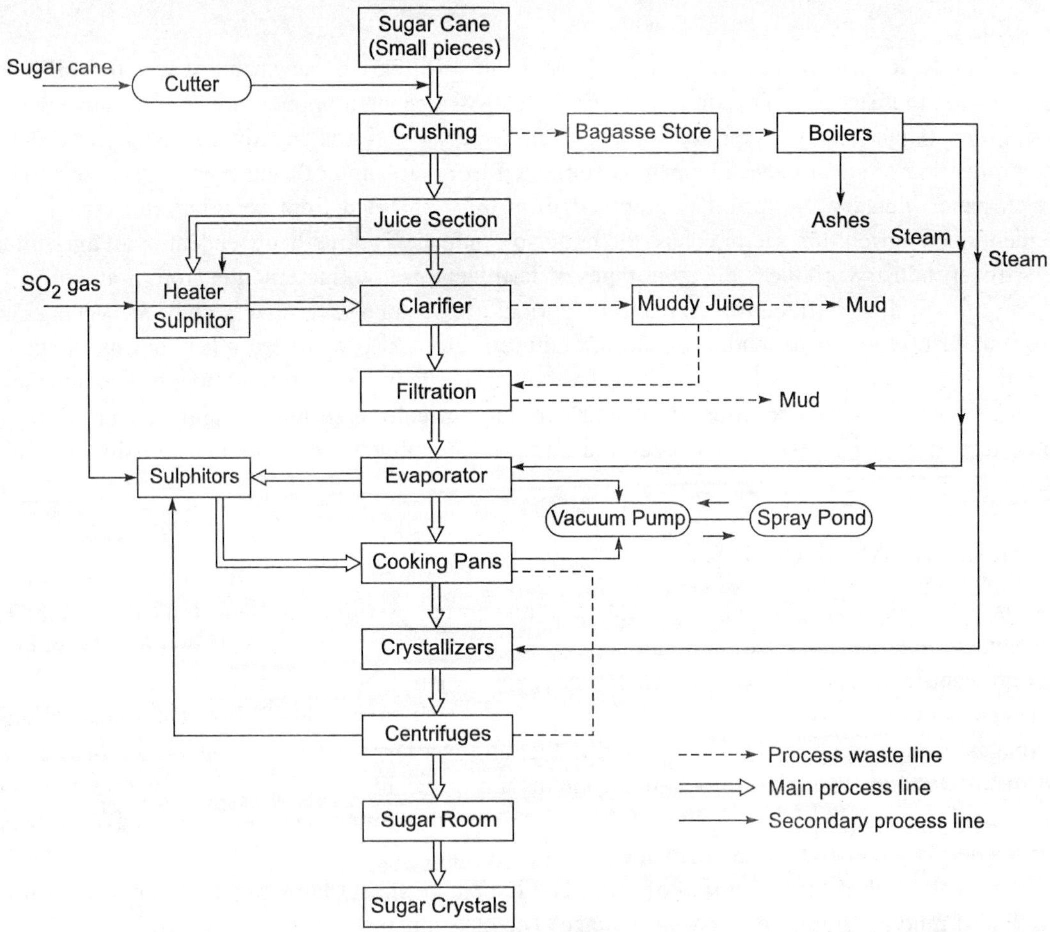

Fig. 7.1

(a) The Feeding System

The system consists of:

(i) Sugar cane supply system A_1 comprises 3 units in series—chain conveyor to carry the sugar cane to cutters, cutter for chopping the sugar cane into small pieces and chain conveyor for carrying small pieces to crushing units. The failure of any one unit causes the failure of feeding system.

(ii) The crushing system A_2 having 5 or 6 units in series. Failure of any unit causes failure of the system. Crusher is equipped with three rollers in a particular configuration to squeeze the sugar cane.

(iii) The bagasse carrying unit A_3 consisting of two or three units having a combination of bucket elevator and conveyor in series. Failure of any one unit causes the piling up of bagasse in front of crushing house, affecting the fuel supply to the boiler and hence the reduction in efficiency of the system, and

(iv) The heat generation system A_4 having three boilers in parallel. The failure of a boiler means the reduction in the efficiency of the system. These boilers generate steam at high pressure and temperature for use in the plant.

Assume that the mean failure and repair rates are constant, failures and repairs are statistically independent, repaired units work like new, each subsystem has separate repair facility to finish any type of waiting in the system. Further, let $p_i(t)$ be the probability density that at time 't' the system is in state 'i', $i = 0, 1, 2, \ldots 11$ and α_j, β_j be the mean constant failure and repair rates, respectively of the system. Taking into consideration the above notations and assumptions, the transition diagram for the feeding system is shown in Fig. 7.2.

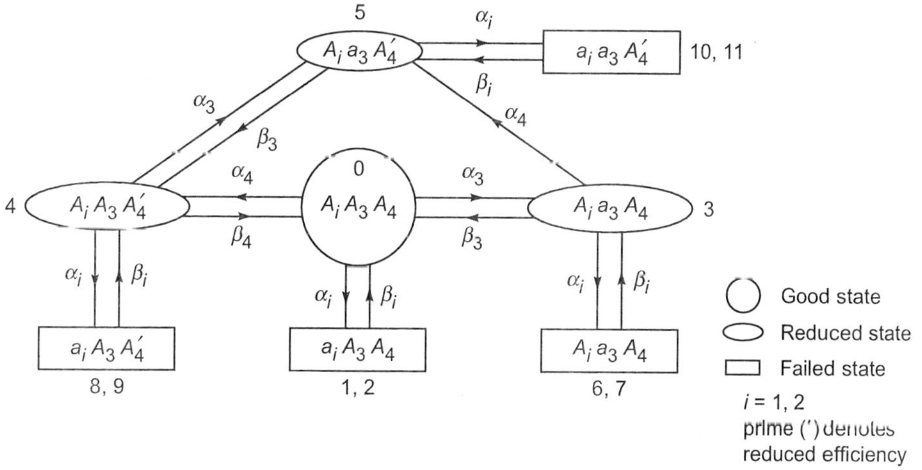

Fig. 7.2

The differential-difference equations governing the system associated with the transition diagram (Fig. 7.2) are

$$\left(\frac{d}{dt} + \sum_{1}^{4} \alpha_i\right) p_0(t) = \sum_{1}^{4} \beta_i p_i(t) \tag{7.1.1}$$

$$\left(\frac{d}{dt}+\beta_i\right)p_i(t) = \alpha_i\, p_0(t), \text{ for } i = 1, 2 \quad (7.1.2)$$

$$\left(\frac{d}{dt}+\alpha_1+\alpha_2+\alpha_4+\beta_3\right)p_3(t) = \beta_1\, p_6(t) + \beta_2\, p_7(t) + \alpha_3\, p_0(t) \quad (7.1.3)$$

$$\left(\frac{d}{dt}+\beta_4+\sum_1^3 \alpha_i\right)p_4(t) = \alpha_4\, p_0(t) + \beta_3\, p_5(t) + \beta_1\, p_8(t) + \beta_2\, p_9(t) \quad (7.1.4)$$

$$\left(\frac{d}{dt}+\sum_1^2 \alpha_i+\beta_3\right)p_5(t) = \alpha_3\, p_4(t) + \alpha_4\, p_3(t) + \beta_1\, p_{10}(t) + \beta_2\, p_{11}(t) \quad (7.1.5)$$

$$\left(\frac{d}{dt}+\beta_i\right)p_{5+i}(t) = \alpha_i\, p_3(t),\ i = 1, 2 \quad (7.1.6)$$

$$\left(\frac{d}{dt}+\beta_i\right)p_{7+i}(t) = \alpha_i\, p_4(t),\ i = 1, 2 \quad (7.1.7)$$

$$\left(\frac{d}{dt}+\beta_i\right)p_{9+i}(t) = \alpha_i\, p_5(t),\ i = 1, 2 \quad (7.1.8)$$

with initial conditions $p_0(0) = 1$ otherwise zero.

Taking Laplace transforms of eqns. (7.1.1)–(7.1.8) and using the initial conditions Kumar et al., [1990] solved the equations getting Laplace transform of availability function, $A_v(s) = L\,A_v(t) = L\{p_0(t) + p_3(t) + p_4(t) + p_5(t)\} = p_0(s) + p_3(s) + p_4(s) + p_5(s)$.

where, $p_0(s) = [s + x_4 - y_4]^{-1}$; $p_3(s) = \alpha_3[s + x_1 - y_1]^{-1}\, p_0(s)$;

$$p_4(s) = \frac{\alpha_4}{(s+x_3-y_3)}\left[1+\frac{\alpha_3\beta_3(s+x_2-y_2)^{-1}}{s+x_1-y_1}\right]p_0(s)\ ;$$

$$p_5(s) = \frac{\alpha_3}{s+x_2-y_2}p_0(s).$$

$$x_1 = \alpha_1 + \alpha_2 + \alpha_4 + \beta_3\ ;\ x_2 = \alpha_1 + \alpha_2 + \beta_3\ ;\ x_3 = \sum_1^3 \alpha_i + \beta_4\ ;\ x_4 = \sum_1^4 \alpha_i$$

$$y_1 = y_2 = \sum_1^2 \frac{\alpha_i\beta_i}{s+\beta_i}\ ;\ y_3 = y_1 + \frac{\alpha_3\beta_3}{s+x_2-y_2}\ ;$$

$$y_4 = y_3 + \frac{\alpha_4}{s+x_3-y_3}\left[1+\frac{\alpha_3\beta_3}{(s+x_1-y_1)(s+x_2-y_2)}\right]$$

Mean time to system failure (MTTF) $= \lim_{s\to 0} s\,R(s)$

$$= \frac{\left(1+\frac{\alpha_3}{\alpha_4+\beta_3}\right)\left(1+\frac{\alpha_4}{\beta_4}+\frac{\alpha_3\alpha_4}{\beta_3\beta_4}\right)}{\left(1+\sum_1^2\frac{\alpha_i}{\beta_i}\right)\left[1+\frac{\alpha_3}{\alpha_4+\beta_3}\left\{\beta_3+\frac{\alpha_4}{\beta_4}\left(\frac{1}{\alpha_4+\beta_3}+\frac{1}{\beta_3}\right)\right\}+\left(1+\frac{\alpha_3}{\beta_3}\right)\left(\frac{\alpha_4}{\beta_4}+\frac{\alpha_3\alpha_4}{\beta_4(\alpha_4+\beta_3)}\right)\right]}$$

For steady-state results, putting $\frac{d}{dt} = 0$, $p_i(t) = p_i$ as $t \to \infty$ and using normalizing condition $\sum_{i=0}^{11} p_i = 1$.

Kumar et al., [1990] obtained availability of the system

$$A = p_0 + p_3 + p_4 + p_5 = \left[1+\sum_1^2 \frac{\alpha_i}{\beta_i}\right]^{-1}.$$

Kumar et al., [1990] and Kumar [1991] analyzed the feeding system in detail and studied the behaviour of the system changing the various effective rates. Kumar et al., [1988] studied the feeding system for general repair time using Lagrange's method for solving partial differential equations.

(b) Refining System

The refining system is comprised of the following four subsystems:

(i) The juice screen B_1 consisting of two units in series to ensure complete removal of small bagasse pieces and impurities from juice. Failure of any one unit causes failure of the system.

(ii) The clarifier B_2 has n units in series. It removes the sugar cane mud from the juice by gravity process. Failure of any unit leads to system failure.

(iii) The sulphonation plant B_3 having m units in parallel. This removes mud from juice using sulphur dioxide. Failure of any unit leads to reduced working capacity of the system. Complete failure occurs only when all m units have failed.

(iv) The heating plant B_4 having k units in parallel. The failure of any unit results into reduced working capacity of the system. Complete failure occurs only when all the k units have failed.

The transition diagram for the refining system is shown in Fig. 7.3.

The differential-difference equations governing the system are (associated with the transition diagram):

$$\left(\frac{d}{dt}+\sum_{j=5}^{6}\alpha_j+\lambda_1+\lambda_2\right)p_0(t) = \sum_{j=5}^{8}\beta_j p_{j-4}(t) \tag{7.1.9}$$

$$\left(\frac{d}{dt}+\sum_{j=5}^{8}\alpha_j+\beta_7\right)p_3(t) = \lambda_2 p_0(t) + \sum_{j=5}^{7}\beta_j p_{j+8}(t) + \beta_8 p_7(t) \tag{7.1.10}$$

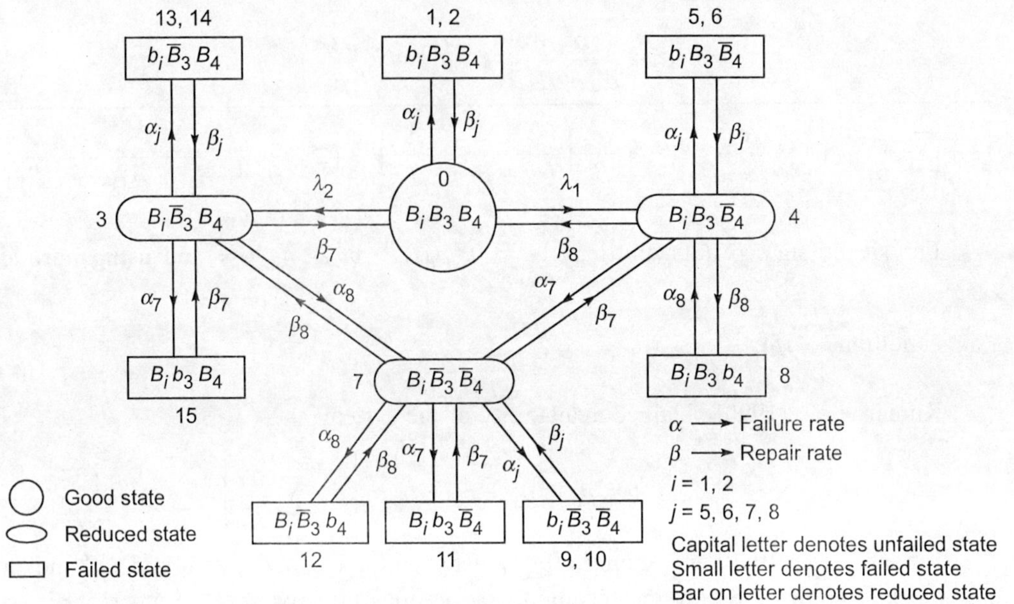

Fig. 7.3

$$\left(\frac{d}{dt}+\sum_{j=5}^{8}\alpha_j+\beta_8\right)p_4(t) = \lambda_1 p_0(t) + \sum_{j=5}^{8}\beta_j p_j(t) \tag{7.1.11}$$

$$\left(\frac{d}{dt}+\sum_{j=5}^{8}\alpha_j+\beta_7+\beta_8\right)p_7(t) = \alpha_7 p_4(t) + \alpha_8 p_3(t) + \sum_{j=5}^{8}\beta_j p_{j+4}(t) \tag{7.1.12}$$

$$\left(\frac{d}{dt}+\beta_j\right)p_i(t) = \alpha_j p_k(t) \tag{7.1.13}$$

with initial conditions $p_0(0) = 1$ otherwise zero and in eqn. (7.1.13) for $k = 0, j = 5, 6, i = j - 4$; $k = 3, j = 5, 6, 7, i = j + 8$; $k = 4, j = 5, 6, 8, i = j$; $k = 7, j = 5, 6, 7, 8, i = j + 4$.

As mentioned in 7.1(a), Laplace transform of availability function $A_v(t)$ is given by

$$L\{A_v(t)\} = A_v(s) = L\{p_0(t) + p_3(t) + p_4(t) + p_7(t)\} = p_0(s) + p_3(s) + p_4(s) + p_7(s).$$

$p_k(s)$, $k = 0, 3, 4, 7$ are obtained from eqns. (7.1.9)–(7.1.13) taking Laplace transforms and solving recursively. Mean time to system failure (MTTF) is given as follows:

$$\text{MTTF} = \left[1+\frac{\lambda_1}{b_1}+\frac{\lambda_2}{b_2}+\frac{1}{b_1+b_2}\left(\frac{\alpha_7\lambda_1}{b_1}+\frac{\alpha_8\lambda_2}{b_2}\right)\left(\frac{b_1}{\beta_7}+\frac{b_2}{\beta_8}+\frac{b_1 b_2}{\beta_7\beta_8}\right)\right] \Bigg/$$

$$\left[1+\sum_{5}^{6}\frac{\lambda_j}{\beta_j}+\frac{\lambda_1 a_1}{b_1}+\frac{\lambda_2 a_2}{b_2}+\frac{1}{b_1+b_2}\left(\frac{\alpha_7\lambda_1}{b_1}\frac{\alpha_8\lambda_2}{b_2}\right)\left(\frac{a_2 b_1}{b_2}+\frac{a_1 b_2}{b_1}+\frac{a_3 b_1 b_2}{\beta_7\beta_8}\right)\right]$$

where, $b_1 = \alpha_7 + \beta_8$; $b_2 = \alpha_8 + \beta_7$; $a_1 = 1 + \sum_{5}^{6} \frac{\alpha_j}{\beta_j} + \frac{\alpha_8}{\beta_8}$;

$$a_2 = 1 + \sum_{5}^{7} \frac{\alpha_j}{\beta_j} \ ; \ a_3 = 1 + \sum_{5}^{8} \frac{\alpha_j}{\beta_j} + \frac{\alpha_7 \beta_7 a_1}{b_1} + \frac{\alpha_8 \beta_8 a_2}{b_2}.$$

For steady state, i.e., when $t \to \infty$, $\frac{d}{dt} \to 0$ and $p_i(t) \to p_i$. The eqns. (7.1.9)–(7.1.13) are solved recursively putting $\frac{d}{dt} = 0$ and using the normalizing condition $\sum_{i=0}^{15} p_i = 1$, obtaining the long-run availability.

$$A_v = p_0 + p_3 + p_4 + p_7$$

Behaviour analysis of refining system is discussed by Kumar et al., [1989, 1990, 1991] in detail giving availability tables, etc. corresponding to changes in repairs and failures.

(c) Crystallization System

It is the most important part of sugar industry. The system is discussed by Kumar et al., [1991, 1992] in detail consists of the following subsystems:

(i) The evaporator D_1, having two units in parallel configuration.
(ii) The cooking pans D_2 ; there are six units in parallel.
(iii) The crystallizer D_3 having 12 units in parallel.
(iv) The centrifuges D_4 comprised 38 units in parallel.
(v) This subsystem (may be called D_5) comprises hopper, elevator, cooler and sugar grader all connected in series. Failure of any one causes complete failure of the system.

The failures in subsystems D_1, D_2, D_3, D_4 reduce only the working capacity and complete failure of these subsystems occurs only when all units of the respective systems fail.

There may be common-cause (cc) failure in the system such as failure of the steam supply to subsystems D_1, D_2, D_3 which stops the plant altogether. Under the above assumptions and following the description of the system (given in the beginning), the transition diagram for the system is given in Fig. 7.4.

The differential-difference equations associated with Fig. 7.4 (using probabilistic arguments) are given as follows:

$$\left(\frac{d}{dt} + \sum_{i,j}^{k,m} \lambda_r + \sigma + \alpha_5\right) p_0(t) = \sum_{1}^{5} \beta_i p_i(t) + \mu\beta_3 p_{3c}(t) + \mu\beta_2 p_{2c}(t) + \mu p_{0c}(t) \qquad (7.1.14)$$

$$\left(\frac{d}{dt} + \alpha_r + \beta_r + \sigma\right) p_r(t) = \beta_r p_{r+5}(t) + \lambda_i p_0(t), \qquad (7.1.15)$$

$n = i$ for $r = 1$; $n = j$ for $r = 2$; $n = k$, for $r = 3$

$$\left(\frac{d}{dt} + \beta_4 + \alpha_4\right) p_4(t) = \beta_4 p_9(t) + \lambda_m p_0(t), \qquad (7.1.16)$$

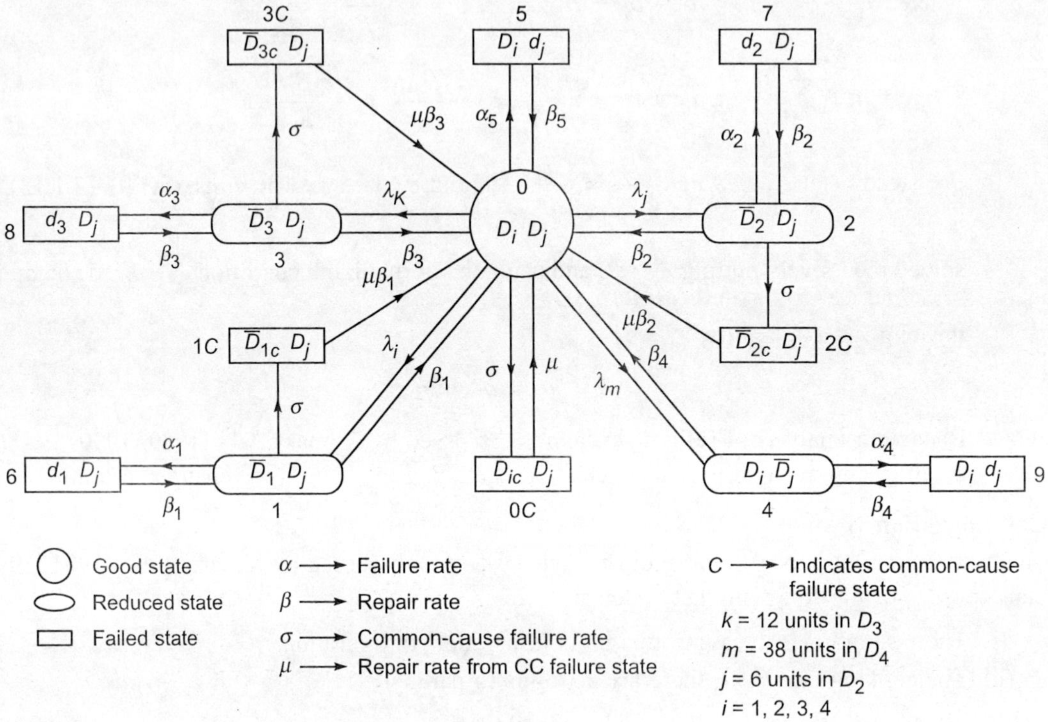

Fig. 7.4

$$\left(\frac{d}{dt}+\beta_5\right)p_5(t) = \alpha_5\, p_0(t) \tag{7.1.17}$$

$$\left(\frac{d}{dt}+\beta_r\right)p_{r+5}(t) = \alpha_r\, p_r(t),\ r = 1,\ 2,\ 3,\ 4 \tag{7.1.18}$$

$$\left(\frac{d}{dt}+\mu\right)p_{0c}(t) = \sigma\, p_0(t) \tag{7.1.19}$$

$$\left(\frac{d}{dt}+\mu\beta_r\right)p_{rc}(t) = \sigma\, p_r(t),\ r = 1,\ 2,\ 3 \tag{7.1.20}$$

with initial condition $p_0(0) = 1$ otherwise zero.

Availability function and MTTF are given by

$$A_v(t) = \sum_{i=0}^{4} p_i(t).$$

$$\text{MTTF} = \left[1+\sum\frac{\lambda_n}{\beta_r+\sigma}+\sum\frac{\lambda_m}{\beta_4}\right]\Bigg/\left[1+\sum\frac{\lambda_n}{\beta_r+\sigma}+\sum\frac{\lambda_n}{(\beta_r+\sigma)^2}\left\{1+\frac{\alpha_r}{\beta_r}+\frac{\beta_r+\sigma}{\beta_r\mu}\right\}\right.$$

$$+\sum \lambda_m\left(1+\frac{\alpha_4}{\beta_4}\right)+\frac{\alpha_5}{\beta_5}+\frac{\sigma}{\mu}\Bigg]$$

In steady state, availability function

$$A_v = \sum_{i=0}^{4} p_i.$$

Equations (7.1.14)–(7.1.20) can be solved using Laplace transform or any other method discussed in Chapter 4. For steady-state results, these equations can be solved recursively taking $\frac{d}{dt} = 0$ as $t \to \infty$, $p_i(t) \to p_i$ and using the normalizing condition $\sum p_i = 1$. Kumar et al., studied the behaviour of the system in detail.

The overall availability of the sugar plant is equal to the product of the availabilities of feeding, refining and crystallization systems.

7.2 THE BUTTER-OIL (GHEE) MANUFACTURING PLANT

The butter-oil plant can be divided into following sections—milk reception and processing, fat separation, cream pasteurisation, butter manufacturing, butter-oil manufacturing, packing.

The milk from collection centres is received on the plant dock. This activity consists of unloading, grading, sampling, weighing and testing of milk. The milk from dump tank is passed through chillers and chilled to 4°C and stored in raw milk silos. The chilled milk is now passed through the milk plate pasteuriser which is a self-contained unit for heating and cooling the milk. The milk is heated to 45°C and is diverted to a cream separator which separates out the cream from the milk. The skimmed milk returns back to the milk pasteuriser where it is further heated to 80°C to make it safe for human consumption. It is then chilled to 4°C to enhance its keeping quality on storage. The cream is heated to 71°C or more. The pasteurised cream is stored in double-jacketed cream storage tank for conversion into butter and butter-oil. Cooling and ageing of cream is done by lowering its temperature and holding for few hours to make churning possible. Cream from the cream storage tank is pumped into the butter churner. The cream is churned into the machine in order to get butter granules. Buttermilk is taken out separately and is pumped back to raw milk silos whereas butter granules are further worked into the machine, so as to get homogeneous mass of butter. The butter is taken out from the machine and is shifted to melting vats by butter trolleys. The butter from melting vats is pumped to the butter-oil boiler where it is heated to 107°C gently so as to evaporate water from the melted butter and is then allowed to remain undisturbed for few hours. The butter-oil is taken out into butter-oil setting tanks through the bottom of the butter-oil boiler and is allowed to settle for a few hours to remove the fine particles of butter-oil residue from the butter-oil. Clarified butter-oil is stored in butter-oil storage tank where it is further cooled to desired temperature (about 30°C) suitable for filling the butter-oil. The butter-oil is then tested for different tests by the quality control staff and is filled into tins. The process flow chart is given in Fig. 7.5.

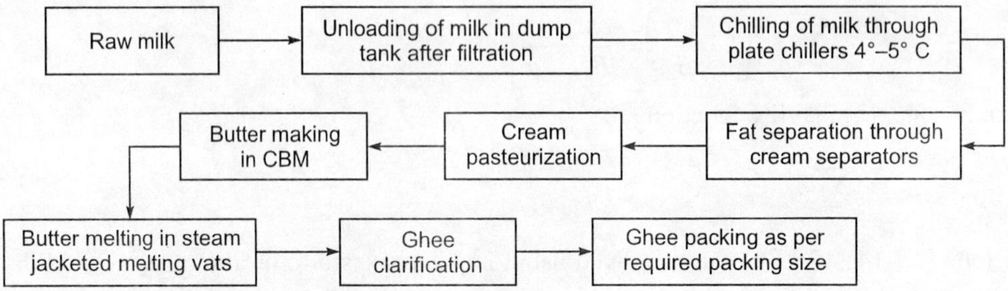

Fig. 7.5

The butter-oil plant consists of eight subsystems, namely:

(i) Pumping system, used for unloading the milk brought from milk collection centres, consists of two pumps one in working and another in standby having perfect switch-over device. This system never fails.

(ii) Chilling system consist of two chillers – one in working and another in standby—with perfect switch-over device. This system never fails.

(iii) Separator (*A*) working on the centrifugal force principle. Chilled milk from chiller is taken to the cream separator where fat is separated from the milk in cream form containing 40 to 50 per cent fat. The skimmed milk is stored in milk silos for preparing milk powder. It consists of three components in series – motor, bearings and high-speed gearbox. Failure of any one component causes failure of this system.

(iv) Pasteuriser (*B*), pasteurises the cream coming from separator (*A*). Here the cream is heated up to 80°C to 82°C for no holding time. The purpose is to destroy pathogenic organism, to destroy undesirable organism, to inactivate the enzymes present and to make possible removal of volatile flavours and to remove the tanning substances. This system can work in reduced state also. It consists of a motor and bearings in series.

(v) Continuous Butter-making Machine (CBM): The cream from cream storage tank is pumped into CBM and churned in this machine in order to get butter granules. The buttermilk produced in this process is pumped back to raw milk silos and the butter granules are further processed in CBM to get homogeneous mass of butter. This butter is shifted to melting vats by butter trolleys. CBM consists of gearbox, motor and bearings in series. Let us denote this subsystem by *C*.

(vi) Melting Vats (*D*), consisting of a double-jacketed storage tank, melt the butter at about 107°C very gently so that the water evaporates from the butter. This melted butter is settled for some time. This system consists of monoblock pumps, motors and bearings in series.

(vii) Butter-oil clarifier (*E*): When the butter-oil from melting vats settled in setting tanks for some time, fine particles of butter-oil residue are removed from butter-oil. It is stored in storage tanks and is cooled upto 30°C. The system consists of motor and gearbox in series.

(viii) Packaging (*F*): A fill, flow and seal automatic machine (called pouch-filling machine) forms packets of the clarified butter-oil. The system consists of printed circuit board and pneumatic cylinder in series.

The pumping and chilling systems, as mentioned above, never fail. So the working of the plant is affected by the rest six systems. In all the above-mentioned subsystems, except subsystem *B*, are subjected to major failures while B can work in reduced state also. Gupta et al., [2005] and Goel et al., [2001] analyzed the butter-oil plant in detail. Let

α_i: represent the respective constant failure rates of subsystems A, C, D, E, F, $\overline{B}$ and B where $i = 1, 2, \ldots, 7$ and $\overline{B}$ denotes the reduced state of system B.

β_i: represent the respective constant repair rates of subsystems A, C, D, E, F, $B(i = 1, 2, \ldots, 6)$

$p_j(t)$, $j = 1, 2, \ldots, 13$: probability density that at time t the system is in jth state.

Small letters a, b, c, d, e and f denote the failed states of the subsystems A, B, C, D, E, F, respectively.

- Repairs and failures are independent of each other and there is no simultaneous failure among the subsystems.
- Switch-over device is perfect and the subsystem B fails only through reduced state.
- Repaired components function like new components.

Taking the above symbols and assumptions into consideration, the transition diagram for the plant is given in Fig. 7.6.

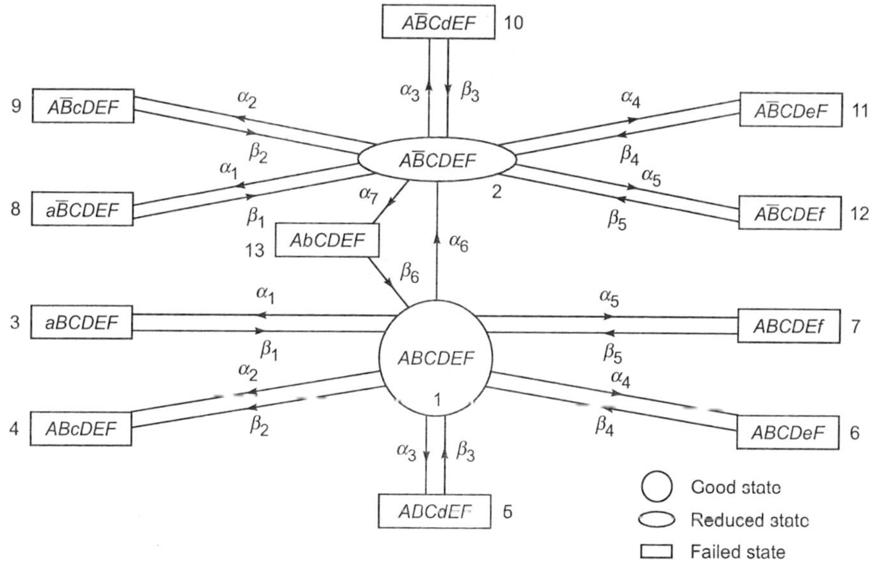

Fig. 7.6

Probability considerations give the following differential-difference equations associated with the transition diagram:

$$\left(\frac{d}{dt} + \sum_{1}^{6} \alpha_i\right) p_1(t) = \sum_{1}^{5} \beta_i p_{2+i}(t) + \beta_6 p_{13}(t) \tag{7.2.1}$$

$$\left(\frac{d}{dt} + \sum_{1}^{5} \alpha_i + \alpha_7\right) p_2(t) = \sum_{1}^{5} \beta_i p_{7+i}(t) + \alpha_6 p_1(t) \tag{7.2.2}$$

$$\left(\frac{d}{dt} + \beta_i\right) p_{2+i}(t) = \alpha_i p_1(t),\ i = 1, 2, \ldots, 5 \tag{7.2.3}$$

$$\left(\frac{d}{dt} + \beta_i\right) p_{7+i}(t) = \alpha_i p_2(t),\ i = 1, 2, \ldots, 5 \tag{7.2.4}$$

$$\left(\frac{d}{dt} + \beta_6\right) p_{13}(t) = \alpha_7 p_2(t) \tag{7.2.5}$$

with initial conditions $p_j(0) = 1$ if $j = 1$ and zero otherwise.

The differential eqns. (7.2.1)–(7.2.5) with initial conditions have been solved by Gupta et al., [2005] using fourth order Runge-Kutta method over the span of one year (360 days) for different parameters. One table for failure rate of CBM, for reference is given here. The availability of the plant is given by

$$A_v(t) = p_1(t) + p_2(t).$$

Effect of failure rate of CBM on availability

$\alpha_2 \rightarrow$ Days ↓	0.0050	0.0052	0.0054	0.0056	0.0058
30	.985743	.958284	.957825	.957366	.956908
60	.958610	.958151	.957692	.957234	.956777
90	.958528	.958069	.957611	.957152	.956695
120	.958477	.958018	.957560	.957102	.956643
150	.958446	.957986	.957528	.957070	.956612
180	.958426	.957967	.957508	.957050	.956592
210	.958414	.957955	.957496	.957038	.956580
240	.958406	.957947	.957489	.957031	.956573
270	.958401	.957943	.957484	.957026	.956568
300	.958399	.957940	.957481	.957023	.956565
330	.958397	.957938	.957479	.957021	.956563
360	.958396	.957937	.957478	.957020	.956562
MTBF	345.468	345.306	345.147	344.985	344.826

Taking $\alpha_1 = .008$, $\alpha_3 = .0027$, $\alpha_4 = .0009$, $\alpha_5 = .0027$, $\alpha_6 = .0055$, $\alpha_7 = .0111$, $\beta_1 = .41$, $\beta_2 = .40$, $\beta_3 = .67$, $\beta_4 = .33$, $\beta_5 = .67$, $\beta_6 = 6$.

Similarly, tables for other failure and repair rates can be prepared to study the effect of these rates upon availability and mean time before failure (MTBF) which helps the management to control the plant running satisfactorily.

Steady state

When $t \rightarrow \infty$, $\frac{d}{dt} \rightarrow 0$ and $p_i(t) \rightarrow p_i$ in steady state, eqns. (7.2.1)–(7.2.5) reduce to the following equations:

$$\sum_{1}^{6} \alpha_i p_1 = \sum_{1}^{5} \beta_i p_{2+i} + \beta_6 p_{13} \tag{7.2.6}$$

$$\left(\sum_{1}^{5} \alpha_i + \alpha_7\right) p_2 = \sum_{1}^{5} \beta_i p_{7+i} + \alpha_6 p_1 \tag{7.2.7}$$

$$\beta_i p_{2+i} = \alpha_i p_1 \ , \ i = 1, 2, \ldots, 5 \tag{7.2.8}$$

$$\beta_i p_{7+i} = \alpha_i p_2 \ , \ i = 1, 2, \ldots, 5 \tag{7.2.9}$$

$$\beta_6 p_{13} = \alpha_7 p_2 \tag{7.2.10}$$

Solving eqns. (7.2.6)–(7.2.10) recursively and using the normalizing condition $\sum_{1}^{13} p_i = 1$, the availability is given by

$$A = p_1 + p_2 = \left(1 + \frac{\alpha_6}{\alpha_7}\right) p_1$$

where,

$$p_1 = \left[\left(1 + \sum_{1}^{5} \frac{\alpha_i}{\beta_i}\right)\left(1 + \frac{\alpha_6}{\alpha_7}\right) + \frac{\alpha_6}{\beta_6}\right]^{-1}.$$

As in transient case (time dependent case) tables for repair rate and failure rate can be formed to study the behaviour of the various subsystems and also the plant as a whole. A table for repair and failure rates (comparative form) is given here as an example.

Availability table for the effect of failure and repair rates of separator

$\alpha_1 \rightarrow$ $\beta_1 \downarrow$	.006	.007	.008	.009	.010
0.30	.956881	.953839	.950816	.947812	.944827
0.35	.959504	.956881	.954272	.951677	.949097
0.40	.961481	.959176	.956881	.954598	.952325
0.45	.963024	.960969	.958920	.957432	.954850
0.50	.964262	.962407	.960558	.958716	.956881

Taking α_2 = .0055, α_3 = .0027, α_4 = .0009, α_5 = .0027, α_6 = .0055, α_7 = .0111, β_2 = .40, β_3 = .67, β_4 = .33 β_5 = .67, β_6 = 6.

Similar tables can be prepared corresponding to other failure and repair rates.

7.3 THE FERTILIZER PLANT

In this section a fertilizer plant is analysed using reliability theory. In agriculture sector, urea fertilizer is required in maximum so we discuss the manufacturing process of urea fertilizer. Carbon dioxide (CO_2) and ammonia (NH_3) are the prime inputs for the production of urea. These gases react at a particular temperature and pressure to form urea. The gaseous urea is cooled down to yield urea crystals. The production process is quite complex, so we divide the whole process in four parts to facilitate the study, namely—Synthesis, Decomposition, Crystallization and Prilling.

The urea synthesis system consists of five subsystems (i) CO_2 booster compressor (A_1), a centrifugal type compressor to raise the pressure of CO_2 from 0.1 atmosphere to 29.5 atm. Its failure leads to complete failure of the system; (ii) CO_2 high pressure compressor (A_2), a reciprocating type compressor to raise the pressure from 29.5 atm to 250 atm. Its failure also leads to complete failure of the system; (iii) ammonia preheater (A_3) consisting of two units in series. The first unit raises the temperature of the gas upto 53.2°C whereas the second unit heats the gas to 82.3°C. Failure of any one causes the system failure; (iv) the liquid ammonia feed pump (B) having reciprocating type pumps. These raise the pressure of ammonia from 16.5 atm to 250 atm. Two pumps work in parallel while two remain in cold standby. Simultaneous failure of three pumps causes the system failure; (v) the recycle solution feed pump (D), a multistage centrifugal pump, to raise the pressure of ammonium carbonate from 17 atm to 250 atm. It has one unit in cold standby. The system fails when both units fail.

The decomposition system comprises four parts—subsystem E_i (i = 1, 2). E_1 the reboiler and E_2 falling film heater. E_1 and E_2 are in series, so failure of any one causes system failure. The subsystem E_j (j = 3, 4) also has two units in series, high pressure absorber E_3 and low pressure absorber E_4. Failure of any one causes system failure. The gas separator system E_5, single unit system, working in series with E_3, E_4. Its failures leads to system failure. The heat exchanger F having one unit in operation and one in cold standby. Failure of both units causes system failure.

The crystallisation system, an important part of the plant comprised five subsystems—(i) the vacuum generator G_1 having two-stage ejector and a barometric condenser. This generates vacuum of 175 mm Hg (abs). Its failure causes the system failure; (ii) the crystallizer G_2 having two parts—upper part called concentrator and the lower part called crystallizer. Failure of any one causes complete failure of the system; (iii) the centrifuge G_3 consisting of five units in series. Failure of any one causes system failure; (iv) crystallizer pump H having one standby unit. Complete failure occurs on failure of both units; (v) slurry feed pump Q having one standby unit. Failure of both units causes system failure.

The prilling system comprises three subsystems : (i) Subsystem U_j consisting four units U_1 cyclone, U_2 screw conveyor, melter U_3 and strainer U_4. Failure of any one unit causes complete failure of the system; (ii) Subsystem V consisting of eleven distributors operating simultaneously having one standby unit. Failure of more than one unit causes system failure (since it causes less inflow of molten urea causing a blockade before distributor); (iii) the belt conveyor W, consisting of one unit, is employed for carrying the product to trommer. Failure of conveyor leads to huge accumulation of urea blocking the flow path of prilled urea.

The schematic diagram of urea production process, transition diagram for urea synthesis system, decomposition system, crystallisation system and prilling system are given in Figs. 7.7, 7.8, 7.9, 7.10, 7.11, respectively. The governing differential equations associated with respective diagrams are only given here. These equations can be solved using any of the methods given in Chapter 4. Detailed analysis with respect to changes in failure and repair rates is given by Kumar et al., [1990, 1991]. Note that in following subsections, capital letters are used for good states while small letters are for failed states.

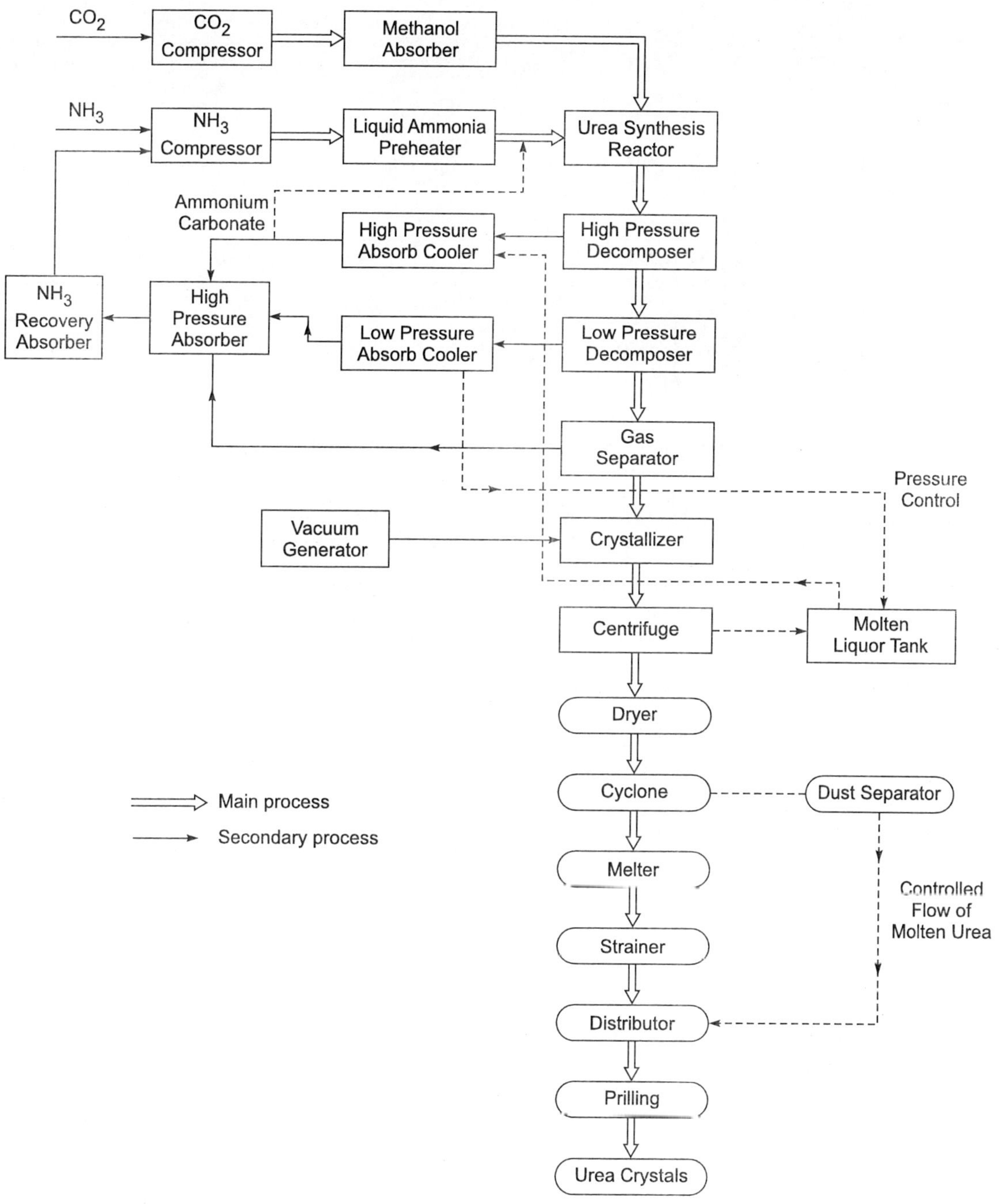

Fig. 7.7

(a) The Urea Synthesis System

Following the notations and other descriptions given above, the governing differential-difference equations for the system are:

$$\left(\frac{d}{dt}+\sum_{1}^{5}\alpha_i\right)p_0(t) = \sum_{1}^{5}\beta_i p_i(t) \tag{7.3.1}$$

$$\left(\frac{d}{dt}+\beta_4+\sum_{1}^{5}\alpha_i\right)p_4(t) = \alpha_4 p_0(t) + \sum_{1}^{5}\beta_i p_{i+5}(t) \tag{7.3.2}$$

$$\left(\frac{d}{dt}+\beta_5+\sum_{1}^{5}\alpha_i\right)p_5(t) = \alpha_5 p_0(t) + \sum_{1}^{3}\beta_i p_{i+15}(t) + \beta_4 p_{10}(t) + \beta_5 p_{19}(t) \tag{7.3.3}$$

$$\left(\frac{d}{dt}+\beta_4+\sum_{1}^{5}\alpha_i\right)p_9(t) = \alpha_4 p_4(t) + \sum_{1}^{5}\beta_i p_{i+10}(t) \tag{7.3.4}$$

$$\left(\frac{d}{dt}+\beta_4+\beta_5+\sum_{1}^{5}\alpha_i\right)p_{10}(t) = \alpha_5 p_4(t) + \alpha_4 p_5(t) + \sum_{1}^{3}\beta_i p_{i+19}(t) + \beta_4 p_{15}(t) + \beta_5 p_{23}(t) \tag{7.3.5}$$

$$\left(\frac{d}{dt}+\beta_5+\beta_4+\sum_{1}^{5}\alpha_i\right)p_{15}(t) = \alpha_5 p_9(t) + \alpha_4 p_{10}(t) + \sum_{1}^{5}\beta_i p_{i+23}(t) \tag{7.3.6}$$

$$\left(\frac{d}{dt}+\beta_i\right)p_j(t) = \alpha_i p_k(t) \tag{7.3.7}$$

with initial conditions $p_0(0) = 1$, otherwise zero. In eqn. (7.3.7), i, j, K are:
for $i = 1, 2, 3, K = 0, j = i$; $K = 4, j = i + 5$; $K = 5, j = i + 15$; $K = 9, j = i + 10$; $K = 10, j = i + 19$; $K = 15, j = i + 23$.
For $i = 4, K = 9, j = 14$; $K = 15, j = 27$.
For $i = 5, K = i, j = 19$; $K = 2i, j = 23$; $K = 3i, j = 28$.

Availability $\quad A_{v_1}(t) = p_0(t) + p_4(t) + p_5(t) + p_9(t) + p_{10}(t) + p_{15}(t)$

For steady state the procedure is same as in previous cases and availability

$$A_v = p_0 + p_4 + p_5 + p_9 + p_{10} + p_{15}.$$

(b) The Decomposition System

The transition diagram using the notations described above is given in Fig. 7.9.
Differential-difference equations associated with Fig. 7.9 are

$$\left(D+\sum_{6}^{11}\alpha_i\right)p_0(t) = \sum_{6}^{11}\beta_i p_{i-5}(t), \tag{7.3.8}$$

$$\left(D+\sum_{6}^{11}\alpha_i+\beta_{11}\right)p_6(t) = \sum_{6}^{11}\beta_i p_{i+1}(t) + \alpha_{11} p_0(t), \tag{7.3.9}$$

$$(D + \beta_i)\, p_j(t) = \alpha_i p_k(t) \tag{7.3.10}$$

with initial conditions $p_0(0) = 1$ otherwise zero.

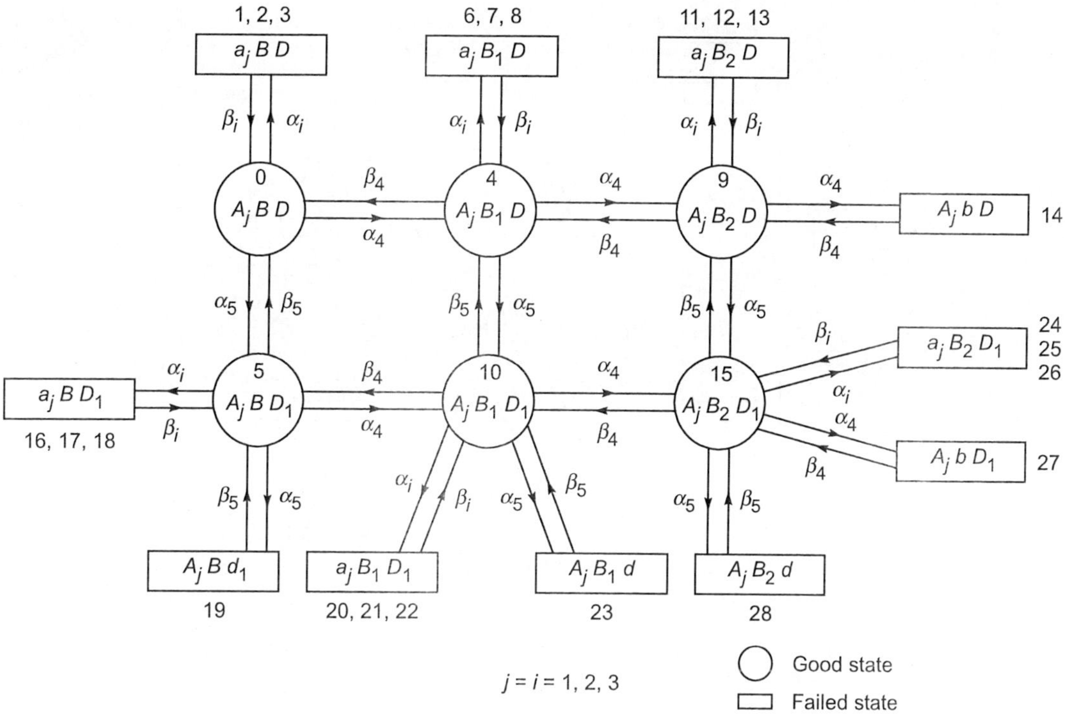

Fig. 7.8

In eqns. (7.3.8)–(7.3.10) $D \equiv d/dt$. In eqn. (7.3.10),
when $K = 0, j = i\ , i = 6, 7, 8, 9, 10\ ;\ K = 6, j = i + 1, i = 6, 7, 8, 9, 10, 11.$
The availability is given by

$$A_{v_2}(t) = p_0(t) + p_6(t).$$

For steady state take $D = 0$, $p_i(t) = p_i$ and use the condition $\sum_{0}^{12} p_i = 1.$

(c) The Crystallization System

The transition diagram and differential-difference equations for the system are:

$$\left(D + \sum_{1}^{5} \lambda_i\right) p_0(t) = \sum_{1}^{5} \mu_i p_i(t) \tag{7.3.11}$$

$$\left(D + \sum_{1}^{5} \lambda_i + \mu_4\right) p_4(t) = \lambda_4 p_0(t) + \sum_{1}^{4} \mu_i p_{i+5}(t) + \mu_5 p_{13}(t), \tag{7.3.12}$$

$$\left(D + \sum_{1}^{5} \lambda_i + \mu_5\right) p_5(t) = \lambda_5 p_0(t) + \sum_{1}^{5} \mu_i p_{i+9}(t) \tag{7.3.13}$$

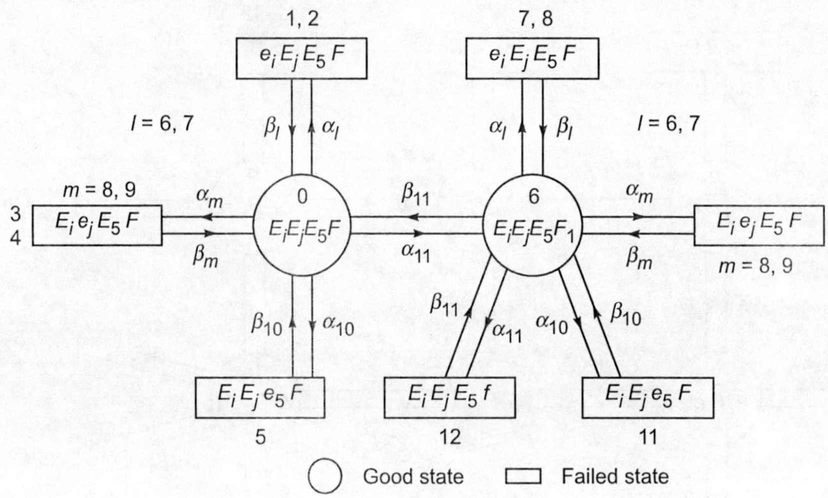

Fig. 7.9

$$\left(D+\sum_{1}^{5} \lambda_i + \mu_4 + \mu_5\right) p_{13}(t) = \sum_{1}^{5} \mu_i p_{i+14}(t) + \lambda_4 p_5(t) + \lambda_5 p_4(t) \tag{7.3.14}$$

$$(D + \mu_i) p_j(t) = \lambda_i p_K(t) \tag{7.3.15}$$

where, $D \equiv d/dt$ and in eqn. (7.3.15) i, j, K are

$i = 1, 2, 3$ $K = 0, j = i$; $K = 4, j = i + 5$; $K = 5, j = i + 9$; $K = 13, j = i + 14$.

$i = 4$, $K = 4, j = i + 5$; $K = 13, j = i + 14$.

$i = 5$, $K = 5, j = i + 9$; $K = 13, j = i + 14$.

The availability is given by

$$A_{v3}(t) = p_0(t) + p_4(t) + p_5(t) + p_{13}(t).$$

In steady state, $D = 0$, $p_i(t) = p_i$ and use the normalizing condition $\sum_{0}^{19} p_i = 1$.

(d) Prilling System

Using the notations given in the description of the system, the transition diagram of the system is The differential-difference equations associated with Fig. 7.11 are as follows:

$$\left(D+\sum_{6}^{11} \lambda_i\right) p_0(t) = \sum_{6}^{9} \mu_i p_i(t) + \sum_{10}^{11} \mu_i p_{i-8}(t) \tag{7.3.16}$$

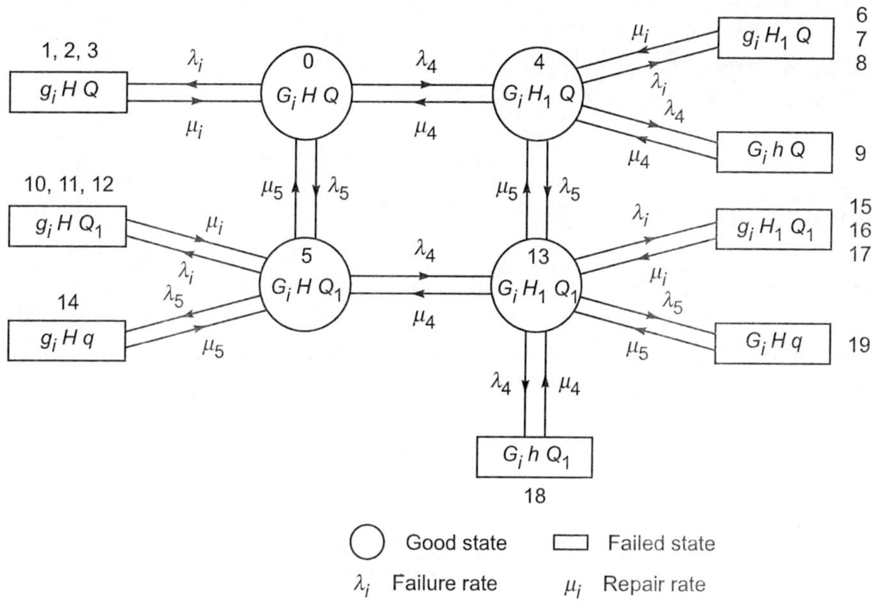

Fig. 7.10

$$\left(D + \sum_{6}^{9} \lambda_i + \mu_{10} + \lambda_{11} + \lambda_{12}\right) p_2(t) = \lambda_{10}\, p_0(t) + \sum_{6}^{9} \mu_i\, p_4(t) + \mu_{10}\, p_5(t) + \mu_{11}\, p_6(t) \quad (7.3.17)$$

$$(D + \mu_i)\, p_j(t) = \lambda_i\, p_K(t) \quad (7.3.18)$$

$$(D + \mu_{10})\, p_5(t) = \lambda_{12}\, p_2(t) \quad (7.3.19)$$

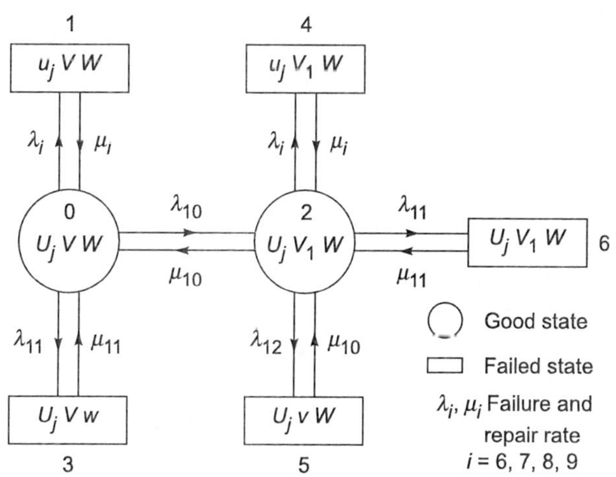

Fig. 7.11

with initial condition $p_0(0) = 1$ otherwise zero.

where, i, j, K in eqn. (7.3.19) are
$i = 6, 7, 8, 9, K = 0, j = 1 \; ; K = 2, j = 4 \; ;$
$i = 11, K = 0, j = i - 8 \; ; K = 2, j = 6.$
$D \equiv d/dt$
The availability $A_{v_4}(t) = p_0(t) + p_2(t)$.

In steady state, $D = 0$, $p_i(t) = p_i$ and use the normalizing condition $\sum_{0}^{6} p_i = 1$.

Since the four systems—urea synthesis system, decomposition system, crystallization system and prilling system are in series, the availability of the whole urea plant is given by

$$A_v(t) = A_{v_1}(t) \times A_{v_2}(t) \times A_{v_3}(t) \times A_{v_4}(t)$$

7.4 AVAILABILITY ANALYSIS OF THE RICE PLANT

7.4.1 Introduction

Rice milling is the most established and the biggest agro preparing industry. Rice is the biggest devoured calorie source among the food grains and it is staple food for 65% of the populace in India. India is the second biggest maker of rice on the planet alongside China, and is additionally one of the main exporters of rice on the planet advertising. Indian Basmati rice has been a most loved among global rice purchasers. Indian rice is winding up much focused and has been distinguished as one of the real products for fare. So reliability engineers, maker and government pioneers must guarantee high profitability, stable task of rice plants. In present part we assessed reliability behaviour of a rice plant which comprises six subsystems to be specific: separator, dryer, grader, husker, polisher and sortex in arrangement design. The numerical model of rice plant has been created utilizing Markov birth-death process. Time subordinate accessibility is determined by utilizing matrix method and steady-state accessibility is determined by recursive technique.

7.4.2 The System

At first dryer dries the crude material (paddy) and places it into the husker. Husker is utilized to strip off husk from paddy. Separator isolates husk and rice. At that point polisher cleans rice, for example, it strips off grain from the rice. Grader is utilized to review the rice, for example, to isolate the long grain and short grain (broken rice). Then with the assistance of sortex dark rice is isolated from reviewed rice. In this way we prepare rice for conclusive conveyance. In like manner the rice plant comprises the accompanying six fundamental subsystems:

(a) **Dryer (A)** comprises one unit. The framework comes up short when this subsystem comes upshort.
(b) **Husker (B)** comprises two units, one working and other in cold standby. Cold standby unit is of lower limit. The framework can chip away at backup unit in diminished limit. Complete disappointment happens when the two units fall flat.
(c) **Separator (C)** comprises one unit. It is exposed to real disappointment as it were.

(d) Polisher (D) comprises two units, one working and the other in cold standby. Cold standby unit is of lower limit. The framework can deal with reserve unit in diminished limit. Complete disappointment happens when the two units come up short.

(e) Grader (E) comprises two units, one working and the other in cold standby. Cold standby unit is of lower limit. The framework can deal with backup unit in diminished limit. Complete disappointment happens when the two units fall flat.

(f) Sortex (F) comprises one unit. It is exposed to real disappointment as it were.

7.4.3 Assumptions and Notations

1. Repair rates and disappointment rates are negative exponential and independent of one another.
2. Not more than one disappointment happens at once.
3. A fixed unit is execution shrewd, all around great.
4. The subsystems B, D and E fail through diminished states.

Notations

A, B, C, D, E, F: Denote the great conditions of subsystem A, B, C, D, E, F individually.

A, b, c, d, e, f: Denote the fizzled conditions.

λ_i / μ_i: Indicate the particular mean failure / repair rates of dryer, husker, separator, polisher, grader, sortex, $I = 1, 2, 3, 4, 5, 6, 7, 8, 9$. Where $I = 3, 6, \& 8$ represent disappointment rates of diminished conditions of B, D, and E individually.

7.4.4 Transition Diagram

The transition diagram of rice plant is given in Fig. 7.12. (p. 173)

By utilizing Markov process, differential equation related to system are developed and are inferred as follows.

$$P_1'(t) = a_1p_1(t) + \mu_9p_{11}(t) + \mu_1p_9(t) + \mu_5p_3(t) + \mu_2p_2(t) + \mu_4p_{10}(t) + \mu_7p_4(t) \tag{7.4.1}$$

$$P_2'(t) = a_2p_2(t) + \mu_9p_{15}(t) + \mu_1p_{12}(t) + \mu_4p_{14}(t) + \mu_3p_{13}(t) + \mu_5p_5(t) + \mu_2p_1(t) \tag{7.4.2}$$

$$P_3'(t) = a_3p_3(t) + \mu_9p_{19}(t) + \mu_4p_{17}(t) + \mu_1p_6(t) + \mu_2p_5(t) + \mu_7p_7(t) + \mu_6p_{18}(t) + \lambda_5p_1(t) \tag{7.4.3}$$

$$P_4'(t) = a_4p_4(t) + \mu_9p_{23}(t) + \mu_1p_{20}(t) + \mu_5p_7(t) + \mu_2p_6(t) + \mu_4p_{21}(t) + \mu_8p_{22}(t) + \lambda_7p_1(t) \tag{7.4.4}$$

$$P_5'(t) = a_5p_5(t) + \mu_7p_8(t) + \mu_6p_{27}(t) + \mu_1p_{24}(t) + \mu_4p_{26}(t) + \mu_9p_{28}(t) + \lambda_2p_3(t) + \mu_3p_{25}(t) + \lambda_5p_2(t) \tag{7.4.5}$$

$$P_6'(t) = a_6p_6(t) + \mu_9p_{33}(t) + \mu_5p_8(t) + \mu_1p_{29}(t) + \mu_3p_{30}(t) + \mu_8p_{32}(t) + \lambda_2p_4(t) + \mu_4p_{31}(t) + \lambda_7p_2(t) \tag{7.4.6}$$

$$P_7'(t) = a_7p_7(t) + \mu_8p_{37}(t) + \mu_4p_{35}(t) + \mu_1p_{34}(t) + \mu_2p_8(t) + \mu_6p_{36}(t) + \lambda_5p_4(t) + \mu_9p_{38}(t) + \lambda_7p_3(t) \tag{7.4.7}$$

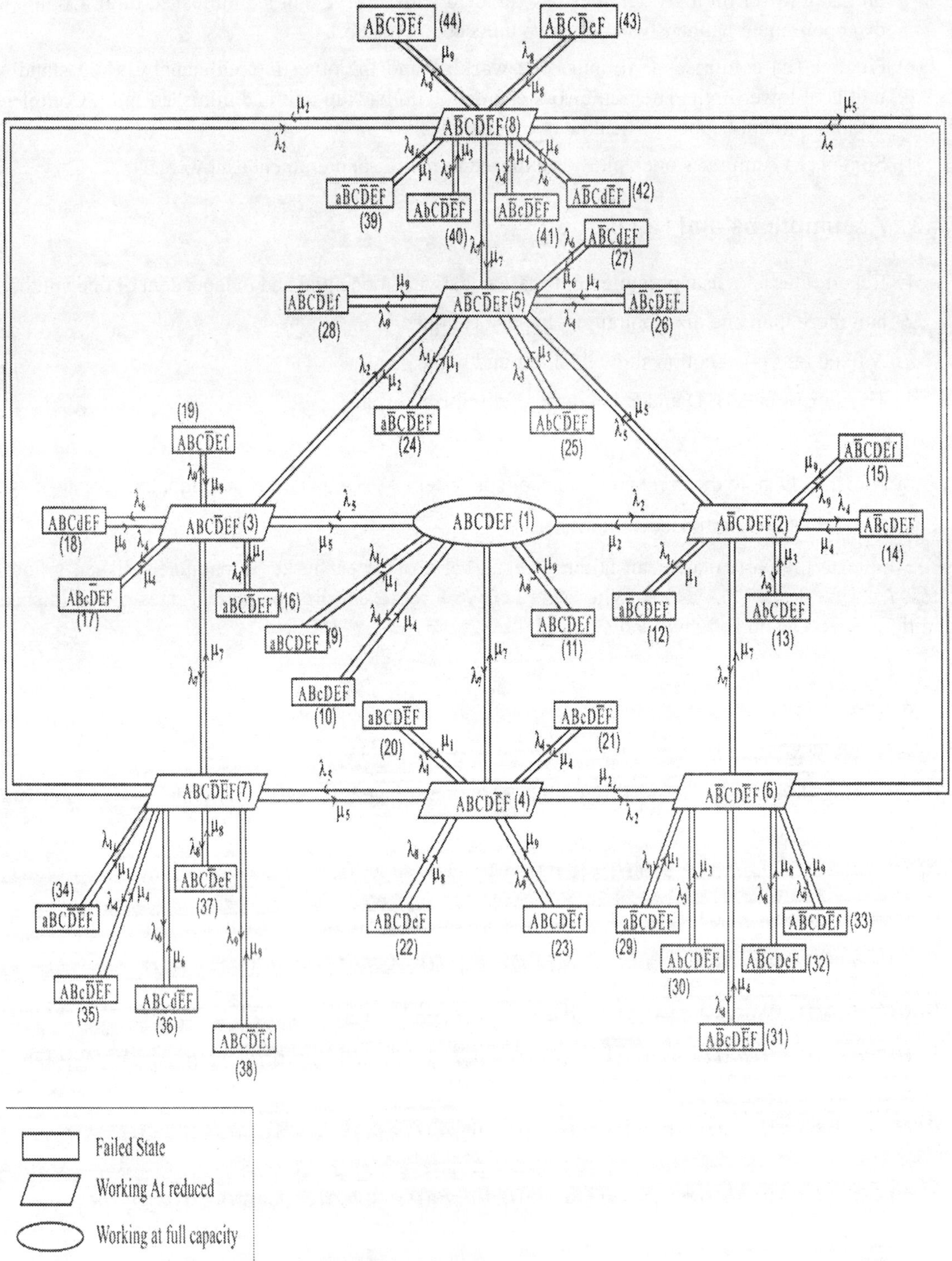

Fig. 7.12 Transition diagram.

$$P_8'(t) = a_8 p_8(t) + \mu_8 p_{43}(t) + \mu_4 p_{41}(t) + \mu_1 p_{39}(t) + \mu_6 p_{42}(t) + \mu_3 p_{40}(t) + \lambda_5 p_6(t) + \mu_9 p_{44}(t) + \lambda_2 p_7(t) + \lambda_7 p_5(t) \quad (7.4.8)$$

$$P_i'(t) + \mu_j p_i(t) = \lambda_j p_1(t)$$
$$i = 9, 11, 10 : j = 1, 9, 4 \quad (7.4.9)$$

$$P_i'(t) + \mu_j p_i(t) = \lambda_j p_2(t)$$
$$i = 13, 12, 15, 14 : j = 1, 4, 3, 9 \quad (7.4.10)$$

$$P_i'(t) + \mu_j p_i(t) = \lambda_j p_3(t)$$
$$i = 19, 17, 16, 18 : j = 1, 9, 6, 4 \quad (7.4.11)$$

$$P_i'(t) + \mu_j p_i(t) = \lambda_j p_4(t)$$
$$i = 23, 20, 22, 21 : j = 1, 8, 4, 9 \quad (7.4.12)$$

$$P_i'(t) + \mu_j p_i(t) = \lambda_j p_5(t)$$
$$i = 27, 24, 26, 25, 28 : j = 1, 6, 3, 9, 4 \quad (7.4.13)$$

$$P_i'(t) + \mu_j p_i(t) = \lambda_j p_6(t)$$
$$i = 31, 33, 30, 29, 32 : j = 1, 9, 3, 8, 4 \quad (7.4.14)$$

$$P_i'(t) + \mu_j p_i(t) = \lambda_j p_7(t)$$
$$i = 36, 34, 38, 35, 37 : j = 4, 8, 1, 6, 9 \quad (7.4.15)$$

$$P_i'(t) + \mu_j p_i(t) = \lambda_j p_8(t)$$
$$i = 40, 42, 44, 39, 41, 43 : j = 1, 6, 9, 3, 8, 4 \quad (7.4.16)$$

With beginning condition $p_1(0) = 1$, generally zero, where

$$a_1 = -(\lambda_2 + \lambda_7 + \lambda_4 + \lambda_1 + \lambda_5 + \lambda_9),\ a_2 = -(\lambda_1 + \lambda_4 + \lambda_7 + \lambda_9 + \lambda_3 + \lambda_5 + \mu_2)$$
$$a_3 = -(\lambda_1 + \lambda_4 + \lambda_7 + \lambda_2 + \lambda_6 + \lambda_9 + \mu_5),\ a_4 = -(\lambda_1 + \lambda_4 + \lambda_8 + \lambda_7 + \lambda_5 + \lambda_9 + \mu_7)$$
$$a_5 = -(\lambda_1 + \lambda_4 + \lambda_7 + \lambda_3 + \lambda_6 + \lambda_9 + \mu_5),\ a_6 = -(\lambda_1 + \lambda_4 + \lambda_8 + \lambda_3 + \lambda_5 + \lambda_9 + \mu_7)$$
$$a_7 = -(\lambda_1 + \lambda_4 + \lambda_8 + \lambda_2 + \lambda_6 + \lambda_9 + \mu_7 + \mu_5)$$
$$a_8 = -(\lambda_1 + \lambda_4 + \lambda_8 + \lambda_3 + \lambda_6 + \lambda_9 + \mu_5 + \mu_2 + \mu_7)$$

7.4.5 Time Dependent Availability Analysis by Matrix Method

Let $p(k, t)$ mean transition probability of occasion that speaks to at time 't' framework is in kth state. Let the quantity of all the conceivable progress conditions of the complex framework are 'n' (number of all out states in the progress outline), then the arrangement of differential contrast conditions related with the plant might be composed as:

$$(\theta I_n - A)\ \overline{P}\ (k, t) = \overline{0}$$

where

$$\theta \equiv d / dt$$

$\overline{0}$ is the null matrix. A is matrix of coefficients of *pi* (t's) in differential contrast equations [7.4.1-7.4.16] for (i = 1, 2… ..n)

$$\overline{P}\ (k, t) = [p_1(t), p_2(t), \ldots\ldots\ldots\ldots\ldots\ldots, p_n(t)]^T$$

What's more, in is the identity matrix of request $n \times 1$.

Give C a chance to be the grid with the end goal that $C^{-1}AC = D$ where $D = (d_1, d_2, \ldots .., d_n)$ be the diagonal matrix of eigen estimations of the network A.

We may compose:

$$C^{-1}[((\theta 1_n - A)\ \overline{P}\ (k, t)] = \overline{0}$$

$$(\theta 1_n - D)\ \mathrm{G}\ (k, t) = \overline{0}$$

where

$$G\ (k, t) = C^{-1}\ \overline{P}\ (k, t)$$

This condition is a direct matrix differential condition in $G(k, t)$ having arrangement

$G\ (k, t)\ e^{-Dt} = K$, for some consistent K, with beginning conditions $p\ (1, 0) = 1$

We get

$$K = C^{-1}\ \overline{P}\ (k, 0),$$

where

$$\overline{P}\ (k, 0) = ([1\ 0\ 0\ 0\ 0\ \ldots\ldots\ldots\ 0]_{1\times n})^T$$

$$G\ (k, t) = e^{Dt}\ C^{-1}\ \overline{P}\ (k, 0)$$

gives,

$$\overline{P}\ (k, t) = C(1 + Dt + D^2t^2 / 2! + \ldots\ldots\ldots\ldots)\ C^{-1}\ \overline{P}\ (k, 0)$$

$$= \overline{P}\ (k, 0) + A\ \overline{P}\ (k, 0)\, t + A^2\ \overline{P}\ (k, 0)\, t^2 / 2! + \ldots\ldots\ldots$$

$$= \overline{P}\ (k, 0) + L_1 t + L_2 t^2 / 2! + \ldots\ldots\ldots\ L_N t^N / N! + \ldots\ldots\ldots$$

where

$$L_N = A^N\ \overline{P}\ (k, 0)$$

Furthermore,

$\overline{P}\ (k, 0)$ is the segment network of request $n \times 1$

The underlying conditions clarify that $\overline{P}\ (k, 0)$ is the column matrix $(1\ 0\ 0\ 0\ \ldots\ldots 0)^T$, $A\ \overline{P}\ (k, 0)$ is only the first column of matrix A. Let us signify this segment grid by

$$A_1 = (a_{11}, a_{12}, \ldots\ldots\ldots\ a_{1,n})^T.$$

$A^2\ \overline{P}\ (k, 0) = AA\ \overline{P}\ (k, 0) = AA_1$ is again a segment network, let us signify it by

$$A_2 = (b_{11}, b_{12}, \ldots\ldots\ldots, b_{1,n})^T.$$

Let $A^{r-1}\ \overline{P}\ (k, 0) = A_{r-1} = (p_{11}, p_{12}, \ldots\ldots\ldots, p_{1,n})^T$

The examination of the condition uncovers that

$$A^r\ \overline{P}\ (k, 0) = A\ A_{r-1} = (q_{11}, q_{12}, \ldots\ldots\ldots, q_{1,n})^T, \text{ say}$$

Change state probabilities of various stages are:

$$p\ (1, t) = 1 + a_{11} t + b_{11} t^2 / 2! + \ldots\ldots\ldots$$

$$\ldots\ldots\ldots\ldots\ldots\ldots$$

$$p\ (n, t) = a_{n1} t + b_{n1} t^2 / 2! + \ldots\ldots\ldots$$

If $p(1, t)$, ……………………… $p\ (i, t)$ are only working states of framework, so

$$AV\ (t) = p\ (1, t) + p\ (2, t) + \ldots\ldots\ldots + p\ (i, t)$$

$$= 1 + (a_{11} + a_{21} + \ldots + a_{i1})\, t + (b_{11} + b_{21} \ldots\ldots + b_{i1})\, t^2/2! + \ldots\ldots\ldots$$

Time Dependent Availability Analysis

In this specific issue $n = 44$, $m = 8$ and estimations of disappointment and repair rates of subsystems are:
$\lambda_1 = 0.001$; $\lambda_2 = 0.003$; $\lambda_3 = 0.003$; $\lambda_4 = 0.0025$; $\lambda_5 = 0.004$; $\lambda_6 = 0.004$; $\lambda_7 = 0.005$; $\lambda_8 = 0.005$; $\lambda_9 = 0.002$;
$\mu_1 = 0.01$; $\mu_2 = 0.03$; $\mu_3 = 0.03$; $\mu_4 = 0.02$; $\mu_5 = 0.04$; $\mu_6 = 0.04$; $\mu_7 = 0.05$; $\mu_8 = 0.05$; $\mu_9 = 0.02$;

Matrix A = Coefficient matrix of differential conditions 7.4.1 to 7.4.16

$$\begin{bmatrix} A_1 & A_2 \\ A_3 & A_4 \end{bmatrix}$$

Where A_1 =

$$\left[\begin{array}{ccccccccccccccccccccccc}
-a_1 & \mu_2 & \mu_5 & \mu_7 & 0 & 0 & 0 & 0 & \mu_1 & \mu_4 & \mu_9 & 0 & 0 & 0 & 0 & 0 & 0 & 0 & 0 & 0 & 0 & 0 & - \\
\lambda_2 & -a_2 & 0 & 0 & \mu_5 & \mu_7 & 0 & 0 & 0 & 0 & 0 & \mu_1 & \mu_3 & \mu_4 & \mu_9 & 0 & 0 & 0 & 0 & 0 & 0 & 0 & - \\
\lambda_5 & 0 & -a_3 & 0 & \mu_2 & 0 & \mu_7 & 0 & 0 & 0 & 0 & 0 & 0 & 0 & 0 & \mu_1 & \mu_4 & \mu_6 & \mu_9 & 0 & 0 & 0 & - \\
\lambda_7 & 0 & 0 & -a_4 & 0 & \mu_2 & \mu_5 & 0 & 0 & 0 & 0 & 0 & 0 & 0 & 0 & 0 & 0 & 0 & 0 & \mu_1 & \mu_4 & \mu_8 & - \\
0 & \lambda_5 & \lambda_2 & 0 & -a_5 & 0 & 0 & \mu_7 & 0 & 0 & 0 & 0 & 0 & 0 & 0 & 0 & 0 & 0 & 0 & 0 & 0 & 0 & - \\
0 & \lambda_7 & 0 & \lambda_2 & 0 & -a_6 & 0 & \mu_5 & 0 & 0 & 0 & 0 & 0 & 0 & 0 & 0 & 0 & 0 & 0 & 0 & 0 & 0 & - \\
0 & 0 & \lambda_7 & \lambda_5 & 0 & 0 & -a_7 & \mu_2 & 0 & 0 & 0 & 0 & 0 & 0 & 0 & 0 & 0 & 0 & 0 & 0 & 0 & 0 & - \\
0 & 0 & 0 & 0 & \lambda_7 & \lambda_5 & \lambda_2 & -a_8 & 0 & 0 & 0 & 0 & 0 & 0 & 0 & 0 & 0 & 0 & 0 & 0 & 0 & 0 & - \\
\lambda_1 & 0 & 0 & 0 & 0 & 0 & 0 & 0 & \mu_1 & 0 & 0 & 0 & 0 & 0 & 0 & 0 & 0 & 0 & 0 & 0 & 0 & 0 & - \\
\lambda_4 & 0 & 0 & 0 & 0 & 0 & 0 & 0 & 0 & \mu_4 & 0 & 0 & 0 & 0 & 0 & 0 & 0 & 0 & 0 & 0 & 0 & 0 & - \\
\lambda_9 & 0 & 0 & 0 & 0 & 0 & 0 & 0 & 0 & 0 & \mu_9 & 0 & 0 & 0 & 0 & 0 & 0 & 0 & 0 & 0 & 0 & 0 & - \\
0 & \lambda_1 & 0 & 0 & 0 & 0 & 0 & 0 & 0 & 0 & 0 & \mu_1 & 0 & 0 & 0 & 0 & 0 & 0 & 0 & 0 & 0 & 0 & - \\
0 & \lambda_3 & 0 & 0 & 0 & 0 & 0 & 0 & 0 & 0 & 0 & 0 & \mu_3 & 0 & 0 & 0 & 0 & 0 & 0 & 0 & 0 & 0 & - \\
0 & \lambda_4 & 0 & 0 & 0 & 0 & 0 & 0 & 0 & 0 & 0 & 0 & 0 & \mu_4 & 0 & 0 & 0 & 0 & 0 & 0 & 0 & 0 & - \\
0 & \lambda_9 & 0 & 0 & 0 & 0 & 0 & 0 & 0 & 0 & 0 & 0 & 0 & 0 & \mu_9 & 0 & 0 & 0 & 0 & 0 & 0 & 0 & - \\
0 & 0 & \lambda_1 & 0 & 0 & 0 & 0 & 0 & 0 & 0 & 0 & 0 & 0 & 0 & 0 & \mu_1 & 0 & 0 & 0 & 0 & 0 & 0 & - \\
0 & 0 & \lambda_4 & 0 & 0 & 0 & 0 & 0 & 0 & 0 & 0 & 0 & 0 & 0 & 0 & 0 & \mu_4 & 0 & 0 & 0 & 0 & 0 & - \\
0 & 0 & \lambda_6 & 0 & 0 & 0 & 0 & 0 & 0 & 0 & 0 & 0 & 0 & 0 & 0 & 0 & 0 & \mu_6 & 0 & 0 & 0 & 0 & - \\
0 & 0 & \lambda_9 & 0 & 0 & 0 & 0 & 0 & 0 & 0 & 0 & 0 & 0 & 0 & 0 & 0 & 0 & 0 & \mu_9 & 0 & 0 & 0 & - \\
0 & 0 & 0 & \lambda_1 & 0 & 0 & 0 & 0 & 0 & 0 & 0 & 0 & 0 & 0 & 0 & 0 & 0 & 0 & 0 & \mu_1 & 0 & 0 & - \\
0 & 0 & 0 & \lambda_4 & 0 & 0 & 0 & 0 & 0 & 0 & 0 & 0 & 0 & 0 & 0 & 0 & 0 & 0 & 0 & 0 & \mu_4 & 0 & - \\
0 & 0 & 0 & \lambda_8 & 0 & 0 & 0 & 0 & 0 & 0 & 0 & 0 & 0 & 0 & 0 & 0 & 0 & 0 & 0 & 0 & 0 & \mu_8 & - \\
- & -
\end{array}\right.$$

$A_2 =$

$$
\begin{bmatrix}
- & 0 & 0 & 0 & 0 & 0 & 0 & 0 & 0 & 0 & 0 & 0 & 0 & 0 & 0 & 0 & 0 & 0 & 0 & 0 & 0 & 0 & 0 \\
- & 0 & 0 & 0 & 0 & 0 & 0 & 0 & 0 & 0 & 0 & 0 & 0 & 0 & 0 & 0 & 0 & 0 & 0 & 0 & 0 & 0 & 0 \\
- & 0 & 0 & 0 & 0 & 0 & 0 & 0 & 0 & 0 & 0 & 0 & 0 & 0 & 0 & 0 & 0 & 0 & 0 & 0 & 0 & 0 & 0 \\
- & \mu_9 & 0 & 0 & 0 & 0 & 0 & 0 & 0 & 0 & 0 & 0 & 0 & 0 & 0 & 0 & 0 & 0 & 0 & 0 & 0 & 0 & 0 \\
- & 0 & \mu_1 & \mu_3 & \mu_4 & \mu_6 & \mu_9 & 0 & 0 & 0 & 0 & 0 & 0 & 0 & 0 & 0 & 0 & 0 & 0 & 0 & 0 & 0 & 0 \\
- & 0 & 0 & 0 & 0 & 0 & 0 & \mu_1 & \mu_3 & \mu_4 & \mu_8 & \mu_9 & 0 & 0 & 0 & 0 & 0 & 0 & 0 & 0 & 0 & 0 & 0 \\
- & 0 & 0 & 0 & 0 & 0 & 0 & 0 & 0 & 0 & 0 & 0 & \mu_1 & \mu_4 & \mu_6 & \mu_8 & \mu_9 & 0 & 0 & 0 & 0 & 0 & 0 \\
- & 0 & 0 & 0 & 0 & 0 & 0 & 0 & 0 & 0 & 0 & 0 & 0 & 0 & 0 & 0 & 0 & 0 & \mu_3 & \mu_4 & \mu_6 & \mu_8 & \mu_9 \\
- & 0 & 0 & 0 & 0 & 0 & 0 & 0 & 0 & 0 & 0 & 0 & 0 & 0 & 0 & 0 & 0 & 0 & 0 & 0 & 0 & 0 & 0 \\
- & 0 & 0 & 0 & 0 & 0 & 0 & 0 & 0 & 0 & 0 & 0 & 0 & 0 & 0 & 0 & 0 & 0 & 0 & 0 & 0 & 0 & 0 \\
- & 0 & 0 & 0 & 0 & 0 & 0 & 0 & 0 & 0 & 0 & 0 & 0 & 0 & 0 & 0 & 0 & 0 & 0 & 0 & 0 & 0 & 0 \\
- & 0 & 0 & 0 & 0 & 0 & 0 & 0 & 0 & 0 & 0 & 0 & 0 & 0 & 0 & 0 & 0 & 0 & 0 & 0 & 0 & 0 & 0 \\
- & 0 & 0 & 0 & 0 & 0 & 0 & 0 & 0 & 0 & 0 & 0 & 0 & 0 & 0 & 0 & 0 & 0 & 0 & 0 & 0 & 0 & 0 \\
- & 0 & 0 & 0 & 0 & 0 & 0 & 0 & 0 & 0 & 0 & 0 & 0 & 0 & 0 & 0 & 0 & 0 & 0 & 0 & 0 & 0 & 0 \\
- & 0 & 0 & 0 & 0 & 0 & 0 & 0 & 0 & 0 & 0 & 0 & 0 & 0 & 0 & 0 & 0 & 0 & 0 & 0 & 0 & 0 & 0 \\
- & 0 & 0 & 0 & 0 & 0 & 0 & 0 & 0 & 0 & 0 & 0 & 0 & 0 & 0 & 0 & 0 & 0 & 0 & 0 & 0 & 0 & 0 \\
- & 0 & 0 & 0 & 0 & 0 & 0 & 0 & 0 & 0 & 0 & 0 & 0 & 0 & 0 & 0 & 0 & 0 & 0 & 0 & 0 & 0 & 0 \\
- & 0 & 0 & 0 & 0 & 0 & 0 & 0 & 0 & 0 & 0 & 0 & 0 & 0 & 0 & 0 & 0 & 0 & 0 & 0 & 0 & 0 & 0 \\
- & 0 & 0 & 0 & 0 & 0 & 0 & 0 & 0 & 0 & 0 & 0 & 0 & 0 & 0 & 0 & 0 & 0 & 0 & 0 & 0 & 0 & 0 \\
- & 0 & 0 & 0 & 0 & 0 & 0 & 0 & 0 & 0 & 0 & 0 & 0 & 0 & 0 & 0 & 0 & 0 & 0 & 0 & 0 & 0 & 0 \\
- & 0 & 0 & 0 & 0 & 0 & 0 & 0 & 0 & 0 & 0 & 0 & 0 & 0 & 0 & 0 & 0 & 0 & 0 & 0 & 0 & 0 & 0 \\
- & 0 & 0 & 0 & 0 & 0 & 0 & 0 & 0 & 0 & 0 & 0 & 0 & 0 & 0 & 0 & 0 & 0 & 0 & 0 & 0 & 0 & 0 \\
- & -
\end{bmatrix}
$$

$$A_3 = \begin{bmatrix}
- & - \\
0 & 0 & 0 & \lambda_9 & 0 & 0 & 0 & 0 & 0 & 0 & 0 & 0 & 0 & 0 & 0 & 0 & 0 & 0 & 0 & 0 & 0 & 0 & 0 & 0 & 0 & - \\
0 & 0 & 0 & 0 & \lambda_1 & 0 & 0 & 0 & 0 & 0 & 0 & 0 & 0 & 0 & 0 & 0 & 0 & 0 & 0 & 0 & 0 & 0 & 0 & 0 & 0 & - \\
0 & 0 & 0 & 0 & \lambda_3 & 0 & 0 & 0 & 0 & 0 & 0 & 0 & 0 & 0 & 0 & 0 & 0 & 0 & 0 & 0 & 0 & 0 & 0 & 0 & 0 & - \\
0 & 0 & 0 & 0 & \lambda_4 & 0 & 0 & 0 & 0 & 0 & 0 & 0 & 0 & 0 & 0 & 0 & 0 & 0 & 0 & 0 & 0 & 0 & 0 & 0 & 0 & - \\
0 & 0 & 0 & 0 & \lambda_6 & 0 & 0 & 0 & 0 & 0 & 0 & 0 & 0 & 0 & 0 & 0 & 0 & 0 & 0 & 0 & 0 & 0 & 0 & 0 & 0 & - \\
0 & 0 & 0 & 0 & \lambda_9 & 0 & 0 & 0 & 0 & 0 & 0 & 0 & 0 & 0 & 0 & 0 & 0 & 0 & 0 & 0 & 0 & 0 & 0 & 0 & 0 & - \\
0 & 0 & 0 & 0 & 0 & \lambda_1 & 0 & 0 & 0 & 0 & 0 & 0 & 0 & 0 & 0 & 0 & 0 & 0 & 0 & 0 & 0 & 0 & 0 & 0 & 0 & - \\
0 & 0 & 0 & 0 & 0 & \lambda_3 & 0 & 0 & 0 & 0 & 0 & 0 & 0 & 0 & 0 & 0 & 0 & 0 & 0 & 0 & 0 & 0 & 0 & 0 & 0 & - \\
0 & 0 & 0 & 0 & 0 & \lambda_4 & 0 & 0 & 0 & 0 & 0 & 0 & 0 & 0 & 0 & 0 & 0 & 0 & 0 & 0 & 0 & 0 & 0 & 0 & 0 & - \\
0 & 0 & 0 & 0 & 0 & \lambda_8 & 0 & 0 & 0 & 0 & 0 & 0 & 0 & 0 & 0 & 0 & 0 & 0 & 0 & 0 & 0 & 0 & 0 & 0 & 0 & - \\
0 & 0 & 0 & 0 & 0 & \lambda_9 & 0 & 0 & 0 & 0 & 0 & 0 & 0 & 0 & 0 & 0 & 0 & 0 & 0 & 0 & 0 & 0 & 0 & 0 & 0 & - \\
0 & 0 & 0 & 0 & 0 & 0 & \lambda_1 & 0 & 0 & 0 & 0 & 0 & 0 & 0 & 0 & 0 & 0 & 0 & 0 & 0 & 0 & 0 & 0 & 0 & 0 & - \\
0 & 0 & 0 & 0 & 0 & 0 & \lambda_4 & 0 & 0 & 0 & 0 & 0 & 0 & 0 & 0 & 0 & 0 & 0 & 0 & 0 & 0 & 0 & 0 & 0 & 0 & - \\
0 & 0 & 0 & 0 & 0 & 0 & \lambda_6 & 0 & 0 & 0 & 0 & 0 & 0 & 0 & 0 & 0 & 0 & 0 & 0 & 0 & 0 & 0 & 0 & 0 & 0 & - \\
0 & 0 & 0 & 0 & 0 & 0 & \lambda_8 & 0 & 0 & 0 & 0 & 0 & 0 & 0 & 0 & 0 & 0 & 0 & 0 & 0 & 0 & 0 & 0 & 0 & 0 & - \\
0 & 0 & 0 & 0 & 0 & 0 & \lambda_9 & 0 & 0 & 0 & 0 & 0 & 0 & 0 & 0 & 0 & 0 & 0 & 0 & 0 & 0 & 0 & 0 & 0 & 0 & - \\
0 & 0 & 0 & 0 & 0 & 0 & 0 & \lambda_1 & 0 & 0 & 0 & 0 & 0 & 0 & 0 & 0 & 0 & 0 & 0 & 0 & 0 & 0 & 0 & 0 & 0 & - \\
0 & 0 & 0 & 0 & 0 & 0 & 0 & \lambda_3 & 0 & 0 & 0 & 0 & 0 & 0 & 0 & 0 & 0 & 0 & 0 & 0 & 0 & 0 & 0 & 0 & 0 & - \\
0 & 0 & 0 & 0 & 0 & 0 & 0 & \lambda_4 & 0 & 0 & 0 & 0 & 0 & 0 & 0 & 0 & 0 & 0 & 0 & 0 & 0 & 0 & 0 & 0 & 0 & - \\
0 & 0 & 0 & 0 & 0 & 0 & 0 & \lambda_6 & 0 & 0 & 0 & 0 & 0 & 0 & 0 & 0 & 0 & 0 & 0 & 0 & 0 & 0 & 0 & 0 & 0 & - \\
0 & 0 & 0 & 0 & 0 & 0 & 0 & \lambda_8 & 0 & 0 & 0 & 0 & 0 & 0 & 0 & 0 & 0 & 0 & 0 & 0 & 0 & 0 & 0 & 0 & 0 & - \\
0 & 0 & 0 & 0 & 0 & 0 & 0 & \lambda_9 & 0 & 0 & 0 & 0 & 0 & 0 & 0 & 0 & 0 & 0 & 0 & 0 & 0 & 0 & 0 & 0 & 0 & -
\end{bmatrix}$$

$A_4 =$

$$
\begin{bmatrix}
- & - \\
- & \mu_9 & 0 & 0 & 0 & 0 & 0 & 0 & 0 & 0 & 0 & 0 & 0 & 0 & 0 & 0 & 0 & 0 & 0 & 0 & 0 & 0 & 0 \\
- & 0 & \mu_1 & 0 & 0 & 0 & 0 & 0 & 0 & 0 & 0 & 0 & 0 & 0 & 0 & 0 & 0 & 0 & 0 & 0 & 0 & 0 & 0 \\
- & 0 & 0 & \mu_3 & 0 & 0 & 0 & 0 & 0 & 0 & 0 & 0 & 0 & 0 & 0 & 0 & 0 & 0 & 0 & 0 & 0 & 0 & 0 \\
- & 0 & 0 & 0 & \mu_4 & 0 & 0 & 0 & 0 & 0 & 0 & 0 & 0 & 0 & 0 & 0 & 0 & 0 & 0 & 0 & 0 & 0 & 0 \\
- & 0 & 0 & 0 & 0 & \mu_6 & 0 & 0 & 0 & 0 & 0 & 0 & 0 & 0 & 0 & 0 & 0 & 0 & 0 & 0 & 0 & 0 & 0 \\
- & 0 & 0 & 0 & 0 & 0 & \mu_9 & 0 & 0 & 0 & 0 & 0 & 0 & 0 & 0 & 0 & 0 & 0 & 0 & 0 & 0 & 0 & 0 \\
- & 0 & 0 & 0 & 0 & 0 & 0 & \mu_1 & 0 & 0 & 0 & 0 & 0 & 0 & 0 & 0 & 0 & 0 & 0 & 0 & 0 & 0 & 0 \\
- & 0 & 0 & 0 & 0 & 0 & 0 & 0 & \mu_3 & 0 & 0 & 0 & 0 & 0 & 0 & 0 & 0 & 0 & 0 & 0 & 0 & 0 & 0 \\
- & 0 & 0 & 0 & 0 & 0 & 0 & 0 & 0 & \mu_4 & 0 & 0 & 0 & 0 & 0 & 0 & 0 & 0 & 0 & 0 & 0 & 0 & 0 \\
- & 0 & 0 & 0 & 0 & 0 & 0 & 0 & 0 & 0 & \mu_8 & 0 & 0 & 0 & 0 & 0 & 0 & 0 & 0 & 0 & 0 & 0 & 0 \\
- & 0 & 0 & 0 & 0 & 0 & 0 & 0 & 0 & 0 & 0 & \mu_9 & 0 & 0 & 0 & 0 & 0 & 0 & 0 & 0 & 0 & 0 & 0 \\
- & 0 & 0 & 0 & 0 & 0 & 0 & 0 & 0 & 0 & 0 & 0 & \mu_1 & 0 & 0 & 0 & 0 & 0 & 0 & 0 & 0 & 0 & 0 \\
- & 0 & 0 & 0 & 0 & 0 & 0 & 0 & 0 & 0 & 0 & 0 & 0 & \mu_4 & 0 & 0 & 0 & 0 & 0 & 0 & 0 & 0 & 0 \\
- & 0 & 0 & 0 & 0 & 0 & 0 & 0 & 0 & 0 & 0 & 0 & 0 & 0 & \mu_6 & 0 & 0 & 0 & 0 & 0 & 0 & 0 & 0 \\
- & 0 & 0 & 0 & 0 & 0 & 0 & 0 & 0 & 0 & 0 & 0 & 0 & 0 & 0 & \mu_8 & 0 & 0 & 0 & 0 & 0 & 0 & 0 \\
- & 0 & 0 & 0 & 0 & 0 & 0 & 0 & 0 & 0 & 0 & 0 & 0 & 0 & 0 & 0 & \mu_9 & 0 & 0 & 0 & 0 & 0 & 0 \\
- & 0 & 0 & 0 & 0 & 0 & 0 & 0 & 0 & 0 & 0 & 0 & 0 & 0 & 0 & 0 & 0 & \mu_1 & 0 & 0 & 0 & 0 & 0 \\
- & 0 & 0 & 0 & 0 & 0 & 0 & 0 & 0 & 0 & 0 & 0 & 0 & 0 & 0 & 0 & 0 & 0 & \mu_3 & 0 & 0 & 0 & 0 \\
- & 0 & 0 & 0 & 0 & 0 & 0 & 0 & 0 & 0 & 0 & 0 & 0 & 0 & 0 & 0 & 0 & 0 & 0 & \mu_4 & 0 & 0 & 0 \\
- & 0 & 0 & 0 & 0 & 0 & 0 & 0 & 0 & 0 & 0 & 0 & 0 & 0 & 0 & 0 & 0 & 0 & 0 & 0 & \mu_6 & 0 & 0 \\
- & 0 & 0 & 0 & 0 & 0 & 0 & 0 & 0 & 0 & 0 & 0 & 0 & 0 & 0 & 0 & 0 & 0 & 0 & 0 & 0 & \mu_8 & 0 \\
- & 0 & 0 & 0 & 0 & 0 & 0 & 0 & 0 & 0 & 0 & 0 & 0 & 0 & 0 & 0 & 0 & 0 & 0 & 0 & 0 & 0 & \mu_9
\end{bmatrix}
$$

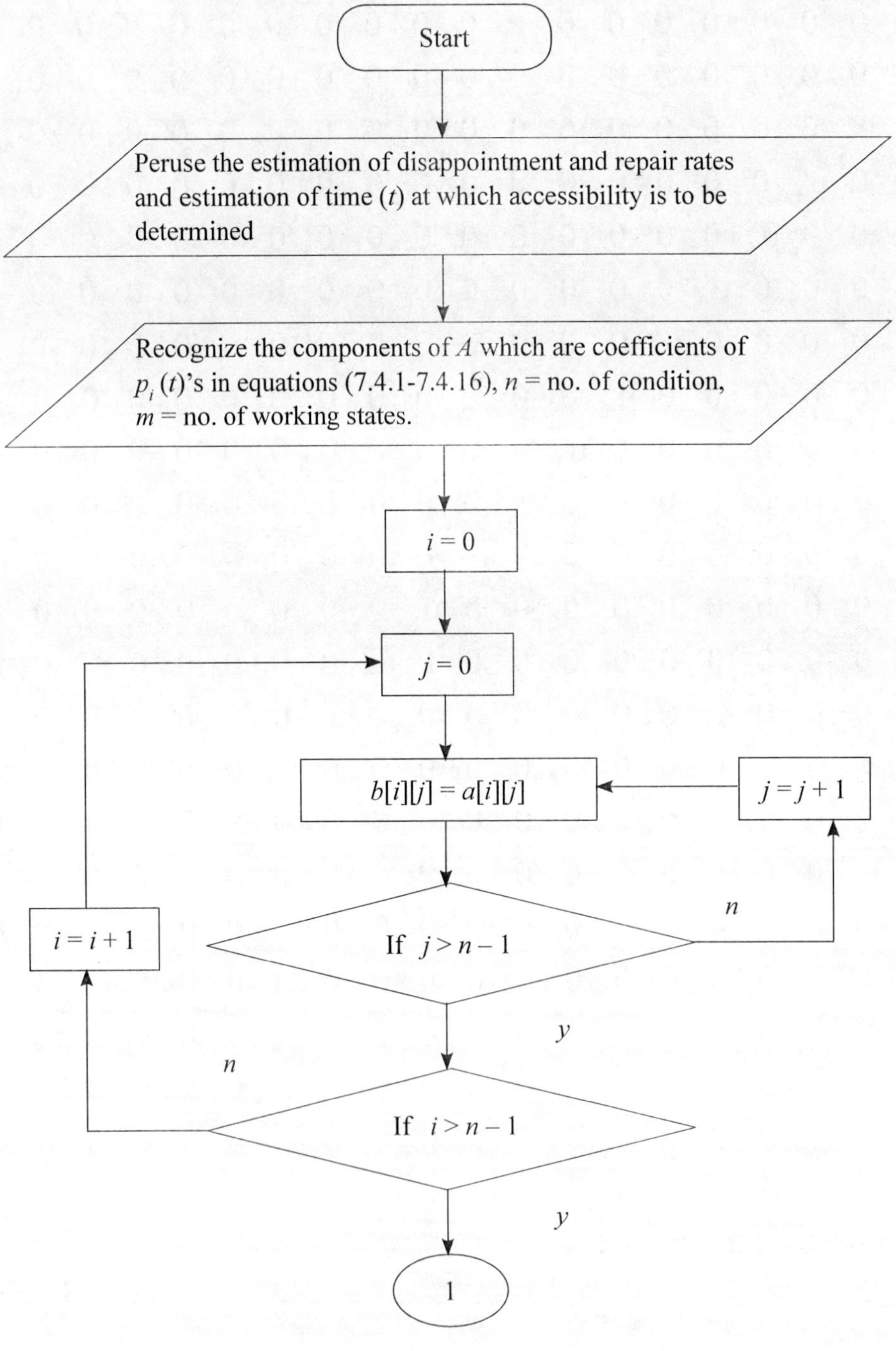

Fig. 7.13 Flowchart representation of matrix method.

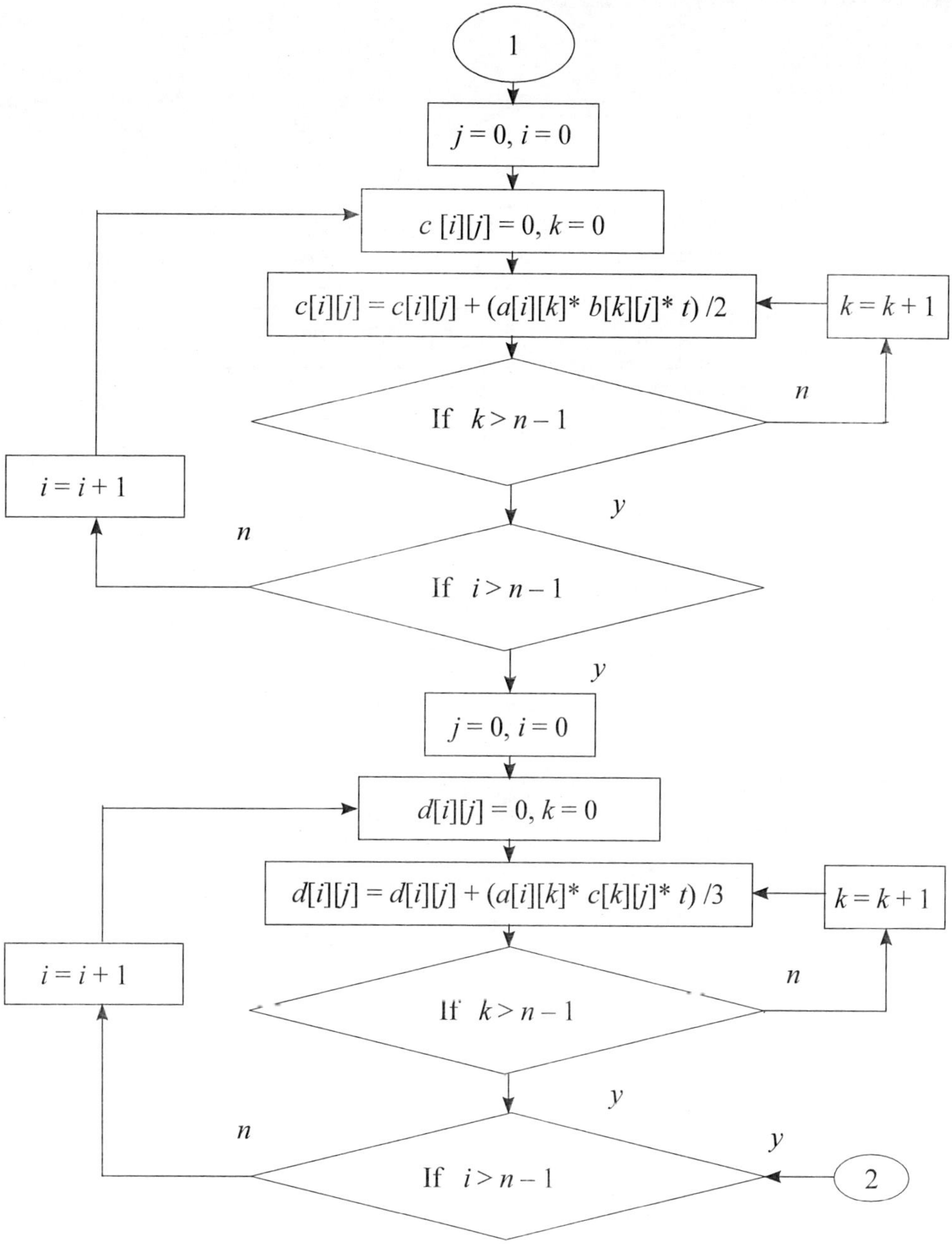

Fig. 7.14 Flowchart representation of matrix method.

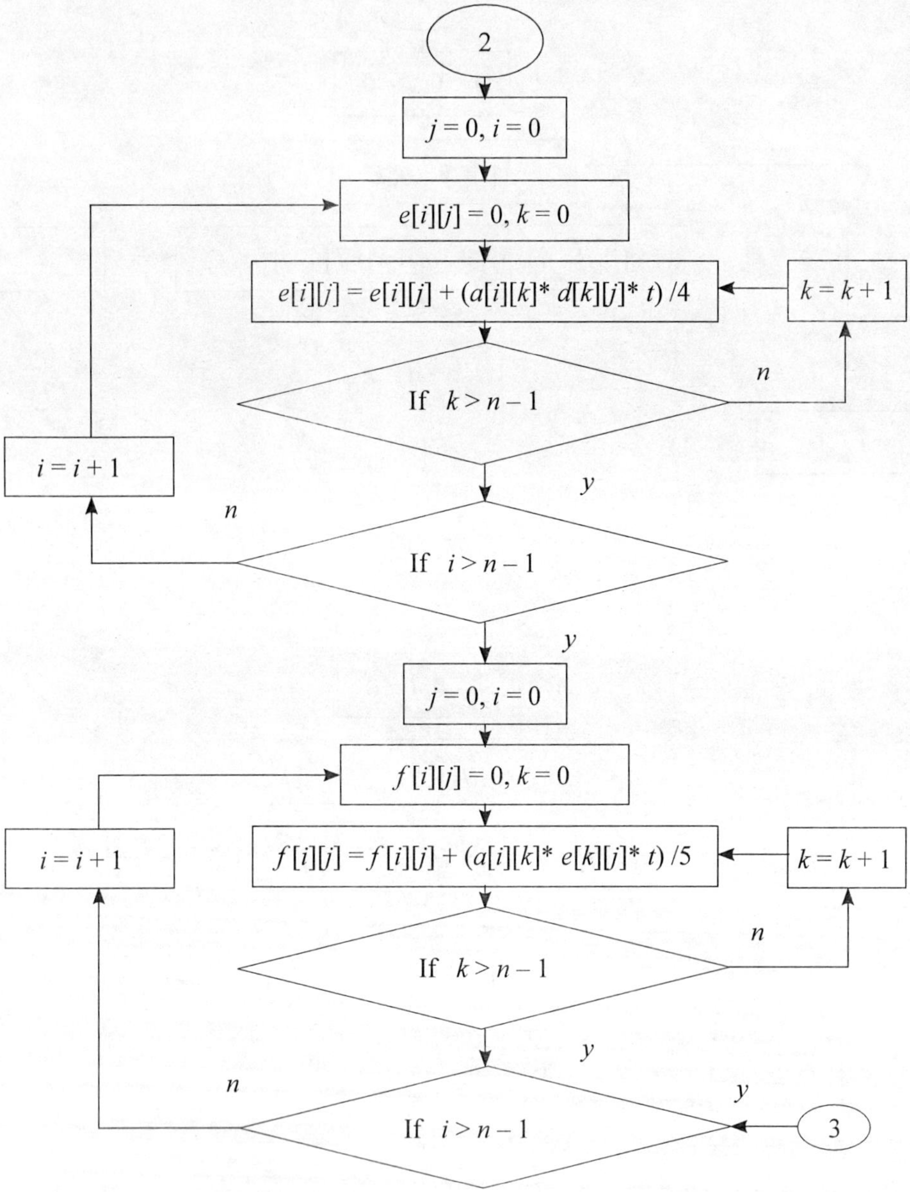

Fig. 7.15 Flowchart representation of matrix method.

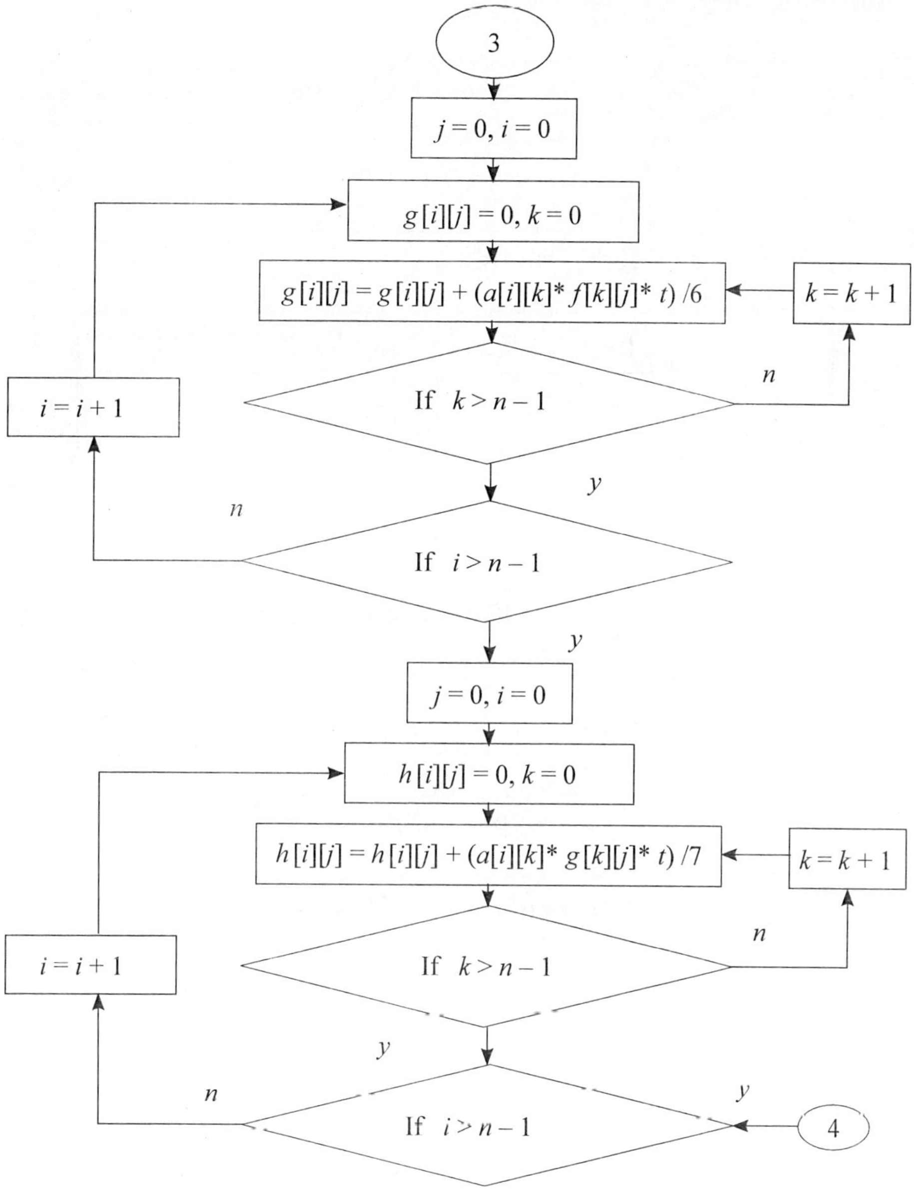

Fig. 7.16 Flowchart representation of matrix method.

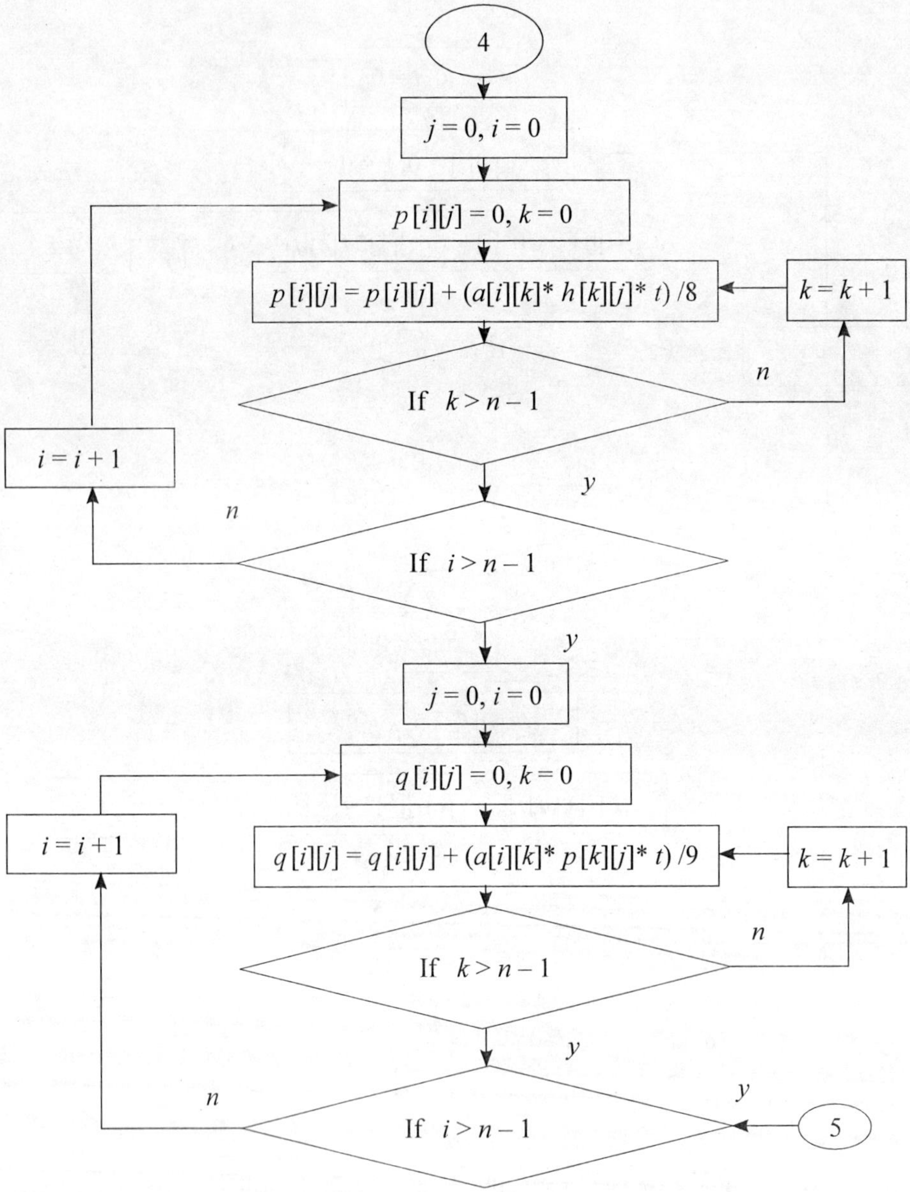

Fig. 7.17 Flowchart representation of matrix method.

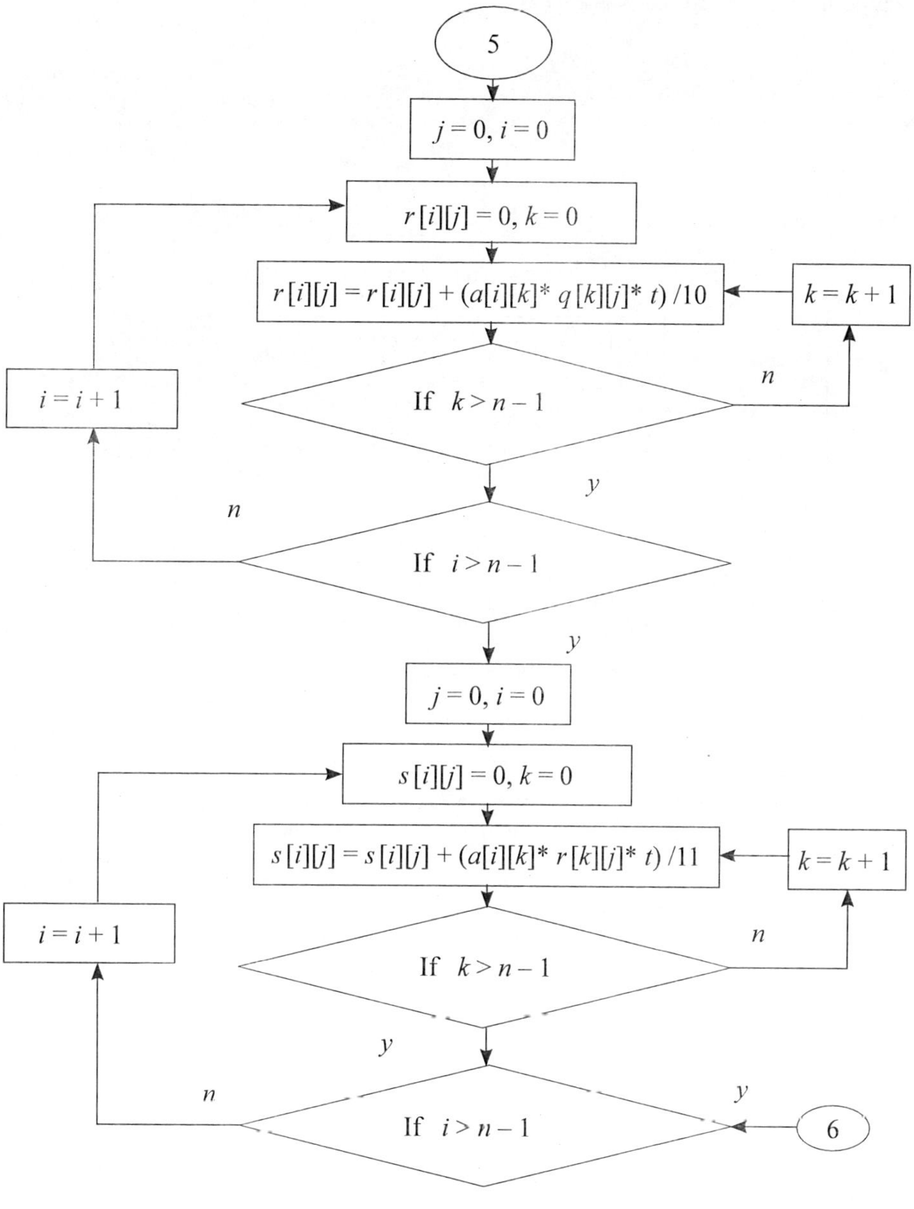

Fig. 7.18 Flowchart representation of matrix method.

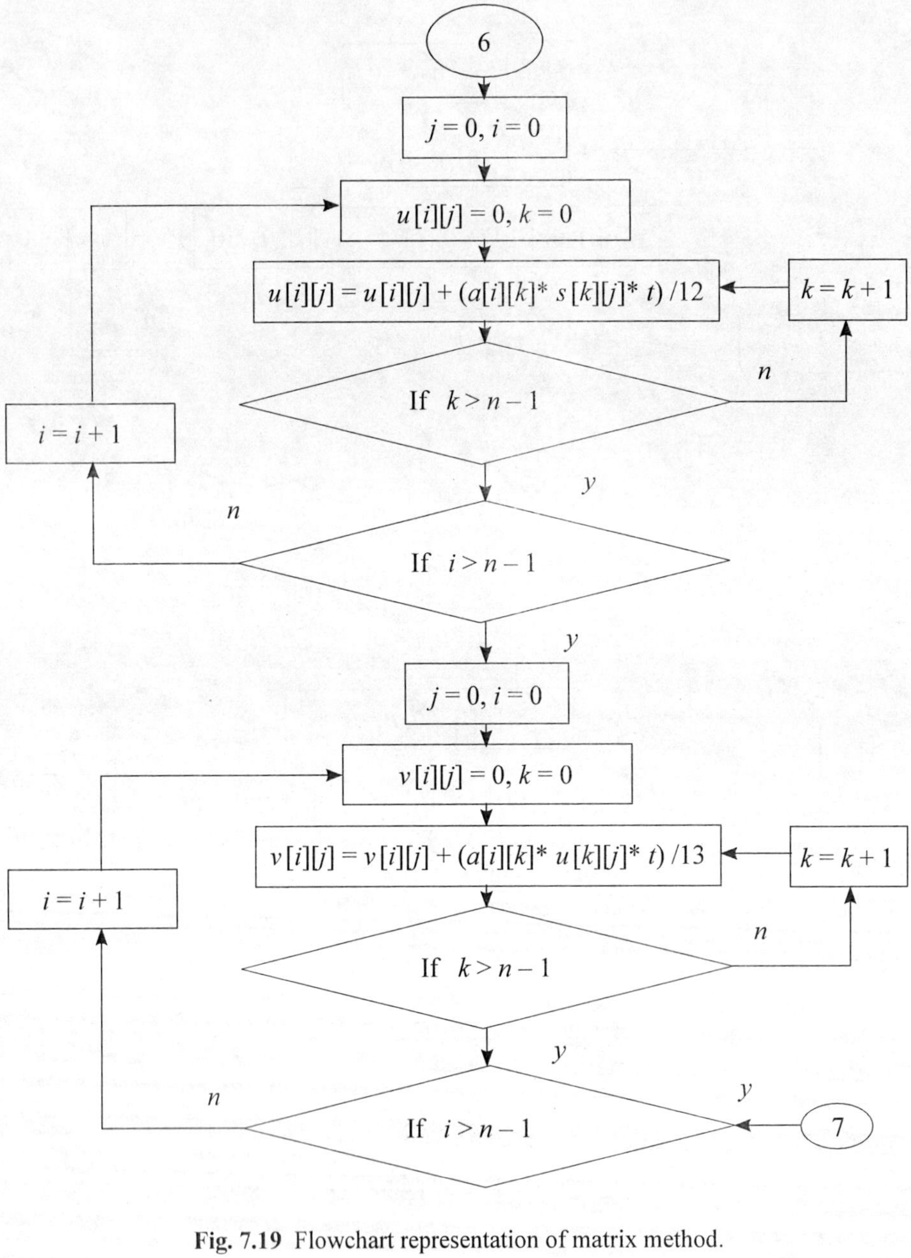

Fig. 7.19 Flowchart representation of matrix method.

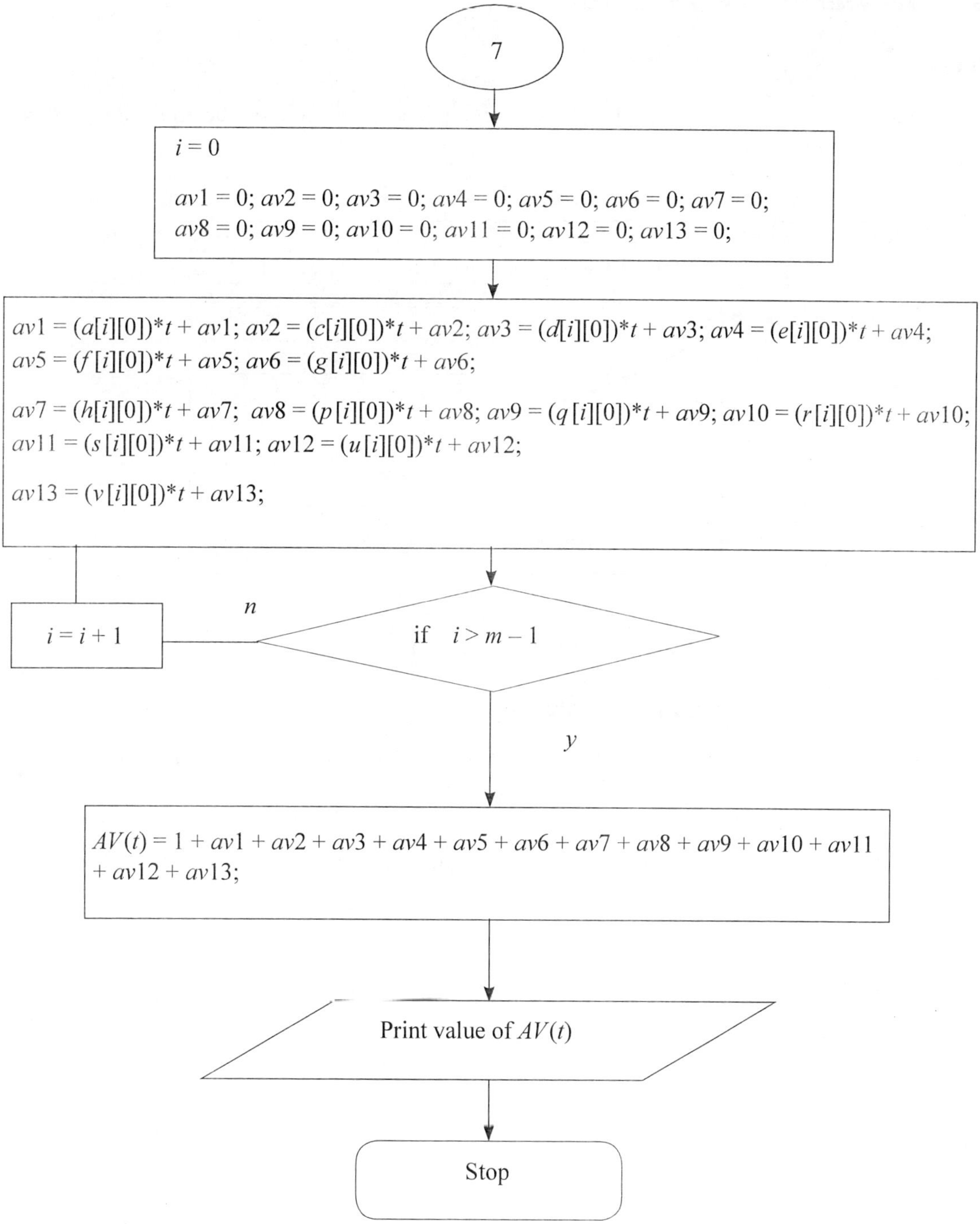

Fig. 7.20 Flowchart representation of matrix.

Results

Estimation of accessibility at various time (t) is given underneath. Results are determined by utilizing C-program (according to flowchart).

Table 7.1 Variation of availability with time

Time (units) (t)	0	10	20	30	40	50	60	70	80	90	100
$AV(t)$	1	0.9492	0.9073	0.8736	0.8468	0.8257	0. 7888	0.7410	0.7	0.66	0.62

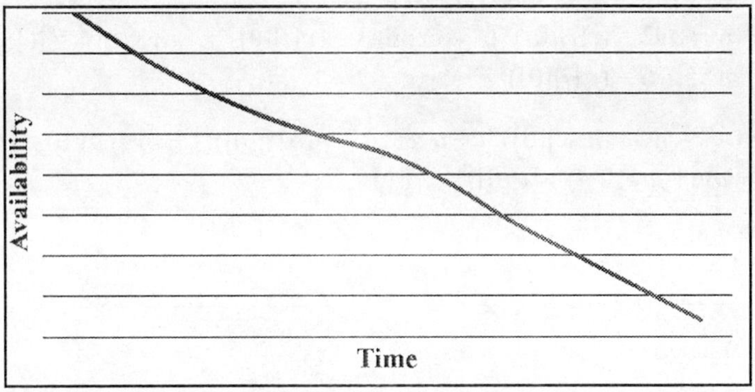

Fig. 7.21 Variation of availability with time.

7.4.6 Steady-State Availability Analysis

Rice plant is required to be accessible for long length of time. Along these lines, the long run or relentless state accessibility of the framework is gotten by putting $d/dt \to 0$ & $p_i(t) \to p_i$ as $t \to \infty$ in the above condition (7.4.1 to 7.4.16). These conditions are understood recursively. The estimations of consistent state probabilities are acquired as far as pursues.

$p_2 = k_2 p_1$ $p_3 = k_3 p_1$ $p_4 = k_4 p_1$ $p_5 = k_5 p_1$

$p_6 = k_6 p_1$ $p_7 = k_7 p_1$ $p_8 = k_8 p_1$ $p_9 = c_1 p_1$

$p_{10} = c_4 p_1$ $p_{11} = c_9 p_1$ $p_{12} = c_1 k_2 p_1$ $p_{13} = c_3 k_2 p_1$

$p_{14} = c_4 k_2 p_1$ $p_{15} = c_9 k_2 p_1$ $p_{16} = c_1 k_3 p_1$ $p_{17} = c_4 k_3 p_1$

$p_{18} = c_6 k_3 p_1$ $p_{19} = c_9 k_3 p_1$ $p_{20} = c_1 k_4 p_1$ $p_{21} = c_4 k_4 p_1$

$p_{22} = c_8 k_4 p_1$ $p_{23} = c_9 k_4 p_1$ $p_{24} = c_1 k_5 p_1$ $p_{25} = c_3 k_5 p_1$

$p_{26} = c_4 k_5 p_1$ $p_{27} = c_6 k_5 p_1$ $p_{28} = c_9 k_5 p_1$ $p_{29} = c_1 k_6 p_1$

$p_{30} = c_3 k_6 p_1$ $p_{31} = c_4 k_6 p_1$ $p_{32} = c_8 k_6 p_1$ $p_{33} = c_9 k_6 p_1$

$p_{34} = c_1 k_7 p_1$ $p_{35} = c_4 k_7 p_1$ $p_{36} = c_6 k_7 p_1$ $p_{37} = c_8 k_7 p_1$

$p_{38} = c_9 k_7 p_1$ $p_{39} = c_1 k_8 p_1$ $p_{40} = c_3 k_8 p_1$ $p_{41} = c_4 k_8 p_1$

$p_{42} = c_6 k_8 p_1$ $p_{43} = c_8 k_8 p_1$ $p_{44} = c_9 k_8 p_1$

where

$c_i = \lambda_i / \mu_i$ $\quad i = 1, 2, 3, 4, 5, 6, 7, 8, 9$ &

$k_2 = a_{21}$ $\qquad k_3 = a_{31} + a_{32}a_{21}$

$k_4 = a_{41} + a_{42}k_2 + a_{43}k_3$ $\qquad k_5 = a_{52}k_2 + a_{53}k_3 + a_{54}k_4$

$k_6 = a_{62}k_2 + a_{63}k_3 + a_{64}k_4 + a_{65}k_5$

$k_7 = a_{73}k_3 + a_{74}k_4 + a_{75}k_5 + a_{76}k_6$

$k_8 = a_{86}k_6 + a_{85}k_5 + a_{87}k_7$

$a_{85} = \lambda_7/a_{88}$ $\qquad a_{86} = \lambda_5/a_{88}$ $\qquad a_{87} = \lambda_2/a_{88}$

$a_{88} = \mu_2 + \mu_5 + \mu_7$ $\qquad a_{73} = \lambda_7/a_{77}$ $\qquad a_{74} = \lambda_5/a_{77}$ $\qquad a_{75} = \mu_2 a_{85}/a_{77}$

$a_{76} = \mu_2 a_{86}/a_{77}$ $\qquad a_{77} = \lambda_2 + \mu_5 + \mu_7 - \mu_2 a_{87}$ $\qquad a_{62} = \lambda_7/a_{66}$

$a_{63} = \mu_5 a_{87} a_{76}/a_{66}$ $\qquad a_{64} = \mu_5 a_{87} a_{74} + \lambda_2/a_{66}$ $\qquad a_{65} = \mu_5 a_{85} + \mu_5 a_{87} a_{75}/a_{66}$

$a_{66} = \lambda_5 + \mu_2 + \mu_7 - (\mu_5 a_{86} + \mu_5 a_{87} a_{76})$ $\qquad a_{52} = \lambda_5/a_{55}$

$a_{53} = \mu_7 a_{87} a_{73} + \lambda_2 + (\mu_7 a_{86} + \mu_7 a_{87} a_{76})\ a_{63}/a_{55}$

$a_{54} = (\mu_7 a_{87} a_{74}) + (\mu_7 a_{86} + \mu_7 a_{87} a_{76})\ a_{64}/a_{55}$

$a_{55} = \lambda_7 + \mu_2 + \mu_5 - (\mu_7 a_{85} + \mu_7 a_{87} a_{75}) + (\mu_7 a_{86} + \mu_7 a_{87} a_{76})\ a_{65}/a_{55}$

$a_{41} = \lambda_7/a_{44}$

$a_{42} = (\mu_2 a_{62} + \mu_5 a_{76} a_{62}) + a_{52}(\mu_5 a_{75} + \mu_2 a_{65} + \mu_5 a_{76} + a_{65})/a_{44}$

$a_{43} = (\mu_5 a_{73} + \mu_2 a_{63} + \mu_5 a_{63} a_{76}) + a_{53}(\mu_5 a_{75} + \mu_2 a_{65} + \mu_5 a_{76} a_{65})/a_{44}$

$a_{44} = \lambda_2 + \lambda_5 + \mu_7 - [(\mu_5 a_{74} + \mu_2 a_{64} + \mu_5 a_{64} a_{76}) + (\mu_5 a_{75} + \mu_2 a_{65} + \mu_5 a_{65} a_{76})(a_{54})]$

$a_{31} = \lambda_5/a_{33}$, $\quad a_{32} = \mu_2 a_{52} + \mu_2 a_{54} a_{42}/a_{33}$ $\qquad a_{33} = \lambda_2 + \lambda_7 + \mu_5 - (\mu_2 a_{53} + \mu_2 a_{54} a_{43})$

$$a_{21} = \begin{array}{l}(\mu_7 a_{64} + (\mu_5 + \mu_7 a_{65})a_{54})\ a_{41} + \lambda_2 \\ +(\mu_7 a_{63} + (\mu_5 + \mu_7 a_{65})a_{53}) + (((\mu_7 a_{64} + (\mu_5 + \mu_7 a_{65})a_{54})a_{43})a_{31})/a_{22}\end{array}$$

$$\begin{aligned}a_{22} = \lambda_5 + \lambda_7 + \mu_2 - \{&(\mu_7 a_{62} + (\mu_5 + \mu_7 a_{65})a_{52}) \\ &+ (\mu_7 a_{64} + (\mu_5 + \mu_7 a_{65})a_{54})\ a_{42} + (\mu_7 a_{63} + (\mu_5 + \mu_7 a_{65})a_{53}) + \\ &(((\mu_7 a_{64} + (\mu_5 + \mu_7 a_{65})\ a_{54})a_{43})a_{32}))\end{aligned}$$

and p_1 is determined by utilizing condition, for example,

$$\sum_{i=1}^{44} p_i = 1$$

or $p_1 (1 + k_2 + k_3 + k_4 + k_5 + k_6 + k_7 + k_8 + c_1 + c_4 + c_9 + c_1k_2 + c_3k_2 + c_4k_2 + c_9k_2 + c_1k_3 + c_4k_3 + c_6k_3 + c_9k_3 + c_1k_4 + c_4k_4 + c_8k_4 + c_9k_4 + c_1k_5 + c_3k_5 + c_4k_5 + c_6k_5 + c_9k_5 + c_1k_6 + c_3k_6 + c_4k_6 + c_8k_6 + c_9k_6 + c_1k_7 + c_4k_7 + c_6k_7 + c_8k_7 + c_9k_6 + c_1k_7 + c_4k_7 + c_6k_7 + c_8k_7 + c_9k_7 + c_1k_8 + c_3k_8 + c_4k_8 + c_6k_8 + c_8k_8 + c_9k_8) = 1$

$p_1 * N = 1$, i.e., $p_1 = {}^{1}\!/_{N}$

Now, steady-state accessibility of rice plant is

$A_v = p_1 + p_2 + p_3 + p_4 + p_5 + p_6 + p_7 + p_8$

Stream outline speaking to the computational method to ascertain consistent state accessibility of concern plant is demonstrated as follows (Figure 7.22).

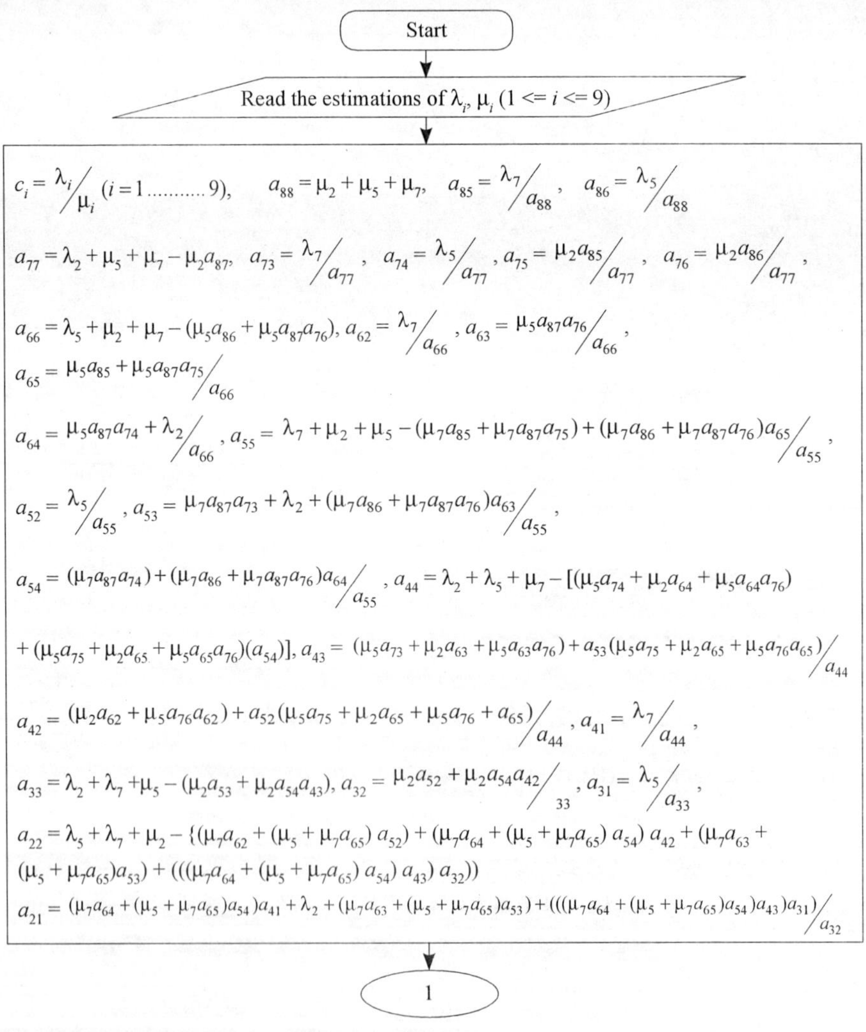

Fig. 7.22 Flowchart.

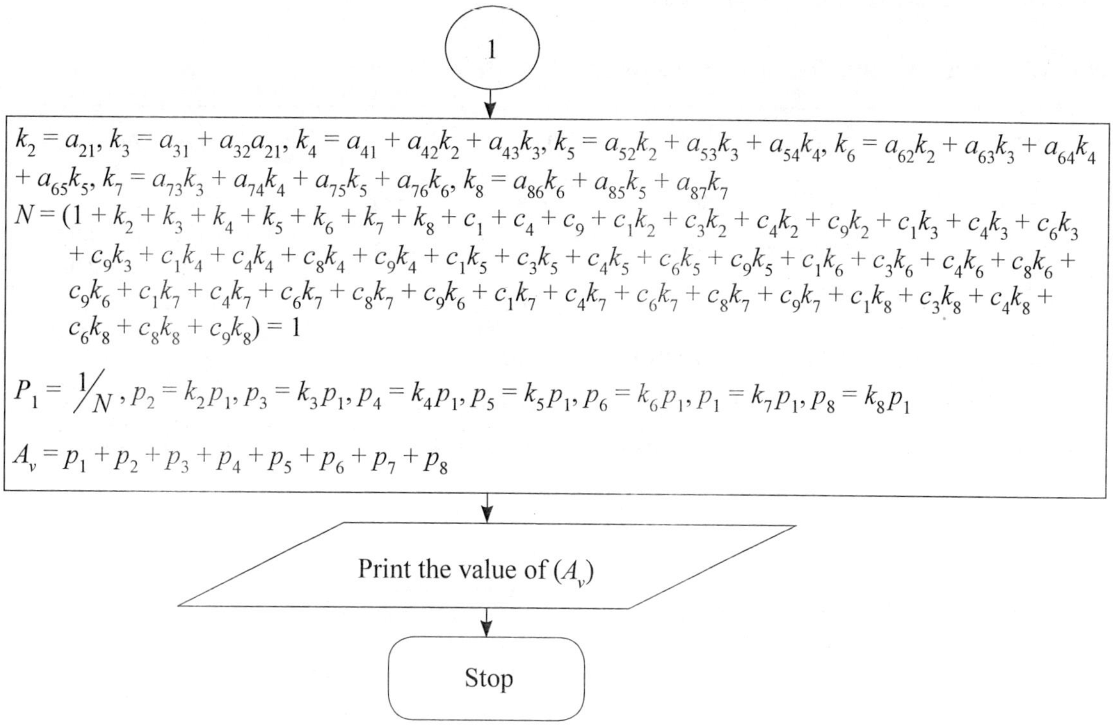

Fig. 7.23 Flowchart for steady-state accessibility.

7.5 DECISION SUPPORT SYSTEM

The choice emotionally supportive network manages the quantitative analysis of a considerable number of elements, viz., upkeep procedures and conditions of nature which impact the support choices related with the rice plant. These choice models are created under the genuine basic leadership condition, for example, probabilistic model. Such models are utilized to actualize the best possible support choices for a rice plant. Tables (7.2-7.7) speak to the choice matrices for different subsystems of rice plant. These matrices essentially uncover the different accessibility levels for various mixes of disappointment and repair rates. Based on choice emotionally supportive network created, we may choose conceivable mix (λ_i, μ_i.) for example, optimal upkeep technique. For instance, consider Table 7.2, decision matrix for the dryer subsystem, for some known estimations of disappointment/repair rates of husker, separator, polisher, grader, sortex, it uncovers the impact of fix rates of dryer sub-framework on the accessibility of rice plant. Additionally, other choice matrices clarify the impact of disappointment/fix rates of husker, separator, polisher, grader, and sortex on the framework accessibility. Here, we may choose the most extreme estimation of accessibility for every subsystem and after that at long last the limit of the maximal would be taken for every subsystem. Thus, the ideal estimations of repair and failure rates might be chosen in like manner for every subsystem of rice plant. Estimations are performed by the C-program which is according to flowchart (Figure 7.22).

Table 7.2 Decision matrix for the dryer subsystem

λ_1 \ μ_1	0.01	0.02	0.03	0.04	0.05	Constant values
0.001	0.6978	0.7230	0.73187	0.7363	0.7390	$\lambda_2 = 0.003$, $\mu_2 = 0.03$; $\lambda_3 = 0.003$, $\mu_3 = 0.03$;
0.002	0.6523	0.6978	0.7144	0.7230	0.7283	$\lambda_4 = 0.0025$, $\mu_4 = 0.02$;
0.003	0.6123	0.6742	0.6978	0.7102	0.7178	$\lambda_5 = 0.004$, $\mu_5 = 0.04$; $\lambda_6 = 0.004$, $\mu_6 = 0.04$;
0.004	0.5770	0.6523	0.6819	0.6978	0.7077	$\lambda_7 = 0.005$, $\mu_7 = 0.05$;
0.005	0.5455	0.6317	0.6668	0.6858	0.6978	$\lambda_8 = 0.005$, $\mu_8 = 0.05$, $\lambda_9 = 0.002$, $\mu_9 = 0.02$;

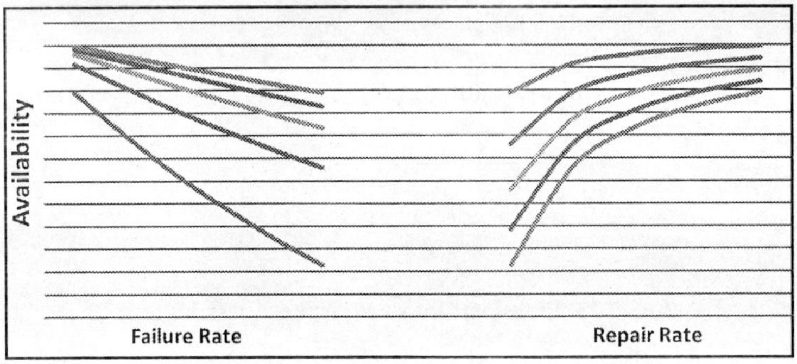

Fig. 7.24 Effect of failure and repair rates of dryer

Table 7.3 Decision matrix for husker subsystem

λ_2 \ μ_2	0.03	0.05	0.07	0.09	0.11	Constant values
0.003	0.6978	0.6981	0.6984	0.6986	0.6987	$\lambda_1 = 0.001$, $\mu_1 = 0.01$; $\lambda_3 = 0.003$, $\mu_3 = 0.03$;
0.004	0.6972	0.6977	0.6980	0.6982	0.6984	$\lambda_4 = 0.0025$, $\mu_4 = 0.02$;
0.005	0.6966	0.6972	0.6976	0.6979	0.6981	$\lambda_5 = 0.004$, $\mu_5 = 0.04$; $\lambda_6 = 0.004$, $\mu_6 = 0.04$;
0.006	0.6961	0.6968	0.6972	0.6976	0.6978	$\lambda_7 = 0.005$, $\mu_7 = 0.05$; $\lambda_8 = 0.005$, $\mu_8 = 0.05$,
0.007	0.6955	0.6963	0.6969	0.6972	0.6975	$\lambda_9 = 0.002$, $\mu_9 = 0.02$;

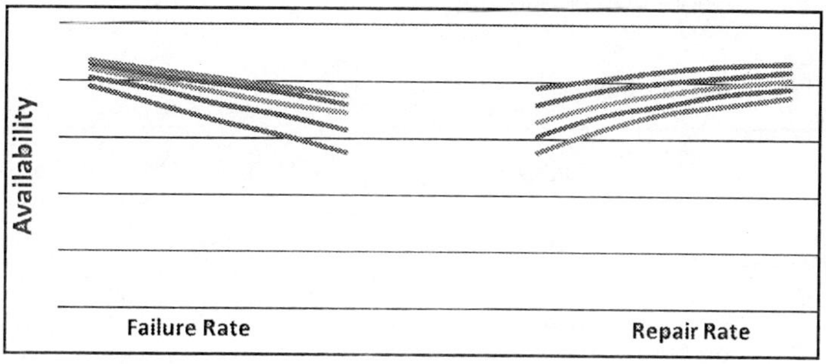

Fig. 7.25 Effect of failure and repair rates of husker.

Table 7.4 Decision matrix for separator subsystem

λ_4 \ μ_4	0.02	0.04	0.06	0.08	0.10	Constant values
0.0025	0.6978	0.7296	0.7409	0.7466	0.7501	$\lambda_1 = 0.001, \mu_1 = 0.01;$ $\lambda_2 = 0.003, \mu_2 = 0.03;$
0.0050	0.6418	0.6978	0.7187	0.7296	0.7363	$\lambda_3 = 0.003, \mu_3 = 0.03;$ $\lambda_5 = 0.004, \mu_5 = 0.04;$
0.0075	0.5941	0.6686	0.6978	0.7133	0.7230	$\lambda_6 = 0.004, \mu_6 = 0.04;$ $\lambda_7 = 0.005, \mu_7 = 0.05;$
0.001	0.5530	0.6418	0.6781	0.6978	0.7102	$\lambda_8 = 0.005, \mu_8 = 0.05,$ $\lambda_9 = 0.002, \mu_9 = 0.02;$

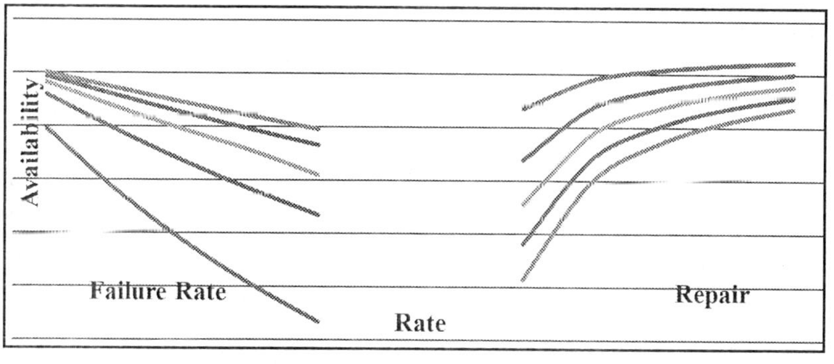

Fig. 7.26 Effect of failure and repair rates of separator.

Table 7.5 Decision matrix for polisher subsystem

λ_5 \ μ_5	**0.04**	**0.06**	**0.08**	**0.10**	**0.12**	**Constant values**
0.004	0.6978	0.6982	0.6984	0.6986	0.6988	$\lambda_1 = 0.001, \ \mu_1 = 0.01;$ $\lambda_2 = 0.003, \ \mu_2 = 0.03;$
0.008	0.6958	0.6965	0.6970	0.6974	0.6977	$\lambda_3 = 0.003, \ \mu_3 = 0.03;$ $\lambda_4 = 0.0025, \mu_4 = 0.04;$
0.012	0.6940	0.6950	0.6957	0.6963	0.6967	$\lambda_6 = 0.004, \ \mu_6 = 0.04;$
0.016	0.6924	0.6936	0.6945	0.6952	0.6957	$\lambda_7 = 0.005, \ \mu_7 = 0.05;$ $\lambda_8 = 0.005, \ \mu_8 = 0.05,$
0.020	0.6909	0.6923	0.6933	0.6941	0.6947	$\lambda_9 = 0.002, \ \mu_9 = 0.02;$

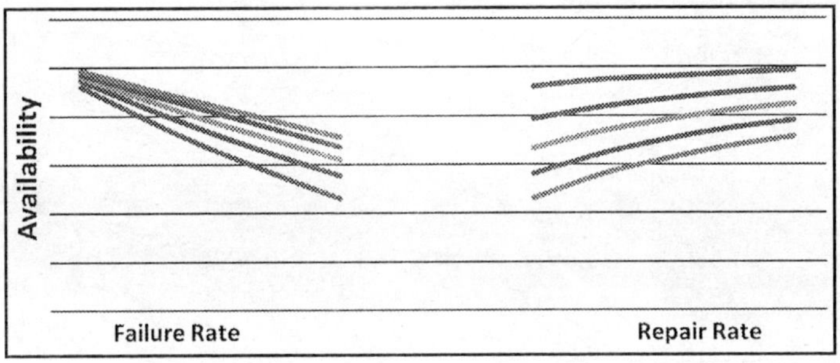

Fig. 7.27 Effect of failure and repair rates of polisher.

Table 7.6 Decision matrix for grader subsystem

λ_7 \ μ_7	**0.05**	**0.55**	**1.05**	**1.55**	**2.05**	**Constant values**
0.005	0.6978	0.6984	0.6988	0.6990	0.6993	$\lambda_1 = 0.001, \ \mu_1 = 0.01;$ $\lambda_2 = 0.003, \ \mu_2 = 0.03;$
0.055	0.6976	0.6981	0.6985	0.6986	0.6988	$\lambda_3 = 0.003, \ \mu_3 = 0.03;$
0.105	0.6970	0.6976	0.6980	0.6983	0.6985	$\lambda_4 = 0.0025, \mu_4 = 0.02;$ $\lambda_5 = 0.004, \ \mu_5 = 0.04;$
0.155	0.6965	0.6972	0.6976	0.6980	0.6982	$\lambda_6 = 0.004, \ \mu_6 = 0.04;$ $\lambda_8 = 0.005, \ \mu_8 = 0.05,$
0.205	0.6959	0.6967	0.6973	0.6976	0.6979	$\lambda_9 = 0.002, \ \mu_9 = 0.02;$

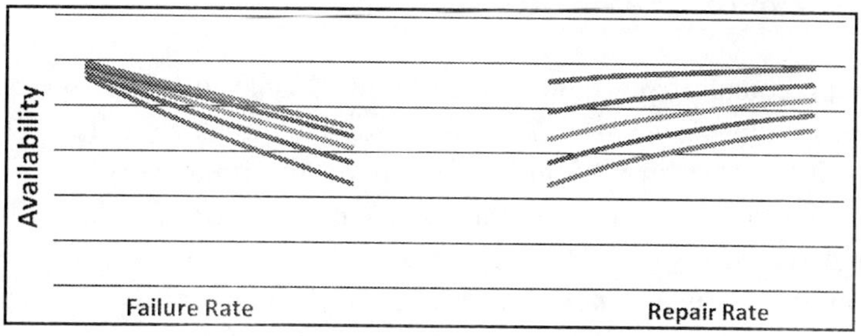

Fig. 7.28 Effect of failure and repair rates of grader.

Table 7.7 Decision matrix for sortex subsystem

μ_9 / λ_9	**0.07**	**0.08**	**0.09**	**0.10**	**0.11**	**Constant values**
0.007	0.645	0.649	0.656	0.659	0.663	$\lambda_1 = 0.001$, $\mu_1 = 0.01$; $\lambda_2 = 0.003$, $\mu_2 = 0.03$;
0.014	0.642	0.647	0.651	0.655	0.659	$\lambda_3 = 0.003$, $\mu_3 = 0.03$;
0.021	0.639	0.645	0.648	0.65	0.657	$\lambda_4 = 0.0025$, $\mu_4 = 0.02$; $\lambda_5 = 0.004$, $\mu_5 = 0.04$;
0.028	0.633	0.641	0.645	0.648	0.652	$\lambda_6 = 0.004$, $\mu_6 = 0.04$; $\lambda_7 = 0.005$, $\mu_7 = 0.05$,
0.035	0.63	0.635	0.641	0.643	0.645	$\lambda_8 = 0.005$, $\mu_8 = 0.05$;

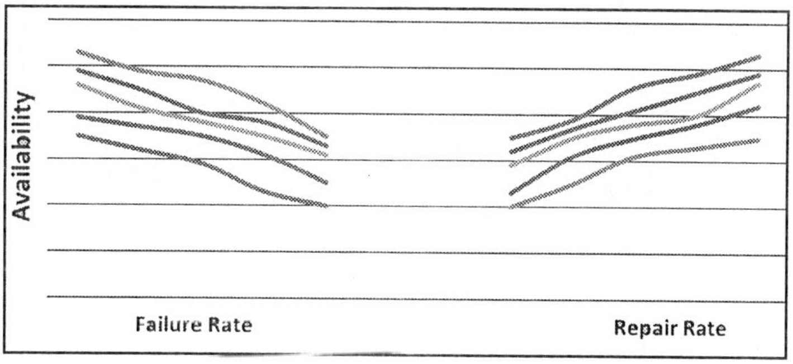

Fig. 7.29 Effect of failure and repair rates of sortex.

Table 7.2 demonstrates the impact of repair and failure rates of dryer on the accessibility of rice plant. It is seen that for some known estimations of repair and failure rates of husker, separator, polisher, grader, sortex ($\lambda_2 = 0.003$, $\mu_2 = 0.03$; $\lambda_3 = 0.003$, $\mu_3 = 0.03$; $\lambda_4 = 0.0025$; $\mu_4 = 0.02$; $\lambda_5 = 0.004$; $\mu_5 = 0.04$; $\lambda_6 = 0.004$, $\mu_6 = 0.04$; $\lambda_7 = 0.005$; $\mu_7 = 0.05$; $\lambda_8 = 0.005$; $\mu_8 = 0.05$; $\lambda_9 = 0.002$; $\mu_9 = 0.02$; as disappointment rate (λ_1) of dryer increments from 0.001 (once in 1000 hr) to 0.005 (once in 200 hr), the framework accessibility diminishes extensively by 15.23%. Likewise, as the fix rate (μ_1) of dryer increments from 0.01 (once in 100 hr) to 0.05 (once in 20 hr), the framework accessibility increment by 4.12%. From Table 7.3, it is additionally seen that as failure rate (λ_2) of husker increments from 0.003 (once in 333.33 hr) to 0.007 (once in 142.85 hr), the framework accessibility diminishes by 0.23%. So also, when repair rate (μ_2) of husker from 0.03 (once in 33.3 hr) to 0.11 (once in 9.09 hr), at that point the framework accessibility increments by 0.09%. Table 7.4 additionally uncovers the variety of framework accessibility with change in failure rates (λ_4) and fix rates (μ_4) of the separator. As failure rate (λ_4) increments from 0.0025 (once in 400 hr) to 0.0125 (once in 80 hr), the framework accessibility lessens by 18.05%. Thus, when fix rate (μ_4) increments from 0.02 (once in 50 hr) to 0.10 (once in 10 hr), at that point the framework accessibility increments by 5.23%. Similarly, Table 7.5 demonstrates the variety in accessibility of rice plant with the adjustment in repair and failure rates of subsystem polisher (λ_5, μ_5). Further, Table 3.5 uncovers that disappointment rate (λ_5) increments from 0.004 (once in 250 hr) to 0.020 (once in 50 hr), the framework accessibility decreases by 0.69%. Also, when repair rate (μ_5) increments from 0.04 (once in 25 hr) to 0.12 (once in 8.33 hr), at that point the framework accessibility increments by about 0.1%. Table 7.6 additionally uncovers the variety of framework accessibility with change in disappointment rates (λ_7) and repair rates (μ_7) of the grader. As disappointment rate (λ_7) increments from 0.005 (once in 200 hr) to 0.205 (once in 4.87 hr), the framework accessibility decreases by 0.19%. Additionally, when repair rate (μ_7) increments from 0.05 (once in 20 hr) to 2.05 (once in 0.487 hr), at that point the framework accessibility increments by about 0.15%. Table 7.7 likewise uncovers the variety of framework accessibility with change in failure rates (λ_9) and repair rates (μ_9) of the sortex. As disappointment rate (λ_9) increments from 0.007 (once in 142.85 hr) to 0.035 (once in 28.57 hr), the framework accessibility lessens by 1.5%. So also, when repair rate (μ_9) increments from 0.07 (once in 14.2 hr) to 0.11 (once in 9.09 hr), at that point the framework accessibility increments by about 1.8%. From Table 7.2, we may choose the most extreme estimation of accessibility of first segment for dryer sub-framework and afterward for other segment moreover. At long last the limit of the maximal (accessibility) would be taken, for example, A_v (max) = 0.7390. Additionally, from Table 7.3 to Table 7.7, husker A_v (max) = 0.6987, separator A_v (max) = 0.7501, polisher A_v (max) = 0.6988, grader A_v (max) = 0.6993 and sortex A_v (max) = 0.663 the comparing ideal estimations of repair and failure rates might be chosen as needs be for every subsystem of rice plant framework and are given in Table 7.8.

Table 7.8 Optimal values of failure and repair rates

Sr. No.	λ	μ	Max. Availability Level
1.	0.001	0.05	0.7390
2.	0.003	0.11	0.6987
3.	0.0025	0.10	0.7501
4.	0.004	0.12	0.6988
5.	0.005	2.05	0.6993
6.	0.007	0.11	0.663

In this way, repair work ought to be done in the accompanying request of execution.

Table 7.9 Order of repair work

Subsystem	Increase in Availability	Rank
Separator	5.23%	1st
Dryer	4.12%	2nd
Sortex	1.8%	3rd
Grader	0.15%	4th
Polisher	0.1%	5th
Husker	0.09%	6th

In this manner, based on repair rates, order of maintenance/repair work must be carried out as per Table 7.9.

CONCLUDING REMARKS

In this chapter, process industries are discussed and their reliability is analyzed. All these industries (sugar industry, butter-oil industries, fertilizer plant, rice plant) are dependent upon agriculture in one or the other way. For each plants, methods for analysis are presented. By varying the failure or repair rates of the subsystems, the behavior analysis of the system(s) can be studied as demonstrated in the same chapter which is helpful for management. In the similar way or with minor changes in methodologies other agriculture based system may be analyzed. Some other industries like plywood, bread manufacturing, urea fertilizer, poly tube industry, washing unit system and cotton industry have been analyzed by Savita et al. (2007), Gyal et al. (2006), Puran Chand et al. (2007), Arun et al. (2019) and Archna Sharma et al. (2009).

EXERCISE

1. What is the example of agro-based industry?
2. What are the agro-based industries?
3. What are the benefits of agro-based industries?
4. What are the agro products?
5. What is the new technology in agriculture?
6. What are the raw materials used in agro industries?
7. What is the importance of agro processing?
8. Define the sugar industry.
9. Write short notes on:
   (a) Fertilizer plant
   (b) Urea synthesis system
   (c) Rice plant
   (d) Refining system

8

Chapter

Availability Analysis of a Biogas Plant and a Plant Pathological System

INTRODUCTION

In Chapters 6 and 7, some industries (paper, cement, utensils, sugar, fertilizer, butter-oil) have been analysed for availability and MTTF to demonstrate that how can the reliability technology be applied to study the behaviour of an industry by changing the effective rates, etc. In this chapter some non-industrial systems directly concerned with agriculture sector are discussed. The first system is a biogas plant producing energy, a non-conventional energy system while the other system is concerned with medicines for plant disease (plant pathology).

8.1 BIOGAS PLANT (NON-CONVENTIONAL ENERGY SYSTEM)

Nowadays people are more or less dependent on electricity and other energy resources. The electricity is produced by coal, atomic destruction, petroleum, water and other conventional means. The energy produced by conventional means is called conventional energy. Other than these, there are resources like sun heat, wind, cow dung which give a good quantity of energy to the people is known as non-conventional energy (in general). The very frequently used energy is biogas produced from cow dung. The biogas produced is generally used for cooking and lighting purposes. The biogas is powerful and it may run even tractors, cutting machines and even crushing systems. About ten years back, the author saw a cane crushing system run by biogas. Also about one year back, an illiterate farmer of U.P. (India) ran a tractor, a grass cutting machine and a water tubewell pump using biogas. In this section, two models analyzed by Singh [1987, 1989] are given.

A biogas system having two closed tanks and using animal dung for gas production is considered. The tanks are similar in design having separate inlets-outlets and different capacities. The two systems considered are–(i) both the units are operating simultaneously and are repairable. When one unit fails due to one or other reason, the system also works but in reduced capacity, (ii) one unit is working and the other unit is in standby. When the working (active) unit fails it is replaced by standby unit (tank), i.e., the gas supply channel from standby unit is open. Thus, the gas supply is continued.

The operating and standby tanks may fail simultaneously. The failures may occur due to leakage from the tanks, due to low atmospheric temperature, due to snowfall. There may be normal failure due to reasons of animal dung, etc. The repair of the system means removal of the above difficulties, e.g., the tanks are covered by mud to reduce or to finish the effect of atmospheric temperature.

Assuming that the system failures occurring are statistically independent, entire system may fail due to sudden critical atmospheric changes and these may occur when either both units are good or when one unit is good. Let λ_1, λ_2 be the constant failure rates of first and second unit, respectively and $\mu_1(x)$, $\mu_2(x)$ be the respective repair rates of first and second unit. Further assume that β_1 is the constant critical failure rate when both units are operating and β_2 is the critical failure rate when one unit is operating.

If $p_i(t)$ is the probability, that at time t the system is in ith state, $i = 0, 1, 2, 3$ and $p_i(x, t)dx$ is the probability that one unit is in repair and the elapsed repair time is x, the transition diagrams are:

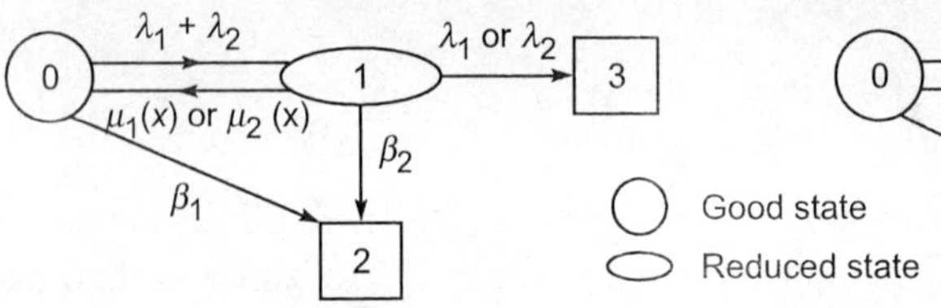

System I

Fig. 8.1

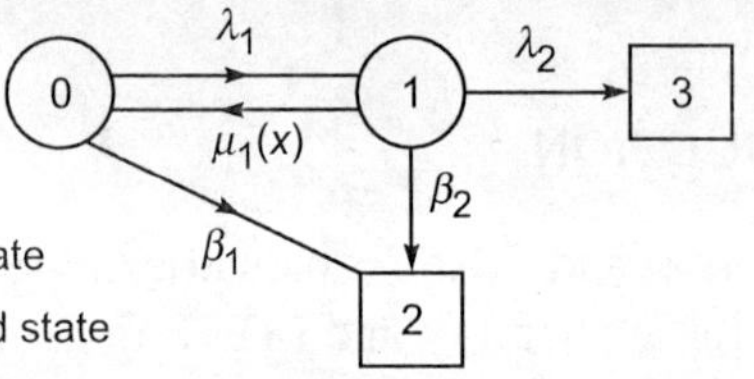

System II

Fig. 8.2

The differential equations for system I (associated with Fig. 8.1) are

$$\left[\frac{d}{dt} + \lambda_1 + \lambda_2 + \beta_1\right] p_0(t) = \int p_1(x, t)\, \mu(x) dx \tag{8.1.1}$$

$$\left[\frac{\partial}{\partial x} + \frac{\partial}{\partial t} + \lambda + \mu(x) + \beta_2\right] p_1(x, t) = (\lambda_1 + \lambda_2) p_0(t) \tag{8.1.2}$$

$$\frac{d}{dt} p_2(t) = \beta_1\, p_0(t) + \beta_2 \int p_1(x, t) dx \tag{8.1.3}$$

$$\frac{d}{dt} p_3(t) = \lambda \int p_1(x, t) dx \tag{8.1.4}$$

with initial and boundary conditions

$$p_1(x, 0) = p_2(0) = p_3(0) = 0; \quad p_0(0) = 1 \tag{8.1.5}$$

$$p_1(0, t) = (\lambda_1 + \lambda_2) p_0(t) \tag{8.1.6}$$

where for λ_1 or λ_2, λ is used and for $\mu_1(x)$ or $\mu_2(x)$, $\mu(x)$ is used whichever is applicable. The limits for integration are from 0 to ∞.

Equations (8.1.1), (8.1.3), and (8.1.4) are first order ordinary differential equations while eqn. (8.1.2) is first order partial differential equation (Lagrange's type). Solving these equations, (using 8.1.5 and 8.1.6) the probabilities are given by

$$p_0(t) = [1 + B(x) + \lambda t \int B(x) dx + \beta_1 + \beta_2 \int B(x) dx]^{-1}$$

$$p_1(x, t) = p_0(t)\, B(x);\ p_2(t) = [\beta_1 + \beta_2 \int B(x) dx]\, p_0(t);\ p_3(t) = [\lambda t \int B(x) dx] p_0(t),$$

where,

$$B(x) = (\lambda_1 + \lambda_2)\ [1 + \int \exp\ (\lambda x + \beta_2 x + \int \mu(x)dx)\, dx]\ \exp\ \{-\lambda x - \beta_2 x - \int \mu(x)dx\}.$$

For system (ii), the differential equations (associated with Fig. 8.2) are

$$\left[\frac{d}{dt} + \lambda_1 + \beta_2\right] p_0(t) = \int p_1(x,\ t)\ \mu_1(x)dx \tag{8.1.7}$$

$$\left[\frac{\partial}{\partial x} + \frac{\partial}{\partial t} + \lambda_2 + \mu_1(x) + \beta_2\right] p_1(x,\ t) = \lambda_1 p_0(t) \tag{8.1.8}$$

$$\frac{dp_2(t)}{dt} = \beta_2\ p_0(t) + \beta_2 \int p_1(x,\ t)dx \tag{8.1.9}$$

$$\frac{dp_3(t)}{dt} = \lambda_2 \int p_1(x,\ t)dx \tag{8.1.10}$$

with boundary condition $p_1(0,\ t) = \lambda_1\ p_0(t)$. Initial conditions and limits are same as for system (i).

The solutions of eqns. (8.1.7)–(8.1.10), using initial and boundary conditions are given by

$$p_0(t) = [1 + B_1(x) + \lambda_2 t \int B_1(x)dx + \beta_2 + \beta_2 \int B_1(x)dx]^{-1};$$

$$p_1(x,\ t) = B_1(x)\ p_0(t); \quad p_2(t) = [1 + \int B_1(x)dx]\beta_2\ p_0(t);$$

$$p_3(t) = [\lambda_2 t \int B_1(x)dx]\ p_0(t),$$

where,

$$B_1(x) = \lambda_1[1 + \int \exp(\lambda_2 x + \beta_2 x + \int \mu_1(x)dx)dx]\ \exp(-\lambda_2 x - \beta_2 x - \int \mu_1(x)dx)$$

The reliability functions and mean time to system failure (MTSF) are given as follows:

$$R_I(t) = p_0(t) + p_1(t) = p_0(t) + \int p_1(x,\ t)dx = [1 + \int B(x)dx]p_0(t)$$

$$R_{II}(t) = [1 + \int B_1(x)dx]p_0(t)$$

$$(\text{MTSF})_I = \int_0^\infty R_I(t)dt \quad ; \quad (\text{MTSF})_{II} = \int_0^\infty R_{II}(t)dt.$$

If the consumer saves C_1 revenue per unit time from system (i) and C_2 revenue per unit time from system (ii) paying C_3 and C_4 service costs for systems (i) and (ii), respectively, then the expected profits for the interval $(0,\ t]$ are given by

$$P_I(t) = C_1 \int_0^t R_I(t)dt - C_3 t$$

$$P_{II}(t) = C_2 \int_0^t R_{II}(t)dt - C_4 t$$

Making profitable, suitable system may be chosen.

8.2 PLANT PATHOLOGY PROBLEM

The pathology to cure diseases from plants is known as plant pathology. The problems of plant pathology may be designed in such a way that we may use known results from reliability technology or the problems may be analyzed as reliability problems. For example, there are diseases in paddy crop when the use of one medicine to cure a particular disease fails, another medicine is used. Similar situations also happen in case of sorghum, bajra (Pennisetum) and maize (Zea) crops. In some cases more than two medicines are used on failure of one after the other. There may be some different type of complex situations in case of plant diseases which may be designed like reliability problems discussed in earlier chapters.

Consider the case of a particular crop. Due to atmospheric and soil effects, the crop may be affected by a number of diseases. Singh [1982] considered a problem of two medicines. Let for a certain disease A, for the crop under consideration, medicine no. 1 is used. If the medicine fails to control the disease, the farmers use another medicine no. 2, say. Medicine no. 2 is used only when medicine no. 1 fails to control the disease. For simplicity, exponential type failure distribution with parameter λ is taken. For example, to control blast disease in paddy crop generally Blitox-50 (0.3%) is used. When it fails to control the disease, Dithane Z-78 (0.2%) is sprayed to control the disease. There are so many diseases in paddy crop which may require more than one treatment. The problem is to find the probability that a particular disease for a particular crop is controlled using medicine no. 1 or no. 2. Success is gained when at least one medicine survives the control system.

Further, let $g_i(t)$, $i = 1, 2$ denotes the density function for t_i, where t_i is the time to failure of ith medicine in controlling the disease and $f(t)$ is the corresponding system failure time density function. The probability density function of the system failure time is equal to the 2-fold convolution of the density function, i.e.,

$$f(t) = \int_0^t g_1(x)\, g_2(t-x)dx \tag{8.2.1}$$

According to our assumption

$$g_i(t) = \lambda_i \exp(-\lambda_i t),\ i = 1, 2$$

$$\therefore \quad f(t) = \int_0^t \lambda_1\lambda_2 \exp(-\lambda_1 x). \exp\{-\lambda_2(t-x)\}dx,$$

$$= \frac{\lambda_1\lambda_2}{\lambda_2-\lambda_1}(e^{-\lambda_1 t} - e^{-\lambda_2 t}) \tag{8.2.2}$$

When $\lambda_1 = \lambda_2 = \lambda,\ f(t) = \lambda^2 t\, e^{-\lambda t}$

Reliability function for the system (i.e., the probability that the control system is quite successful at time t) is given by

$R(t) = R_1(t) + R_2(t) =$ Pr{medicine 1 is working successfully at time t}

+ Pr{medicine 1 is failed prior to t and medicine 2 is working successfully at time t}

$$= e^{-\lambda_1 t} + \int_0^t \lambda_1 e^{-\lambda_1 x}\, e^{-\lambda_2(t-x)}dx$$

$$= \frac{\lambda_2 e^{-\lambda_1 t} - \lambda_1 e^{-\lambda_2 t}}{\lambda_2 - \lambda_1} \tag{8.2.3}$$

For $\lambda_1 = \lambda_2 = \lambda, \quad R(t) = e^{-\lambda t} + \lambda t\, e^{-\lambda t}$

In general, for n medicines,

$$R(t) = \sum_{i=1}^{n} \frac{\lambda_1 \lambda_2 \ldots \lambda_{i-1} \lambda_{i+1} \ldots \lambda_n e^{-\lambda_i t}}{\prod_{\substack{j=1 \\ j \neq i}}^{n} (\lambda_j - \lambda_i)}, \; \lambda_j \neq \lambda_i \text{ for } i \neq j \tag{8.2.4}$$

If $\lambda_i = \lambda$ for $1 \leq i \leq n$,

$$R(t) = e^{-\lambda t} \sum_{i=0}^{n-1} \frac{(\lambda t)^i}{i!}$$

The mean time to failure of the disease control system is given by

$$\text{MTBF} = \sum_{1}^{n} \frac{1}{\lambda_i}$$

8.3 RELIABILITY OPTIMIZATION

An important part of reliability technology is the optimization of reliability for a given system under given conditions. Here a formulation for reliability optimization is given.

Since the spray of the medicines on crops is subject to percentage of the chemical contents, etc., the use of medicines varies from crop to crop and also with time. Also if the medicine used exceeds the prescribed quantity per square area for a crop, there are chances of harming the crop; so it should be taken according to the directions of plant pathologists.

Let for a general problem $Q_i\{w(t)\}$, $i = 1, 2 \ldots, n$ be the weight function depending upon the area to be used and the time (i.e., season) for the use of medicines, the weight constraint is

$$\sum_{i=1}^{n} Q_i\{w(t)\} \leq w \tag{8.3.1}$$

Now the problem is

Maximize $$R(t) = \sum_{i-1}^{n} \phi_i(t)$$

subject to

$$\sum_{i=1}^{n} Q_i\{w(t)\} \leq w \tag{8.3.2}$$

$w(t) \geq 0$ for $t \geq 0$,

where, $R(t)$ is given by (8.2.4) and $\phi_i(t)$ is the function under the summation sign in (8.2.4).

The other constraints may be taken such as cost constraint for some particular type of medicines or, in general, by the farmers according to their financial position. Other generalization of the problem is possible taking a number of crops considered simultaneously treated by some similar or dissimilar medicines.

The problem given by (8.3.2) may be solved easily using dynamic programming. The recurrence relation for (8.3.2) is

$$f_k(\beta) = \max_{w_k} [\phi_k(t) + f_{k-1} (\beta - Q_k(w_k))], \; k = 2, 3 \ldots, n$$

and,

$$f_1(\beta) = \phi_1(t), \; Q_1\{w_1(t)\} \leq w_1$$

where w_1 is the weight taken for first stage calculation.

Singh [1982] analyzed another problem on plant pathology (considering cost constraint also) using Differential Dynamic Programming (Jacobson and Mayne 1970) after formulating the problem as a control problem. He also provides an algorithm for the problem.

EXERCISE

1. How can comparison and analyses of biogas production systems be realized at different system levels?
2. How can organic waste from the food industry be treated in a resource efficient way?
3. How to treat organic waste from the food industry in a resource efficient way?
4. Explain the reliability optimization.
5. Given a budget of Rs. 700 and the following data on three components that must operate in series. Determine, using marginal analysis, the optimum number of redundant units. Compute the achieved reliability.

Components	Reliability	Unit Cost
1	0.80	Rs. 200
2	0.90	Rs. 100
3	0.95	Rs. 75

Bibliography

1. Archna Sharma, Jai Singh; and Kuldip Kumar — 'Availability estimation of certain parts of human body'; Int. J. of Contemporary Mathematical Sciences, 4(25), 2009 pp. 1219-1235.
2. Barlow, R.E. and Proschan, F.; — Mathematical Theory of Reliability, John Wiley & Sons, 1967.
3. Christiaanse, W.R.; — 'A new technique for reliability calculations'; IEEE Trans. Power Appar. Systems, 89, 1970 pp. 1836-1847.
4. Dhillon, B.S. and Rayapati, S.N.; — 'Reliability modeling of on-surface transit systems with human errors'; Micro. and Reliab. 25(6), 1985, pp. 1087-1098.
5. Deepika, Jai Singh & Kuldip Kumar; — 'Redundancy allocation in, pharmaceutical plant; [Int. J. of Engg. Sc. and Tech., 2(4), 2009 pp. 1088-1097]
6. Elsayed A. Elsayed; — Reliability Engineering; Addison Wesley Longman, INC. N.Y. 1996
7. Feller, W.; — An Introduction to Probability Theory and its Applications Vol. (I, II) John Wiley & Sons. 1957.
8. Fukuta, Jiro and Kodama, Masanori; — 'Mission reliability for a redundant repairable system with two dissimilar units; IEEE Trans. On Reliab. R-23(4), 1974, pp. 280-282.
9. Gupta, S.M., Jaiswal, N.K. and Goel, L.R.; — 'Switch failure in a two unit standby redundant system'; Micro. and Reliab. 23(1), 1983, pp. 129-132.
10. Gupta, Pardip, Singh, Jai and Singh, I.P.; — Behavioural Analysis of Duplex Casting system under preemptive resume priority repair discipline–A Case Study (Industrial Engg. Journal, Vol. (XXXIII) No. 12 Dec. 2004, pp. 4-13)

11. Gupta, Pardip, Singh, Jai and Singh, I.P.; 'Maintenance planning bread on performance analysis of 7-out-of-14:G chemical system: A Case Study'; Int. J. of Industrial Engg. 12(3), 2005 pp. 264-274.

12. Gupta, Pardip, Singh, Jai and Singh, I.P.; 'Mission reliability and availability prediction of flexible polymer power production system; OPSEARCH, 42(2), 2005 pp. 152-167.

13. Gupta, Pawan, Lal A.K., Sharma, R.K, and Singh, Jai; 'Behavioural study of the cement manufacturing plant–A numerical approach'; J. of Mathematics and Systems Sciences (JMASS), 1(1), 2005 pp. 50-70.

14. Gupta, Pawan, Lal A.K., Sharma, R.K. and Singh, Jai; 'Numerical analysis of reliability and availability of the serial processes in butter-oil processing plant'; Int. J. of Quality and Reliability Management, (IJQRM) 22(3), 2005, pp. 303-316

15. Gupta, Pawan, Lal A.K., Sharma, R.K. and Singh, Jai; 'Analysis of reliability and availability of serial processes of plastic pipe manufacturing plant', I JQRM, 24(4-5), 2007, pp. 404-419.

16. Gupta, Pawan; Mathematical Analysis of Reliability and Availability of Some Process Industries; Ph.D. Thesis, Thapar Instt. of Engg. and Technology, Patiala (India), 2003.

17. Goel, Pardip and Singh, Jai; 'Availability analysis of butter manufacturing system in a dairy plant; Recent Developments in Operations Research, Narosa Publications, India 2001, pp. 109-116

18. Gupta, V.K. and Singh, Jai; The analysis of a PLCs system—A new approach JMASS, 5(2), 2009 pp. 39-57

19. Gupta, V.K. and Singh, Jai; Profit analysis of a single unit operating system with a capacity factor and undergoing degradation. JMASS, 5(2), 2009 pp. 129-142

20. Jacobson, D.H. and Mayne, D.Q.; Differential Dynamic Programming; American Elsevier Publishing Co. Inc. 1970.

21. Kendall, M.G.; Advanced Theory of Statistics; Charles Grifin, London 1943.

22. Kodama, M. and Takamatsu, S.; 'Mission reliability for a complex redundant system with allowed downtime'; Technical Report No. 1156, Osaka University Japan 1973.

23. Kumar, Dinesh; Singh, Jai and Singh, I.P.; 'Availability of the feeding system in sugar industry'; Micro. and Reliab. 28(5), 1988, pp. 867-71.

24. Kumar, Dinesh; Singh, Jai and Singh, I.P.; 'Reliability analysis of the feeding system in paper industry'; Micro. and Reliab. 28(2), 1988 pp. 213-15.

25. Kumar, Dinesh; Singh, Jai and Pandey, P.C.; 'Availability of the feeding system in paper industry with general repair time'; Proc. of ICMMST Vol. 2, 1988, World Scientific Singapore.

26. Kumar, Dinesh; Singh, Jai and Pandey, P.C.; 'Reliability and cost analysis of bleaching system in paper industry; Proc. of Asian Congress on Quality and Reliability, Wiley Pub. 1989.

27. Kumar, Dinesh; Singh, Jai and Pandey, P.C.; 'Maintenance planning for pulping system in paper industry'; Reliab. Engg. and System Safely, 25(4), 1989 pp. 293-302.

28. Kumar, Dinesh; Singh, Jai and Pandey, P.C.; 'Availability of washing system in paper industry'; Micro. and Rel. 29(5), 1989 pp. 779-81.

29. Kumar, Dinesh; Singh, Jai and Pandey, P.C.; 'Cost analysis of a multicomponent screening system in the paper industry'; Micro. and Rel. 30(3), 1990 pp. 457-61.

30. Kumar, Dinesh; Singh, Jai and Pandey, P.C.; 'Behavioral analysis of a paper production system with different repair policies'; Micro. and Rel. 31(1), 1991 pp. 47-51.

31. Kumar, Dinesh; Singh, Jai and Pandey, P.C.; 'Operational behavior and profit function for bleaching and screening system in paper industry'; Micro. and Rel. 33(8), 1993 pp. 1101-1105.

32. Kumar, Dinesh; Singh, Jai and Pandey, P.C.; 'Availability of the crystallization system in the sugar industry under common-cause failure'; Trans. of IEEE on Rel. 41(1), 1992 pp. 85-91.

33. Kumar, Dinesh; Singh, Jai and Pandey, P.C.; 'Operational behavior of feeding system in sugar plants; Institution of Engineers Journal, Vol. 70, 1990 pp. 113-115.

34. Kumar, Dinesh; Singh, Jai and Pandey, P.C.; 'Availability of the feeding system in sugar industry' ; Micro. and Rel. 28(6), 1988 pp. 867-71.

35. Kumar, Dinesh; Singh, Jai and Pandey, P.C.; 'Behavior analysis of refining system in sugar industry'; Proc. of Int. Symposium on 'Stochastic Models', 1989.

36. Kumar, Dinesh; Singh, Jai and Pandey, P.C.; 'Design and cost analysis of refining system in sugar industry'; Micro. and Rel. 30(6), 1990, pp. 1025-28.

37. Kumar, Dinesh; Singh, Jai and Pandey, P.C.; 'Maintenance planning for refining system in sugar industry'; Int. J. of Quality and Reliability Mgt. 1991.

38. Kumar, Dinesh; Singh, Jai and Pandey, P.C.; 'Designing for reliable operation of urea synthesis in fertilizer industry'; Micro. and Rel. 30(6), 1990 pp. 1021-24.

39. Kumar, Dinesh; Singh, Jai and Pandey, P.C.; 'Behaviour analysis of urea decomposition system with general repair policy in fertilizer industry; Micro. and Rel. 31(5), 1991 pp. 851-54.

40. Kumar, Dinesh; Singh, Jai and Pandey, P.C.; 'Process design for crystallization system in the urea fertilizer industry'; Micro. and Rel. 31(5), 1991 pp. 855-59.

41. Kumar, D.; Analysis and optimization of Systems Availability in Sugar, Paper and Fertilizer Industries; Ph.D. Thesis, I.I.T. Roorkee (India), 1991.

42. Linton, G. Darrell; 'Generalized reliability results for 1-out-of-n: G repairable systems'; IEEE Trans. on Rel. 38(4), 1989.

43. Morse, P.M.; Queues, Inventories and Maintenance, John Wiley & Sons, N.Y. 1958.

44. Nakaguwa, H.; 'A decomposition method for computing system reliability by a Boolean expression'; IEEE Trans. Rel. 26, 1977 pp. 250-52.

45. Parzen, E.; Modern Probability Theory and its Applications; John Wiley & Sons, 1960

46. Prabhu, N.U.; Stochastic Processes; The Macmillan Co. N.Y. 1965.

47. Premo Jr., A.F.; 'The use of Boolean algebra and truth table in the formulation of a mathematical model of success'; IEEE Trans. Rel. 12, 1963 pp. 45-49.

48. Puran Chand, Singh, Jai and S.C. Malik; Availability Analysis of Cotton Mill, [JMASS, 3(1), 2007 pp. 50-59]

49. Rau, J.G.; Optimization and Probability in Systems Engineering; Van Nostrand Reinhold Co., N.Y., 1970.

50. Rao, S. Subba and Jaiswal, N.K.; 'On a Class of queueing problems and discrete transforms'; Operations Research, 17(6), 1969 pp. 1062-76.

51. Shooman, M.L.; Probabilistic Reliability: An Engineering Approach; McGraw-Hill, 1968.

52. Shao, Jiajun and Lamberson, R. Leonard; 'Modelling a shared-load k-out-of-n: G system'; IEEE Trans. Rel.40(2), 1991 pp. 205-9.

53. Singh, Jai; Some Problems on Queues and Reliability; Ph.D. Thesis, Kurukshetra University (India), 1977.

54. Singh, Jai; 'A warm standby redundant system with common-cause failures'; Reliability Engg. and System Safety, 26(2), 1989 pp. 135-42.

55. Singh, Jai; 'Effect of switch failure on 2-redundant systems'; IEEE Trans. On Rel., 29(1), 1980 pp. 82-83.

56. Singh, Jai; 'Reliability analysis of a biogas plant having two dissimilar units'; Micro. and Reliab. 29(5), 1989 pp. 779-81.

57. Singh, Jai; 'Reliability and cost analysis of a bioconversion system'; Proc. of Int. Conf. on Structural Failure, Product Liability and Technical Insurance, Technical University Vienna 1989.

58. Singh, Jai; 'An application of reliability technology to energy modeling'; Proc. of IFORS Argentina 1987.

59. Singh, Jai; 'Reliability technology applied to plant pathology'; Proc. of Operational Research Society of India, (ORSI) 1982.

60. Singh, Jai; 'An O.R. approach to plant pathology'; Proc. of ORSI, 1982.

61. Singh, Jai; 'On M/M/C queueing system with finite number of sources'; The Mathematics Education, 7(3), 1973 pp. 70-73.

62. Singh, Jai and Dayal, B.; '1-out-of-N:G system with common-cause failures and critical human errors'; Micro. and Rel. 31(5), 1991 pp. 847-49.

63. Singh, Jai and Mahajan, Pardip; 'Availability analysis of Container manufacturing plant'; Proc. of Int. Conference on Combinotorics, Statistics and Pattern Recognition and Related Areas, 1998.

64. Singh, Jai and Mahajan, Pardip; 'Reliability of utensils manufacturing plant—A case study'; OPSEARCH, 36(3), 1999 pp. 260-69.

65 Singh, Jai and Goyal, Yogesh; 'Availability and behavior analysis of bread manufacturing plant'; JMASS, 2(2), 2006 pp. 35-45.

66. Singh, I.P.; Singh, Jai and Gupta, Pardeep; 'Behavioral analysis of duplex casting system under preemptive resume priority repair discipline—A case study'; Industrial Engg. Journal 33(12), 2004, pp. 4-13.

67. Singh, Jai and Savita; 'Availability analysis of core Veneer manufacturing system in plywood industry'; Proc. of Int. Conf. on Reliability and Safety Engineering, 2005 (IIT Kharagpur, India).

68. Savita and Jai Singh; Behaviour analysis of plywood industry under preventive and corrective maintenance. (JMASS, 3, 2007, pp. 60-73)

69. Aravendan, M. and Paneerselvam, R. (2016) "Implementation of hybrid genetic algorithm for CLSC network design problem: A case study on fashion leather goods industry", American Journal of Operations Research, ISSN (Copy), Vol. 6, pp. 300-316.

70. A. Kumar, D. Garg (2019) "Reliability Technology Theory and Application", Lap Lambert Academic Publishing in Germany, ISBN: 978- 613-9-47665-7.

71. Kumar A., Goel P., Garg D., and Sahu A. (2017) "System Behavior Analysis in the Urea Fertilizer Industry", Book: Data and Analysis [978-981-10-8526-0, REDSET 2017, CCIS 799, Paper Approval (464880-1-En, Chapter no:1).

72. Zhang, X. and Zeng (2016) "Optimal maintenance modeling for systems with multiple non-identical units using extended DSSP method", American Journal of Operations Research, ISSN (copy from old), Vol. 6, pp. 275-295.

73. Adlakha N., Taneja G. and Shilpi (2017) "Reliability and cost benefit analysis of two unit standby communication system with assembling of one unit at a time", AJMI, 9(2), pp. 281-288.

74. Cruz, F.R.B., Duarte, A.R. and Woense Van, T. (2008), "BuffersAllocation in general single server Queuing Networks", Computers and Operation Research, 35(11), ISSN https://doi.org./10.1016/j.cor2007.3.004, pp. 3581-3598.

75. A. Kumar, D. Garg and P. Goel (2017) "Mathematical Modeling and profit analysis of an edible oil refinery industry", Airo International Research Journal, ISSN 2320-3714, Vol. XIII, pp. 1-14.

76. Linmin Hu, Dequan Yue and Jiandong Li (2012) "Availability analysis and design optimization for a repairable series parallel system with failure dependencies", International Journal of Innovative Computing, Information & Control, Vol. 8 (10 A), ISSN 1349-4198, pp. 6693-6705.

77. Tseng-Chang Yen, Wu-Lin-Chen and Jia-Yu-Chen (2016) "Reliability & Sensitivity Analysis of the Controllable Repair System with Warm Standbys and Working Breakdown, Vol. 97, https://doi.org/10-1016/j.cie.2016.04.019, pp. 84-92.

78. Sridharan, V. and Jayashree, P. J. (1996) "Some Characteristics on a finite queue with normal partial & total failures", Micro Electronics Reliability, 36(2), ISBN 0026-2714 (95)-00088-J, pp. 265-267.

79. Rade, L. (1993) "Reliability Equilance", Micro Electronics Reliability, 33(3), ISBN 0026-2714 (93)-90020-Y, pp. 323-325.

80. Muthukani, M. (2018) "A study on analytical solution or acoustic waves propagation in fluids", International Journal of Statistical and Applied Mathematics, 3(1), ISSN 2456-1452, pp. 54-59.

81. Zoramawa, A.B., Gueuembe S.U., Kantam, RRL. and Musa Y. (2018) "Developing double acceptance sampling plans for percentiles based on the inverse Rayleigh distribution", International Journal of Statistics and Applied Mathematics, 3(1), ISSN 2456-1452, pp. 39-44.

82. Garg D., Kumar K. and Meenu (2010) "Availability optimization for screw plant based on genetic algorithm", International Journal of Engineering Science and Technology, ISSN 0975-5462, 2(4), pp. 658-668.

83. Sarita D. and Garg D. (2017) "Redundancy allocation in neel metal products limited, Indian Journal of Science and Technology, ISSN 0974-5645, 10(30), pp. 1-5.

84. A. Kumar, P. Goel and D. Garg (2017) "Sensitivity analysis of a distillery industry using RPGT", presented a paper 3rd International conference on communication systems , B.K. Birla Institute of Engineering & Technology, Pilani.

85. Rajbala, Kumar A. and Garg D. (2019) "System modeling and analysis: A case study EAEP manufacturing plant", International Journal of Advanced Science and Technology, 28(14), pp. 08-18.

86 Kumar A., Garg D. and Goel P., (2019) "Mathematical modeling and behavioral analysis of a washing unit in paper mill", International Journal of System Assurance Engineering and Management, 1(6), ISSN 0976-4348, pp. 1639-1645.

87. A. Kumar, D. Garg, P. Goel and O. Ozer (2018) "Sensitivity analysis of 3 : 4 :: Good system", International Journal of Advance Research in Science and Engineering, ISSN 2319-8354, 7(2), pp. 851-862.

88. Garg D. and Kumar K. (2011) "Reliability analysis and optimization of some mechanical systems", Ph.D. thesis of National Institute of Technology (NIT), Kurukshetra, Haryana, India.

89. Garg D. and Kumar K. (2009) "Reliability analysis of pharmaceutical plant using Matlab-Tool", International Journal of Electronic Engineering research, ISSN 0975-6450, 1(2), pp. 127-133.

90. Kumar R., Poonia M. and Goel P. (2016) "Availability analysis of two unit system with warm standby having imperfect switch-over device using RPGT", International Journal of Applied Research, ISSN Print: 2394-7500, ISSN Online: 2394-5869; 2(6), pp. 280-288.

91. Nandal J. and Anand J. (2018) "Profit comparison of a cold standby system with priority to repair over preventive maintenance and server failure during repair", International Journal of Statistics and Applied Mathematics, ISSN: 2456-1452, 3(1), pp. 221-228.

92. J. Fukuta and Kodama M. (1974) "Mission reliability for a redundant repairable system with two dissimilar units", IEEE Transactions on Reliability, Vol. 4, pp. 280-282.

93. Sharma A., Singh J. and Kumar K. (2009) "Availability estimation of certain parts of human body", International Journal of Contemporary Mathematics Sciences, 4(25), pp. 1219-1235.

94. R.E. Barlow and L.C. Hunter (1961) "Reliability analysis of a one-unit system," Operations Research, Vol. 9, pp. 200-208.

95. Garg D., Sing J. and Kumar K. (2009) "Redundancy allocation in, pharmaceutical plant", International journal of Engg. Sci. and Technology, 2(4), pp. 1088-1097.

96. J. Cao and Wu, Yanhong (1989) "Reliability analysis of a two-unit cold standby system with replaceable repair facility", Microelectronics, Reliab, 29(02), pp. 145-150.

97. Charistiaanse, W.R. (1970) "A new technique for reliability calculations", IEEE Trans. Power Appar. System, Vol. 89, pp.1836-1847.

98. A. Kumar, P. Goel, D. Garg and A. Sahu (2017) "Sensitivity analysis of a urea fertilizer industry application", presented a paper in international conference Recent Trends in Operations Research and Statistics (RTORS-2017), Dep. of Mathematics Indian Institute of Technology Roorkee.

99. Kumar D., Jai S. and P.C. Pandey (1989) "Reliability and cost analysis of bleaching system in paper industry", proc. of Asian congress on quality and reliability, Wiley.

100. Asthana A. and Okumoto K. (2012) "Integrative software design for reliability: Beyond models and defect prediction", Bell Labs Technical Journal, Vol. 17, pp. 37-59.

101. Kumar M. (2018) "Comparative study of reliability models in respect of profit and availability of single unit systems with different repair policies", International Journal of Advanced Research in Computer Science, ISSN 0976- 5697, 9(1), pp. 310-314.

102. Kanso A., Toeroe M., and Khendek F. (2014) "Comparing redundancy models for high availability middleware," Computing, Vol. 96, pp. 975-993.

103. A. Kumar, D. Garg and P. Goel (2018) "Behaviour analysis of a bread making system", International Journal of Statistics and Applied Mathematics, ISSN 2456-1452, 3(6), pp. 56-61.

104. Gupta R. (2017) "Stochastic analysis of a repairable model for one unit system with three types of repair policy", International Journal of Statistics and Applied Mathematics, ISSN 2456-1452, 2(6), pp. 126-130.

105. A. Kumar, D. Garg and P. Goel (2019) "Sensitivity analysis of a cold standby system with priority for preventive Maintenance", Journal od Advance and Scholarly Researches in Allied Education, ISSN 2230-7540, 16(4), pp. 253-258.

Index

A

Accelerated life testing, purpose of 48
Accelerated life tests 49
Accelerated tests, types of 47
Active redundancy 4
Availability 3
Availability analysis of a biogas plant 203

B

Bay's theorem 13
Bernoulli trial 19
Beta distribution 22
Binomial distribution 17
Biogas plant 203
Birth and death process 26
Boolean function technique 68
Boolean table method 57
Busy period of the server 74
Butter-oil (ghee) manufacturing plant 163

C

Cattle feed plant, availability analysis 141
Cauchy distribution 22
Cement industry 131
Chi-squared distribution 23
Cold standby redundancy 39
Complex systems 7
Conditional probability 10
Container manufacturing 111
Continuous distribution function 16
Cut-set method 55

D

Data analysis procedures 4
Data analysis, issues in 40
Data collection 39
Data recovery 40
Data reporting system 45
Decision support system 195
Decomposition method 54
Discrete distribution function 16
Discrete transform 66
Discrete transform method 65
Down time ratio 33
Downtime 4

E

Environmental stress screening 50
Ergodic chain 25
Erlang distribution 23
Event space method 56
Exponential distribution 23
Extremal (smallest) distribution 7

F

Failure, types of 7
Fault tree analysis method 41
Force of mortality 5
Fundamental renewal equation 60

G

Gamma distribution 22
Geometric distribution 19
Gumbel distribution 7
Gupta-Singh technique 72

H

Hazard rate 5
Hot standby redundancy 39
Hyper geometric distribution 20

I

Inherent availability 33
Intensity function 5
Internal consistency reliability 45
Inter-rater reliability 44
I-out-of-N : G)special) 94
I-out-of-n: G system 88

J

Joint probability density function 17
Joint probability mass function 16

K

(K, m) systems 7
(k, n) systems 36
K-out-of-(N + M) : G system 91
K-out-of-m:F system 4
K-out-of-m:G system 4
k-out-of-n : G system 89
k-Out-OF-n systems 87

M

Maintainability 3
Maintained systems 6
Mann-Whitney U test 83
Markov chain 25
Markov method 61
Markov process 25
Mathematical modeling 54
Matrix method 68
Mean downtime 32
Mean time
  between failures 32
  to failure 3
  to first failure 31
  to repair 32
  to system failure 31
  to system failure 73, 79
Mill's ratio 5
Mission reliability 106
Mission reliability function 34
Mixed distribution function 16
Multinomial distribution 21

N

Negative binomial distribution 19, 20
Normal distribution 21
Normal random variable 21
Number of replacements 75
Number of server's visits 75

O

One sample runs test 85
On-surface transit system 96

P

Paired-sample sign test 82
Paper industry 119, 120
Parallel forms reliability 45
Parallel redundant systems 6
Parallel systems 34
Parallel-series systems 7, 35
Partial redundancy 4
Pascal's distribution 20
Path-tracing method 56
Plant pathological system203
Plant pathology problems 206
Poisson distribution 18
Poisson experiment 18
Polya's distribution 20
Preventive maintenance 39
Priority systems 98
Probability, definitions of 9
  derivative of 54

Q

Quality 3
Quality control 50

R

Random variable 15
Rank sum test 83
Reduction method 58
Redundancy 4, 39
Reliability allocation 37
Reliability, requirement of 50
Reliability enhancement testing 49
Reliability function 4
Reliability group 1
Reliability optimization 207
Reliability technology 2
  applied to agro-based industries 155
  in process industries 119
Reliability testing 43

Reliability, definition of 2
Renewal density 60
Renewal density equation 61
Renewal function 59
Renewal theoretic approach 58
Reparability 3
Rice plant 174
Risk analysis and management 45

S

Series systems 6, 34
Series-parallel systems 7, 35
Serviceability 3
Shell construction system 111
Sign test 81
Standby redundancy 4, 39
Standby redundant systems 6
Standby systems 35
Steady-state availability 73
Stochastic process 24
Structural redundancy 45
Sugar industry 155
Supplementary variables technique 63
System availability 33
System effectiveness 3
System mean time before failure 31
System reliability 34
System
  analyses of 7
  description of 53
  with arbitrary repair 36
  with imperfect switch and waiting time 101
  with parallel repair 36

T

Test-planning 45'
Test-retest reliability 44
Theorems of probability 10
Theory of reliability 1
Tie-set method 56
Time-dependent reliability 61
Transition matrix 25
Two-unit system 98

U

Uniform probability distribution 17
Uptime ratio 33
Uptime 3
Utensils industry 136

W

Warm redundancy 46
Warm standby redundancy 39
Weibull distribution 23